Pollution: Causes, Effects, and Control

Second Edition

Edited by
R. M. Harrison
University of Essex

ROYAL
SOCIETY OF
CHEMISTRY

British Library Cataloguing in Publication Data
Pollution: causes, effects and control – 2nd ed
  1. Pollution
  I. Harrison, Roy M. *1948–*
363.7'3

ISBN 0-85186-283-7

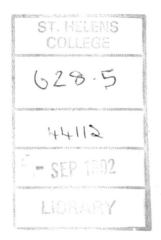

© The Royal Society of Chemistry 1990

Reprinted 1992

Published by The Royal Society of Chemistry,
Thomas Graham House, Cambridge CB4 4WF

Filmset by Bath Typesetting Ltd., Bath
and printed by Henry Ling Ltd., at the Dorset Press, Dorchester.

Made in Great Britain

Pollution:
Causes, Effects, and Control
Second Edition

Preface

The first edition of this book, published in 1983 arose from collation of the course notes from a Residential School held at the University of Lancaster in 1982, supplemented with additional chapters to give a more complete overview of the field. The aim was to provide a basic textbook covering the more important concepts.

The second edition has been considerably expanded, both by including totally new subject areas and by giving each author a greater length in which to cover his topic. The level of treatment is again essentially introductory, although some authors have sought to go into rather greater depth in some aspects. Encyclopaedic coverage of the entire field is impossible, but the contributions combine to give a broad overview, touching on most of the important areas, and delving deeper into many of them. Acknowledging the importance now attached to these topics, wholly new chapters deal with radioactive pollution and chemistry and pollution of the stratosphere.

The authors have been selected on the basis of their established reputations in the field and their ability to write with clarity of presentation. A high proportion of them wrote for the first edition, but in the case of this edition have had the advantage of preparing their material specifically for a book, rather than for lecture notes as before.

Much has happened in the pollution field since 1982 when the first edition was written. Thus much updating has been carried out, and changes in emphasis, reflecting changing perceptions of a fast moving field have been incorporated. The environment is now high on the political agenda, heightening the need for authoritative scientific information. I hope that this book goes some way towards meeting that need.

Roy M. Harrison
Colchester

Contents

Contributors

G. J. K. Acres, *Johnson Matthey plc, New Garden House, 78, Hatton Garden, London EC1N 8JP, UK.*

J. R. Ashby, *School of Chemistry, Leicester Polytechnic, PO Box 143, Leicester LE1 9BH, UK.*

I. Colbeck, *Institute of Aerosol Science, University of Essex, Wivenhoe Park, Colchester CO4 3SQ, UK.*

P. J. Craig, *School of Chemistry, Leicester Polytechnic, PO Box 143, Leicester LE1 9BH, UK.*

B. Crathorne, *Water Research Centre plc, Medmenham Laboratory, PO Box 16, Henley Road, Medmenham, Buckinghamshire SL7 2HD, UK.*

A. J. Dobbs, *Water Research Centre plc, Medmenham Laboratory, PO Box 16, Henley Road, Medmenham, Buckinghamshire SL7 2HD, UK.*

J. C. Farman, *British Antarctic Survey, High Cross, Madingley Road, Cambridge CB3 0ET, UK.*

R. M. Harrison, *Institute of Aerosol Science, University of Essex, Wivenhoe Park, Colchester CO4 3SQ, UK.*

C. N. Hewitt, *Institute of Environmental and Biological Sciences, University of Lancaster, Bailrigg, Lancaster LA1 4YQ, UK.*

A. James, *Department of Civil Engineering, University of Newcastle-upon-Tyne, Newcastle-upon-Tyne NE1 7RU, UK.*

J. N. Lester, *Public Health Engineering, Department of Civil Engineering, Imperial College of Science, Technology and Medicine, London SW7 2BU, UK.*

W. L. Linton, *RTZ Limited, York House, Bond Street, Bristol BS1 3PE, UK.*

P. W. Lucas, *Division of Biological Sciences, Institute of Environmental and Biological Sciences, University of Lancaster, Bailrigg, Lancaster LA1 4YQ, UK.*

R. Macrory, *Barrister, Lecturer in Environmental Law and Policy, Imperial College Centre for Environmental Technology, 48, Prince's Gardens, London SW7 1LU, UK.*

T. A. Mansfield, *Division of Biological Sciences, Institute of Environmental and Biological Sciences, University of Lancaster, Bailrigg, Lancaster LA1 4YQ, UK.*

C. F. Mason, *Department of Biology, University of Essex, Wivenhoe Park, Colchester CO4 3SQ, UK.*

R. F. Packham, *22, Harwood Road, Marlow, Buckinghamshire SL7 2AS, UK.*

D. R. Turner, *Plymouth Marine Laboratory, Citadel Hill, Plymouth PL1 2PB, UK.*

H. A. Waldron, *Occupational Health Department, St. Mary's Hospital, Praed Street, London W2 1NY, UK.*

M. L. Williams, *Warren Spring Laboratory, Department of Trade and Industry, Gunnels Wood Road, Stevenage, Hertfordshire SG1 2BX, UK.*

P. C. Young, *Centre for Research on Environmental Systems, University of Lancaster, Bailrigg, Lancaster LA1 4YQ, UK.*

CHAPTER 1

Chemical Pollution of the Aquatic Environment by Priority Pollutants and its Control

B. CRATHORNE AND A. J. DOBBS

1 INTRODUCTION

A clumsy but essentially accurate definition of pollution is 'too much of something in the wrong place.' This definition, when applied to chemical pollutants in the aquatic environment recognizes that any chemical can become a 'pollutant' if it is present at a high enough concentration. This was shown by the recent example of a pollution incident caused by sugar.[1] Indeed in truth there is no such thing as a 'pollutant chemical' only a given chemical present at too high a concentration. Nonetheless some chemicals can be selected as being of high priority for control in the aquatic environment; because they are frequently found there, because they are capable of exerting adverse effects at low concentrations, because they remain unchanged in the aquatic environment for long periods, or because they biomagnify in food chains. Often the priority chemicals selected display more than one of these characteristics. Priority chemicals and their control will be the subject of this chapter, but before discussing them it may be useful to consider briefly the subjects of 'gross pollution' and 'effluent consenting' since these are also critical subjects for aquatic pollution and its control.

The majority of pollution incidents in the UK continue to be due to gross organic pollution, frequently originating from silage effluents and slurries from farms and from sewage treatment plant effluents. These effluents generally act as pollutants not because they contain chemicals at concentrations that are toxic but rather the reverse. They contain chemicals that provide food for micro-organisms which multiply rapidly. The micro-organisms oxidize the organic chemi-

[1] J. McAngus and P. Garrett, Water Bulletin, 21 November, 1986.

1

cals in the effluent and use up the oxygen rapidly thereby, in some cases, killing higher organisms. This was the mechanism by which sugar, quoted above, became a pollutant. Organic pollution of this kind is fairly well understood and its control is mainly a matter of investment, better treatment plant operation, and/or better storage and control during use, though doubtless there are improvements that can be made in the technology used at various stages.

Consents for effluents have historically been set in terms of the gross pollution parameters identified in the Royal Commission Report of 1912 namely a suspended solids concentration and a five day biochemical oxygen demand and these still form the basis of most consents in the UK. With a much increased and slowly growing array of synthetic organic chemicals being made and used, these two simple measurements cannot provide sufficient environmental protection in all cases. On the other hand, extensive analytical investigation is an impractical proposition for all effluents even if there was sufficient knowledge to (a) identify all the necessary components and (b) to be able to assess the environmental impact of those components identified. The use of toxicity tests for effluents consenting and monitoring is being increasingly considered as an answer to this problem (*e.g.*, see ref. 2 and 3).

Most of the priority pollutants identified and discussed in this chapter have been in production and use for several years and yet in many cases there are significant gaps in the data on their environmental behaviour and effects. This should not happen in the future because there is now a New Chemicals Testing Scheme which requires 'new' chemicals or old chemicals being used in new applications, to be tested and the results of the test submitted for assessment prior to marketing. Although the 'Base set' data requirements do not provide sufficient information for the setting of EQSs, it is possible for additional data to be required by the competent authority. The details of the New Chemicals scheme are presented elsewhere.[4]

As indicated previously this chapter is concerned with specific 'priority' chemicals as pollutants of the aquatic environment. The first section deals with chemicals of concern in direct discharges and their control, while the following section discusses pollutants from diffuse sources. The next section summarizes information on the sources, fates, and analytical considerations of some selected priority substances and finally some concluding comments are made.

2 DIRECT DISCHARGES

The environmental policy of the UK is now being profoundly influenced by developments within the Commission for the European Community and a number of European Community (EC) Directives concerning water now exist.[5,6] The most important one for this chapter is the Directive on pollution

[2] T. M. Wall and R. W. Hanmer, *J. Wat. Poll. Control Fed.*, 1987, **59**, 7–12.
[3] OECD, Environmental Monographs No. 11, 'The use of biological tests for water pollution assessment and control,' OECD Paris, 1987.
[4] Health and Safety Executive, 'A guide to the notification of new substances regulations HS(R)14(Rev),' HMSO, ISBN 0-11-885454-2, 1988.
[5] J. Gardiner and G. Mance, Water Research Centre, Technical Report TR 205, 1984.
[6] N. Haigh, 'EEC Environmental Policy and Britain,' Environmental Data Services Limited, 1984.

caused by certain dangerous substances discharged into the aquatic environment of the community (76/464/EEC published 4 May, 1976). This is sometimes known as the Dangerous Substances Directive. It applies to all water bodies to which discharges of the substances covered by the Directive are made, *i.e.*, fresh, estuarine, marine, and ground waters. This Directive has been the origin of the debate involving the relative merits of the EQO/EQS and the uniform emission standards approaches to pollution control. Both are recognized as ways of controlling substances of concern and it is worth briefly summarizing the elements of each.

The EQO/EQS approach seeks to define the use that is to be made of a given water body, examples would be 'use for drinking water abstraction' or 'use for the support of salmonid fish populations' which defines an Environmental Quality Objective (EQO). In order to secure this objective in the presence of dangerous chemicals, Environmental Quality Standards (EQS) are needed. These are concentrations of the chemicals concerned below which there is expected to be no impact on the EQO. Emission limits for effluents can then be established by taking account of dilution capacity within the receiving waters on the basis that the EQS limits must not be exceeded outside the immediate impact zone also called the mixing zone or zone of non-compliance.

The uniform emission standard (UES) approach on the other hand sets limits for the concentration of the dangerous substance concerned in the effluent without taking specific account of the available dilution capacity or the presence of other inputs. These UES limits usually act as effluent concentrations expressed as monthly flow-weighted averages with additional limits on daily values, the daily values being usually a factor of 2 to 4 higher. Values are also expressed as a total amount of substances per unit of production or use, *e.g.*, 40 g CCl_4 per tonne of production.

The relative merits of the two approaches have been debated long and hard and there seems little point in repeating these arguments here since there is now a growing recognition that the most sensible approach to pollution control is a fusion of the two approaches as has been proposed in the 'Red List' Discussion Document which will be discussed later.

2.1 Chemicals covered by the Dangerous Substances Directive – Black and Grey List Chemicals

In this Directive selected chemicals are either placed on List I which has come to be known as the 'Black' List or on List II the 'Grey' List. Different control procedures are to be applied to chemicals on these lists. Those on List I have limit values and EQSs agreed at Community level, these appear in daughter Directives, *e.g.*, Directive of 26 September, 1983 on limit values and quality objectives for cadmium discharges (83/513 EEC). (Confusingly, in Directives the Commission uses 'quality objective' to mean the same as EQS in the discussion above!) The List I chemical categories listed in the Directive are given in Table 1 and specific chemicals agreed for control as List I chemicals are given in Table 2. Substances which are on List I but which have no daughter directive are to be treated as List II. In 1982 the Commission also published a list of 129 potential List I chemicals and this is given in Table 3. These chemicals were selected by the Commission on the advice of consultants on the basis of their production volume and estimates of their toxicity, persistence, and bioaccumulation.

Table 1 *List I substances*

List I contains certain individual substances which belong to the following families and groups of substances, selected mainly on the basis of their toxicity, persistence and bioaccumulation, with the exception of those which are biologically harmless or which are rapidly converted into substances which are biologically harmless:

1 organohalogen compounds and substances which may form such compounds in the aquatic environment
2 organophosphorus compounds
3 organotin compounds
4 substances in respect of which it has been proved that they possess carcinogenic properties in or via the aquatic environment
5 mercury and its compounds
6 cadmium and its compounds
7 persistent mineral oils and hydrocarbons of petroleum origin
8 persistent synthetic substances which may float, remain in suspension or sink and which may interfere with any use of the waters

Table 2 *Chemicals selected for control as List I chemicals*

Mercury	Pentachlorophenol
Cadmium	'Drins' (aldrin, dieldrin, endrin, and isodrin)
Hexachlorocyclohexane	Hexachlorobenzene
Carbon tetrachloride	Hexachlorobutadiene
DDT	Chloroform

Table 3 *Candidate chemicals for List I*

Chlorinated hydrocarbons	—aldrin, dieldrin, chlordane, chlorobenzene, dichlorobenzenes, chloronaphthalene, chloroprene, chloropropene, chlorotoluenes, chlorotoluidene, endosulfan, endrin, heptachlor, hexachlorobenzene, hexachlorobutadiene, hexachlorocyclohexane, hexachloroethane, PCBs, tetrachlorobenzenes, trichlorobenzenes.
Chlorophenols	—monochlorophenols, 2,4-dichlorophenol, 2-amino-4-chlorophenol, pentachlorophenol, 4-chloro-3-methylphenol, trichlorophenol.
Chloroanilines and nitrobenzenes	—monochloroanilines, 1-chloro-2,4-dinitrobenzene, dichloroanilines, 4-chloro-2-nitrobenzene, chloronitrobenzenes, chloronitrotoluenes, dichloronitrobenzenes.
Polycyclic aromatic hydrocarbons	—anthracene, biphenyl, naphthalene, PAH.
Inorganic chemicals	—arsenic and compounds, cadmium and compounds, dibutyl tin compounds, mercury and compounds, tetrabutyltin.

Solvents	—benzene, carbon tetrachloride, chloroform, dichloroethane, dichloroethylene, dichloromethane, dichloropropane, dichloropropanol, dichloropropene, ethylbenzene, toluene, tetrachloroethylene, trichloroethane, trichloroethylene.
Others	—benzidine, benzyl chloride, benzylidene chloride, chloral hydrate, chloroacetic acid, chloroethanol, dibromomethane, dichlorobenzidine, dichloro diisopropyl ether, diethylamine, dimethylamine, epichlorohydrin, isopropylbenzene, tributyl phosphate, trichlorotrifluoroethane, vinyl chloride, xylenes.
Pesticides	—azinphos ethyl and methyl, coumaphos, cyanuric chloride, 2,4-D and derivatives, 2,4,5-T and derivatives, DDT, demeton, dichlorprop, dichlorvos, dimethoate, disulfoton, fenitrothion, fenthion, linuron, malathion, MCPA, mecoprop, methamidiophos, mevinphos, monolinuron, omethoate, oxydemeton-methyl, parathion, phoxim, propanil, pyrazon, simazine, triazophos, tributyl tin oxide, trichlorofon, trifluralin, triphenyltin compounds.

Table 4   *List II substances*

List II contains

— substances belonging to the families and groups of substances in List I for which the limit values referred to in Article 6 of the Directive have not been determined
— certain individual substances and categories of substances belonging to the families and groups of substances listed below and which have a deleterious effect on the aquatic environment, which can, however, be confined to a given area and which depend on the characteristics and location of the water into which they are discharged.

Families and groups of substances referred to

1   The following metalloids and metals and their compounds:

1	zinc	6	selenium	11	tin	16	vanadium
2	copper	7	arsenic	12	barium	17	cobalt
3	nickel	8	antimony	13	beryllium	18	thallium
4	chromium	9	molybdenum	14	boron	19	tellurium
5	lead	10	titanium	15	uranium	20	silver

2   Biocides and their derivatives not appearing in List I
3   Substances which have a deleterious effect on the taste and/or smell of the products for human consumption derived from the aquatic environment, and compounds
4   Toxic or persistent organic compounds of silicon and substances which may give rise to such compounds in water, excluding those which are biologically harmless or are rapidly converted in water into harmless substances
5   Inorganic compounds of phosphorus and elemental phosphorus
6   Non-persistent mineral oils and hydrocarbons of petroleum origin
7   Cyanides, fluorides
8   Substances which have an adverse effect on the oxygen balance, particularly: ammonia, nitrites

List II chemicals are to be controlled by using the EQO approach using quality standards set nationally. Member states are also required by the Directive to establish programmes to reduce pollution by these substances. The families of chemicals identified as List II are given in Table 4 and those for which National Quality Standards have been set are given in Table 5.

Table 5   *List II chemicals for which UK National Standards have been set*

Lead
Chromium
Zinc
Copper
Nickel
Arsenic
Boron
Iron*
pH*
Vanadium
Tributyltin compounds
Triphenyltin compounds
Cyfluthrin
Sulcofuron
Flucofuron
Permethrin
Polychlorochloromethylsulphonamidodiphenylether mothproofers

* Neither of these falls within the scope of the substances listed in Table 4 but DoE Circular 7/89 requires that they be treated in the same way as List II substances.[7]

2.2   Red List Chemicals

In July, 1988 the Department of the Environment and the Welsh Office issued a consultation paper entitled 'Inputs of Dangerous Substances to Water: Proposals for a Unified System of Control.'[8] This paper set out the Government's proposals for tightening controls over the input to water, of dangerous substances. Dangerous substances were defined as those which represented the greatest threat to the aquatic environment due to their persistence, toxicity, their ability to bioaccumulate, and their likely presence in the aquatic environment. The aim of the proposals in the consultation paper were stated as:

(i)   To reduce inputs to the aquatic environment of those substances which represent the greatest potential hazards.

---

[7] Department of the Environment, 'Water and the Environment,' Circular 7/89, HMSO, London, 1989.
[8] Department of the Environment, 'Inputs of dangerous substances to water: proposals for a unified system of control. The Government's consultative proposals for tighter controls over the most dangerous substances entering the aquatic environment,' 'The Red List,' July, 1988.

(ii) To improve the scientific basis of the system for identifying the most dangerous substances.

(iii) To develop a more integrated approach to pollution control.

(iv) In controlling dangerous substances in water, to take full account not only of 'point source' discharges from production plants, but also of 'diffuse sources' of entry into the aquatic environment.

A further important objective was to propose a unified system for controlling the discharge of dangerous substances and reconcile the approach favoured in the UK, the EQO/EQS system, and the approach favoured by other states in the EC, that of uniform emission standard (UES) as noted above. The more stringent of the two being applied as a control measure in any given case, for example:

— Substances which have a large input from diffuse sources as well as point discharges, cannot be effectively controlled using a UES. In such cases, control can be only achieved using a system of EQSs.

— It is not yet possible to set safe environmental quality standards for some substances, because not enough is known about their long-term risks to the aquatic environment (and to man). In these circumstances, a precautionary principle would be adopted and steps taken to minimize their input to the environment from all sources, taking into account best-available pollution control technology.

With these considerations in mind, the Government set out proposals for a unified approach to controlling the discharge of dangerous substances. The essential points of these proposals are as follows:

(i) Identification of a limited range of the most dangerous substances selected according to clear scientific criteria, whose discharge to water should be minimized as far as possible—the 'Red List';

(ii) The setting of strict environmental quality standards for all 'Red List' substances;

(iii) The introduction of a system of 'scheduling' of industrial processes discharging significant amounts of 'Red List' substances, and the progressive application of technology-based emission standards, based on the concept of best available technology not entailing excessive costs ('BATNEEC');

(iv) Measures designed, where possible, to reduce inputs of 'Red List' substances from diffuse sources.

The control regime for the 'Red List' chemicals is more stringent than that for 'Black List' substances and marks a change in philosophy by becoming more 'precautionary' and requiring 'batneec' for all 'Red List' substances. The proposals also imply that because there is always some uncertainty about the long term environmental effects of 'Red List' chemicals we should aim to reduce their environmental concentrations as much as is possible—regardless of the current environmental concentrations or whether or not that there is any evidence that they are causing environmental damage. As a first stage for selection of 'Red List' chemicals, the priority chemicals from the 'Black List' and the List of 129 were studied.

The selection procedure adopted for the 'Red List' is described in more detail in the consultation paper and will not be discussed further here. However, it is intended that selection procedures are refined, for example to take into account data on carcinogenicity. In addition it is envisaged that the 'Red List' will be modified on the basis of new data being generated on particular substances. A provisional 'Red List' of 26 priority substances was published in the consultation paper. A list of substances considered as priority candidates for inclusion in the 'Red List' was also given.

As a result of consultation and the consideration of new and corrected data, the provisional 'Red List' has now been slightly amended. This amended list has been issued as the initial priority 'Red List' and is shown in Table 6. The Government has initiated a monitoring programme for these priority chemicals and the need for specific action will be considered on an individual basis. Ministers have also given a commitment,[9] to reduce the input of 'Red List' substances to the North Sea by 50% in the period from 1985 to 1995. While as individual citizens we may welcome this as a desirable objective, as scientists we must recognize that it will be very difficult to establish the loads of these chemicals with sufficient accuracy to assess whether this objective has been achieved or not.

Table 6   *The initial priority UK Red List*

Mercury and its compounds
Cadmium and its compounds
Gamma-Hexachlorocyclohexane
DDT
Pentachlorophenol
Hexachlorobenzene
Hexachlorobutadiene
Aldrin
Dieldrin
Endrin
Polychlorinated biphenyls
Dichlorvos
1,2-Dichloroethane
Trichlorobenzene
Atrazine
Simazine
Tributyltin compounds
Triphenyltin compounds
Trifluralin
Fenitrothion
Azinphos-methyl
Malathion
Endosulfan

[9] Department of the Environment, Ministerial Declaration Second International Conference on the Protection of the North Sea, London, 24–25 November, 1987.

3 DIFFUSE SOURCES

Pollution from diffuse sources is of course more difficult to control than that from point sources and generally can only be exerted by control of chemical use. However, for many pollutants it is likely that diffuse sources are of equal or greater importance than point sources. A good example of this is the estimation of loads of heavy metals to the North Sea where despite the low volatility of heavy metals, deposition from the atmosphere makes a significant contribution to the total pollution load.[10]

The 'Red List' discussion document undertakes to attempt 'where possible' to reduce diffuse inputs but it is not presently clear what specific control measures will be used. In the rest of this section we have selected three pollutants where diffuse sources are the major ones and we outline the current position and the control measures introduced.

3.1 Acidity

It is now accepted that the oxides of sulphur and nitrogen produced and released to the atmosphere during the combustion of coal, gas, and oil can, in areas with particular geologies, bring about the acidification of lakes and streams. In remote areas of the world average rainfall acidity is typically about pH 5.0, with individual episodes ranging from pH 4.0 to 5.6.[11] The term 'acid rain' usually refers to rainfall events which are significantly more acidic than those in remote areas. With this definition, measurements made during the last two decades have shown that 'acid rain' occurs throughout Europe and over much of Eastern North America. The acidity of rainfall in these areas is typically about pH 4.0 to 4.5.

Contemporary interest in the subject of acid rain began in Scandinavia about 20 years ago.[12] This arose from concern regarding freshwater acidification and an observed reduction in fish populations and diversity of species in areas with a low capacity to neutralize acidic inputs. Similar concerns have now been raised about freshwater acidification in parts of the UK,[13] and regions where surface and groundwaters are potentially susceptible to acidification have been identified in reports from the UK Acid Waters Review Group.[14,15] This group was established specifically to assess the effects of acid deposition in the UK and their reports provide a more complete assessment of the effects on the aqueous environment.

[10] Department of the Environment, 'Quality status of the North Sea,' summary, Septemter, 1987.
[11] J. N. Galloway, G. E. Likens, W. C. Keene, and J. M. Miller, *J. Geophys. Res.*, 1982, **87**, 8771–8786.
[12] S. Oden, 'Nederbordens och luftens fursurning dess orsaker, forlopp och verken i olika miljoer. Statens naturvenetenskapliga forskningrad,' Ekologikommitteen Bull. No. 1, Stockholm, 1968.
[13] R. Harriman and B. Morrison, 'Ecology of acid streams draining forested and non-forested catchments in Scotland,' in 'Ecological Impact of Acid Precipitation,' Proc. Int. Conf. Sandenfjord, Norway, 1980, ed. D. Drablos and A. Tollan, Oslo-As: SNSF 312–313, 1980.
[14] UK Acid Waters Review Group, 'Acidity in United Kingdom fresh waters,' Interim report, Department of the Environment, London, 1986.
[15] UK Acid Waters Review Group, 'Acidity in United Kingdom fresh waters,' Second report, HMSO, London, ISBN 0-11-752158-2, 1988.

The only practicable way to reduce acid depositions is to control emissions from the sources of acidic oxides. This is a long-term solution, it may not reduce the acidity of already acid waters and it is possible that some waters would never recover. In any event, it has been estimated that a reduction of about 90% from 1985 emission levels would be required to return most surface waters to near-pristine condition.[15] This would call for concerted action in Europe and possibly even more widespread action. To maintain current conditions a deposition reduction of around 30% from 1985 levels is necessary.[15]

In order to achieve rapid improvements, techniques other than a reduction in deposition have to be used. There is already a considerable experience in the use of liming to improve water quality at least to the point where fisheries can be sustained. However any benefits from this process should be weighed against its other environmental impact, since catchment liming will change the chemistry of the area with potentially adverse effects on natural, acid-loving species. In addition, liming must be continued indefinitely if the benefits are to be maintained and this can be a costly process, particularly in remote areas.

The scope for remedial action is therefore limited and further reductions in emissions would appear to be necessary to sustain an improvement in the situation. However the role of afforestation in the increase in acid loading is now accepted and future plantings should be planned to avoid catchments sensitive to acid inputs.

3.2 Nitrates

Nitrate is naturally occurring in soil and water and is required for the production of chlorophyll and amino acids, *etc.*, by plants. Yields of crops and the productivity of livestock have increased considerably over the last 30 years due, to a large part, to the use of nitrogen-containing fertilizers. Nitrate is not bound to soil and therefore moves through soil with water and eventually percolates into lower water-bearing rocks or runs into rivers and other surface waters.

There has been a marked increase in the nitrate content of both groundwaters and surface waters over the last 50 years. By far the major contribution to this has been from changes in agricultural practices, with increased use of nitrogen fertilizers and the release of nitrate following the ploughing of permanent pasture being the prime sources.[16]

There are implications for human health arising from the presence of high nitrate levels in drinking water with the occurrence of methaemoglobinemia in infants the prime concern.[16] In addition, nitrate in lakes, rivers, estuaries, and coastal waters has contributed to high nutrient levels leading to the formation of algal blooms and to oxygen depletion.

In the UK, it has been found that rivers show a marked seasonal pattern of nitrate concentration; these are higher in the winter when drainage from the land occurs during autumn and winter rains.[17] Peak winter levels can often exceed $100 \, mg \, NO_3 \, l^{-1}$ in some lowland rivers with extensive arable farming

[16] ECETOC, Technical Report No. 27, 'Nitrate and Drinking Water,' ISBN 0-773-8072-27, 1988.
[17] Department of the Environment, 'Nitrate in water,' Pollution Paper 26, HMSO, London, 1986.

catchments. In groundwaters, elevated nitrate levels are found where arable land overlies aquifers and there is no intervening layer of clay. Water containing high nitrate levels percolates downwards and, depending on the thickness of the overlying rock, may take between 5 and 40 years to reach the water table. The extent of nitrate accumulation depends on both climatic and geological factors. Mathematical models have predicted that, for the most severely affected areas in the UK, levels may reach about 150 mg NO_3 1^{-1} assuming a continuation of current agricultural practice.[18] However there is now some evidence from the UK and from the Federal Republic of Germany that the rate of increase in the nitrate content in river water is slowing down,[16] possibly reflecting a reduction in nitrate fertilizer usage.

The EC have recently proposed a Directive concerning the protection of fresh and coastal marine waters against pollution by nitrates from diffuse and point sources and in particular the application to land of chemical fertilizers and manure. The Directive would require Member States to designate areas as 'vulnerable zones' within which controls are placed on the application to land of chemical fertilizers and the application and storage of livestock manure are to be set by national governments. Controls would also be imposed on discharges of sewage effluents to vulnerable zones. Member States would also have to introduce regular monitoring and reporting procedures, including the keeping of records of the total quantity of nitrogen from fertilizers and manure applied annually in vulnerable zones.

Comments have been requested on these wide-reaching proposals by March, 1989 and it is likely that the draft Directive will be considered by the Council of Ministers in June, 1989. There is likely to be some criticism of the Commission's proposals, particularly as they are directed solely at one approach to reducing nitrate levels. As has been described in several publications (*e.g.*, see ref. 16), a reduction in the application of nitrogenous fertilizers would not, in itself, have an immediate significant impact on nitrate levels in water, particularly groundwaters. This is largely because the nitrate leached from the soil over the last few years has not yet reached the water table and, in some areas, may not do so for several years.

For this reason, other forms of control may be necessary to reduce nitrate concentrations in drinking water. Possible actions were listed in a recent report.[16]

— Controlled blending of high and low nitrate waters. However, this would only be a short-term measure if levels of nitrate in water continue to increase.
— Maximize natural denitrification by long storage in large reservoirs.
— Provide additional water treatment, *e.g.*, use of ion exchange resins or microbiological denitrification.
— Provision of bottled water to populations identified at risk, particularly infants under 6 months of age.

[18] S. S. D. Foster, L. R. Bridge, A. K. Geok, A. R. Lawrence, and J. H. Parker, 'The groundwater nitrate problem: a summary of research on the impact of agricultural land use practices on groundwater quality between 1976 and 1985,' *Hydrogeol. Rep. Brit. Geol. Survey*, No. 86/2, 1986.

3.3 Pesticides

Pesticides feature prominently in the priority lists discussed previously. Since pesticides are designed to control living organisms they can be expected to have an environmental impact and many examples of effects in the aquatic environment can be cited. However most of these were due to accidents or deliberate misuse and there is little evidence to suggest that when used properly, and thereby acting as a diffuse source, pesticides cause any adverse environmental effects. Recent studies in the UK and other countries have demonstrated, however, that low concentrations of a wide range of pesticides can be detected in the aquatic environment. More work is needed on the environmental fate of some pesticides and on the impact of low concentrations of pesticides on the aquatic environment.

The EC Directive Relating to Water Intended for Human Consumption (80/778/EEC) has, however, significantly affected our reaction to the presence of low levels of pesticides in the aquatic environment by stipulating a maximum admissible concentration (MAC) of $0.1\ \mu g\ l^{-1}$ for individual pesticides, regardless of toxicity. Since some pesticides are not efficiently removed by conventional drinking water treatment, $0.1\ \mu g\ l^{-1}$ is now one level of concern for pesticides in water. A further level of concern is that needed to protect the aquatic environment although for most pesticides $0.1\ \mu g\ l^{-1}$ tends to be the lower value.

There are several difficulties associated with monitoring pesticide levels in the aquatic environment:

— A large number of pesticides are approved for use and, in theory, any of these could enter the aquatic environment. It is impractical to monitor all water sources for all pesticides.
— If present at all, most pesticides would be found only at extremely low concentrations and, in some instances, sensitive analytical techniques are not available to monitor such low levels.
— Little information is generally available on the environmental fate and distribution of pesticides. Such data could be useful in trying to assess which pesticides to monitor in a given area.
— Although some information is available on the range and quantity of pesticides applied for agricultural purposes, it is very difficult to obtain specific information relating to a particular water catchment. Again this information is essential in developing a pesticide monitoring programme in a given area.
— Very little recent information is available on the non-agricultural use of pesticides and it is now accepted that such sources can have a significant impact on pesticide levels in the aquatic environment.

Due to a combination of these factors, there has been little systematic monitoring of surface waters and groundwaters for pesticides. In recent years, however, water authorities and water companies in the UK have started to develop extensive pesticide monitoring programmes.

Protection of the aquatic environment should ultimately be afforded by the development of new pesticides able to 'target' pests more accurately, rapidly break down in the environment, and ensure minimum leaching from soil, *etc.* Pesticides are now being designed with such criteria in mind and new compounds have to go through a rigorous testing procedure before approval is granted for widespread use.

In the UK, the use of pesticides is covered generally by the Food and Environment Protection Act (FEPA), 1985, and, more specifically, by the Control of Pesticide Regulations (COPR), 1986. Under these regulations pesticides are approved for specific uses and if an approval is revoked, the pesticide is effectively banned. Under the Control of Pollution Act (COPA), 1974, Water Authorities, as pollution control authorities, have powers to set up areas in order to protect water resources and these powers could be applied to pesticide usage. Under the provisions of the Water Act, 1989, these powers are transferred to the National Rivers Authority.

A variety of actions would be possible under these powers; for example pesticide users could be advised, wherever possible, to reduce pesticide usage in specified, sensitive areas. This action would probably have little effect on the agricultural use of pesticides but might reduce non-agricultural use, for example by local authorities, British Rail, golf clubs, *etc.* More formal restrictions on pesticide use would be achieved by setting up water protection zones (WPZs), for example to protect water sources used for drinking water supplies. However it might be difficult to identify such zones and the resource costs for policing and monitoring pesticide use in a WPZ would be high.

An alternative approach would be to control pesticide use nationally via the pesticide registration scheme, under the auspices of the Advisory Committee on Pesticides (ACP). Approval for the use of a pesticide could be withdrawn following a review by the ACP of its toxicity and pattern of use. This approach to controlling pesticide use would be simpler and probably more effective than setting up WPZs.

It is impossible to assess, at the moment, how long it would take for any pollution control measures to reduce the concentration of pesticides in the aquatic environment, particularly in groundwaters.

4 SOURCES, FATES, AND ANALYSIS

In this section we have summarized information on some of the priority chemicals identified in the earlier sections in order to illustrate the types of chemicals involved and the analytical techniques used to solve the problems that monitoring requirements pose. The basis for selection was the desire to cover a range of different chemical types and there was no attempt to select on the basis of relative harm or priority. Table 7 shows the chemical structures of the chemicals discussed.

It will be seen that the priority chemicals have very little in common in terms of use, properties, or behaviour. This would also have been apparent if we had extended these brief reviews to cover toxicological and ecotoxicological properties. We have not included the special cases (mercury and cadmium) of the two elements identified as priority chemicals.

Table 7 *Chemical structures of selected chemicals*

Atrazine

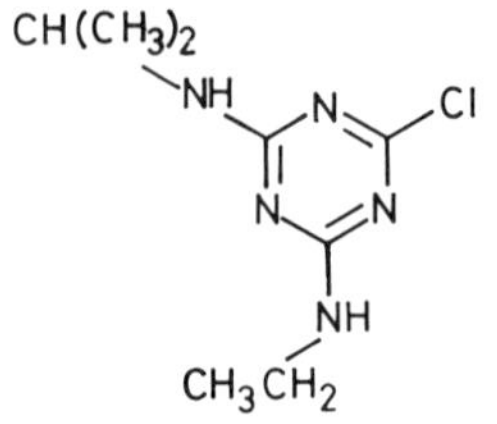

Azinphosmethyl

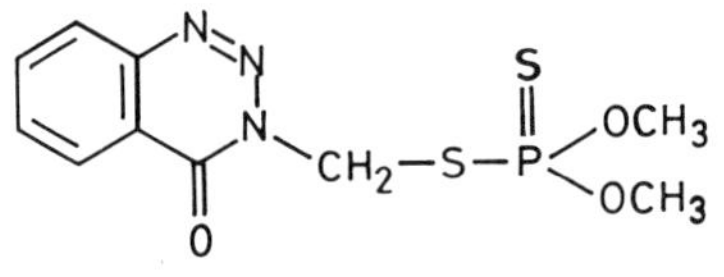

1,2-Dichloroethane CH_2Cl——CH_2Cl

Endosulfan

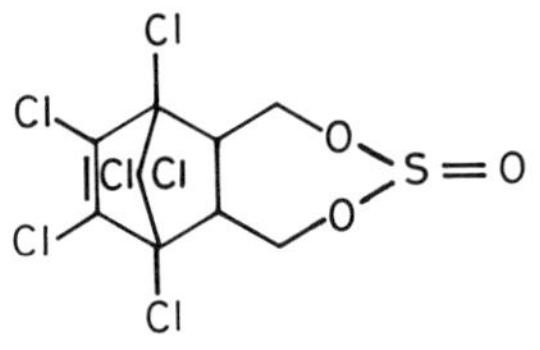

Tributyltin oxide $(C_4H_9)_3$ Sn—O—Sn $(C_4H_9)_3$

4.1 Atrazine

Atrazine is a systemic herbicide which acts by inhibiting photosynthesis in the target plants. Typical agricultural application rates are in the range 0.5 to 2.5 kg per hectare although higher non-agricultural application rates are frequently used for total plant control on railway tracks and industrial sites. In 1987 it was estimated that 50–200 tonnes of atrazine were applied agriculturally and up to 5000 tonnes used non-agriculturally.[19] Atrazine is not now manufactured in the UK and although it is used at many formulating sites and these could be point

[19] D. Beaton, *Farmers Weekly*, 7 October, 1988.

sources, the major release will undoubtedly be from diffuse sources. Atrazine is moderately water soluble (approx 0.2 mmol l^{-1}, 33 mg l^{-1}) and loss from storm run-off can be quite high, *e.g.*, 5–20%.[20] Residues of atrazine are widely found in ground and surface waters in both the UK and the rest of Europe.

Atrazine can be degraded by micro-organisms although, under typical use conditions, degradation rates are not high and half-lives in the range 1 to 12 months have been reported in top soils. Atrazine does not react with water at an appreciable rate and it is not particularly photolytically labile. It seems likely therefore that atrazine will only slowly be removed from the aquatic environment and it is likely to persist for long periods in groundwaters where there are few micro-organisms and no sunlight.

Analysis for atrazine in water samples is relatively straightforward involving, for example, solvent extraction followed by either capillary column gas chromatography or high performance liquid chromatography.[21] Nitrogen specific detectors for GC greatly improve the detection limits.

4.2 Azinphosmethyl

Azinphosmethyl is a broad spectrum organophosphorus insecticide. It does not appear to be manufactured in the UK. It is generally applied at rates in range 0.1–0.4 kg ha^{-1} and is most widely used in the UK for insect control on fruit and vegetables. Total usage in the UK is probably in the range 1–10 tonnes per year.

Azinphosmethyl is moderately soluble in water (0.1 mmol l^{-1}, 33 mg l^{-1}) and in common with many other organophosphorus insecticides is fairly rapidly hydrolysed. A half-life of approximately 2 days has been reported in a farm pond.[22] This is typical of results from other laboratory and field studies, although field studies have generally indicated higher breakdown rates, probably due to a contribution from photolytic degradation. Half-lives in soil and crops in a similar range (3–5 days) have been reported.[23,24] Therefore the environmental fate appears to be fairly rapid degradation, possibly assisted by photodegradation.

Analysis of water samples can be undertaken by gas chromatography with phosphorus specific detection following solvent extraction.[25]

[20] D. E. Glotfelty, A. W. Taylor, A. R. Isensee, J. Jersey, and S. Glenn, *J. Environ. Quality*, 1984, **13**, 115–121.
[21] Standing Committee of Analysts, 'Chlorophenoxy acidic herbicides, trichlorobenzoic acid, chlorophenols, triazines, and glyphosate in water.' Methods for the examination of waters and associated materials, HMSO, London, 1985.
[22] F. P. Meyer, *Trans. Am. Fish Soc.*, 1965, **94**, 203.
[23] C. A. Anderson, J. C. Cavagnol, C. J. Cohen, A. D. Cohick, R. T. Evans, L. J. Everett, J. Hensel, P. Honeycut, E. R. Levy, W. W. Loeffler, D. L. Nelson, T. Parr, T. B. Waggoner, and J. W. Young, *Residue Rev.*, 1974, **51**, 213.
[24] S. Smith, T. E. Reagen, J. L. Flynn, and G. H. Willis, *J. Env. Qual.*, 1983, **12**, 534–537.
[25] Standing Committee of Analysts, 'Organophosphorus pesticides in sewage sludge, river water, and drinking water,' Methods for the examination of waters and associated materials, HMSO, London, 1985.

4.3   1,2-Dichloroethane

1,2-Dichloroethane is manufactured in very large quantities (~ 5 million tonnes per year). It is mainly used as an intermediate, particularly in vinyl chloride manufacture, so production volume in this case can be misleading as an indicator of environmental release. Other uses for the chemical are as a solvent and as a fumigant for stored grain and these are likely to be the major diffuse sources.

The major properties determining its enviromental fate are its high vapour pressure, 8 kPa at 20 °C, and its moderate water solubility (90 mmol l^{-1}, 8 g l^{-1}) which give it a high volatilization rate from water. Once in the atmosphere it will be broken down by oxidation with hydroxyl radicals. Biodegradation is slow and soil adsorption is low so the chemical is likely to persist in groundwaters, although hydrolysis does occur slowly and this may be the major degradation route in groundwaters.

Analysis of samples of water for dichloroethane can be undertaken by solvent extraction and gas chromatography but its high volatility can cause problems with the extraction and concentration stages. Better approaches are to exploit its high volatility and utilize head space or purge and trap techniques coupled with analysis with gas chromatography.

4.4   Endosulfan

Endosulfan is a broad spectrum insecticide similar in many respects to other, better known organochlorine pesticides such as dieldrin, but of lower persistence. It is not manufactured in the UK and most sources are likely to be diffuse but figures for current or recent UK use are not available. The commercial product is a mixture of two isomers both of which have low aqueous solubilities (~ 0.2 µmol l^{-1}, 80 µg l^{-1}) and low vapour pressures (~ 1 mPa). The sulphate group appears to be a centre that allows the chemical to be degraded, albeit slowly, both abiotically and by micro and higher organisms. Slow degradation in water bodies would therefore be expected. The other major loss from water would be expected to be due to particulate adsorption and sedimentation. As expected with a low water solubility chemical, particle adsorption is very significant and it was found that $>75\%$ of the endosulfan in a River Rhine sample was bound to particulates.[26]

Analysis of endosulfan as with other organochlorines is relatively straightforward involving solvent extraction, concentration, clean-up if necessary using column chromatography, and capillary column gas chromatography with an electron capture detector.[27]

4.5   Tributyltin Oxide

Tributyltin oxide (TBTO) and other tributyltin (TBT) compounds are mainly

---

[26] P. A. Greve and S. L. Wit, *J. Wat. Poll. Control Fed.*, 1971, **43**, 2338–2348.
[27] Standing Committee of Analysts, 'Organochlorine insecticides and polychlorinated biphenyls in waters. Methods for the examination of waters and associated materials,' HMSO, London, 1978.

used as a microbiocide, *e.g.*, as a fungicide for wood treatment and as an anti-fouling additive for paints applied to boats. It is manufactured at several sites in the EC including the UK and the total annual production is in the range 1000–10 000 tonnes per year. Because it is manufactured and formulated in the UK it is likely to occur in direct discharges but diffuse sources will also be important for example from its use in wood preservatives and in anti-fouling paints, although its use in anti-fouling paints is now restricted.[28]

Tributyltin oxide is not very soluble in water ($\sim 30 \, \mu mol \, l^{-1}$, $18 \, mg \, l^{-1}$); however the aqueous chemistry of these organotin compounds is quite complex and, probably, a large number of dissociated forms exist in water. The use of the term 'solubility' may therefore be misleading. None the less the low apparent solubility is consistent with the tendency of TBTO to adsorb strongly to particulate material in the water column and to sediments. The major abiotic and biotically mediated degradation mechanisms appear to involve oxidative debutylation in discrete stages to give dibutyltin, butyltin, and eventually inorganic tin compounds. The major abiotic degradation route is photolytic degradation and half-lives of ~ 10 days having been reported. Biodegradation appears to occur only slowly and is even slower under anaerobic conditions which is probably important given the tendency of TBT to adsorb strongly to particulate material.

Analysis of organotins is complicated by the variety of forms that can exist. The more specific the analytical technique the more time consuming and expensive it becomes. Total tin determination by atomic absorption is not generally considered to be specific enough and a determination based on 'organic solvent extractable tin' is often the minimum level of speciation acceptable. Most of the organotin species partition strongly into organic solvents such as toluene and therefore 'organic solvent extractable total tin' can be roughly equated with total organotin.[29] Atomic absorption techniques can be used to quantify the tin in the organic solvent extract. For a higher level of speciation it is necessary to distinguish between the different oganotin forms. One approach to this is to convert the organotin species into non-polar volatile forms, extract them, and separate them by GC. Borohydride reduction, with purge and trapping of the volatile organotin hydrides and analysis by capillary gas chromatography with flame photometric detection has been particularly successful.[30] This technique allows mono, di, and tributyltin compound to be separately quantified.

A review on the sources, fate, and determination of tributyltin in the environment has recently been published[31] and the data used to derive the values and the recommended EQS have been published.[32]

[28] Department of the Environment, Pollution Paper No. 25, 'Organotin in antifouling paints—environmental considerations,' 1986.
[29] S. C. Apte and M. J. Gardner, *Talanta*, 1988, **35**, 539–544.
[30] M. O. Andreae and J. T. Bird, *Anal. Chim. Acta*, 1984, **156**, 147.
[31] Commission of the European Communities, Water Research Report 8, 'Tributyltin in the environment — sources, fate, and determination,' Cost 641, EUR 11562, 1988.
[32] T. F. Zabel, J. Seager, and S. D. Oakley, Water Research Centre, Technical Report, TR 255, 1988.

5 CONCLUSIONS

As chemists we must recognize that following synthesis of a new chemical which is moderately kinetically stable, no matter how carefully it is controlled, it will eventually become quite widely dispersed in the environment. This is a natural consequence of the second law of thermodynamics which ensures that eventually, traces of that chemical may be found in any environmental sample taken, provided the analytical detection limit is low enough. The concentrations detected may be very low but many synthetic and naturally occurring chemicals are designed to exert physical and biological effects at low concentrations and it is not uncommon nowadays to quote detection limits in terms of ppt (which almost always means one part per 10^{12} expressed as a mass : mass basis). This almost unimaginable analytical power is a tribute to both the scientists that synthesize chemicals which are effective at such low concentrations and the scientists that devise and use such exquisitely sensitive analytical methodologies. The power to analyse samples at such levels does however present a perceptual problem for the man in the street who does not recognize any difference between 1 ppm and 1 ppt. He may feel that 'toxic' and 'pollutant' chemicals should simply not be present in given environmental samples. This is a particular problem with identifying 'priority' chemicals because our selection procedures are crude. The danger is that public pressure may focus attention on the occurrence of trace levels of these priority chemicals, which may not be the cause of the most significant environmental impact.

On the other hand with many tens of thousands of synthetic chemicals in use, many of which have the potential to cause environmental damage and some of which have fulfilled this potential, action is clearly necessary. Since it is not possible to control everything at once, some prioritization is required. This chapter has attempted to summarize the current state of pollutant prioritization in the UK and has illustrated the kinds of chemicals involved with some specific examples.

CHAPTER 2

The Chemistry of Metal Pollutants in Water

D. R. TURNER

1 INTRODUCTION

Increased awareness of pollution and its effects has emphasized the importance of water quality management in maintaining our natural waters in a fit state for various purposes (*e.g.*, for use as drinking water, for recreation, or to assure the viability of the native biota). Effective water quality management requires that the concentrations and effects of pollutants can be accurately assessed, and that the consequences of proposed discharges can be predicted. This chapter will consider the chemistry of pollutant metals and its consequences for their dispersal and biological effects. The discussion will concentrate on general principles rather than individual case studies, although the elements copper and plutonium and the chemical context of sea and estuarine waters, will be used to illustrate many of the points in this discussion. Copper and plutonium are two widely studied elements with very different chemistries and very different source terms. Copper is a micronutrient element which becomes toxic at elevated concentrations, and which is in widespread industrial use. Plutonium, in contrast, is a synthetic element whose presence in natural waters arises from civil and military uses of nuclear technology. Its chemistry is extremely complex, with four oxidation states possible in aqueous solution, but modelling this chemistry is becoming increasingly urgent in connection with the assessment of radioactive waste disposal options. Similarly, saline waters provide chemically complex and widely studied media to illustrate the discussion, but the same principles apply equally to fresh waters.

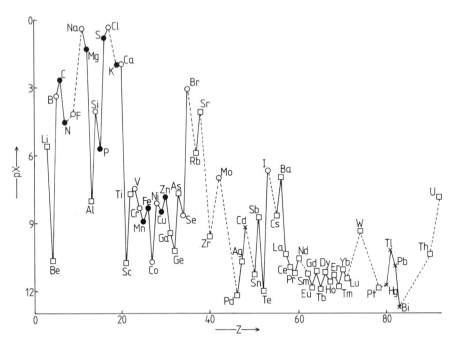

Figure 1 *Seawater concentrations of the elements expressed as* $pX = -\log_{10}[X]$ *plotted as a function of atomic number* (Z). ● *elements essential in all species;* ○ *elements essential in some species;* □ *non-essential elements;* × *toxic elements. Reproduced with permission from D. R. Turner and M. Whitfield, Ecol. Bull. (Stockholm), 1983, **35**, 9–37; updated with information from M. Whitfield and D. R. Turner, in 'Aquatic Surface Chemistry: Chemical Processes at the Particle–Water Interface,' ed. W. Stumm, John Wiley, New York, 1987, pp. 457–493*

2 METAL POLLUTANTS IN PERSPECTIVE

It is important to realize that with the exception of the synthetic elements and nuclides produced by nuclear installations (*e.g.*, Pu, ^{60}Co) all pollutant metals are naturally present in the aquatic environment and it is the presence of higher than usual concentrations which present a threat to the biota. Figure 1 shows the dissolved concentrations of the stable elements in the world's largest aquatic reservoir, seawater, as a function of atomic number. Life began in the sea, and its tolerance to, and usage of, trace metals reflect seawater concentrations. Thus all the essential elements are present at concentrations of 1 nM or greater, while strongly toxic elements are present at lower concentrations. Cobalt appears as an exception to this rule since, due to redox changes, its concentration in the contemporary oxygen-rich ocean is lower than in the primordial anoxic ocean in which life began. From the perspective of pollution the important conclusion to be drawn from this distribution is that the mere presence in a water body of a strongly toxic element such as lead or cadmium is not sufficient to indicate pollution: significant pollution arises when biota are unable to protect themselves against increased concentrations. Similar problems arise with many of the micronutrient elements with concentrations between 1 nM and 1 μM and which become toxic at elevated levels; in some cases (*e.g.*, copper) relatively small increases in concentration are sufficient to induce toxicity.

The seawater concentrations summarized in Figure 1 are the result of intricate biogeochemical cycles operating over timescales of thousands to millions of years. To a first approximation, mean seawater concentrations are controlled by geochemical processes such as adsorption on settling particle surfaces, while the distributions of many trace elements within the oceans are controlled by biological processes. Within other aquatic reservoirs (lakes, rivers, estuaries, *etc.*) residence times are shorter, and concentrations tend to be more variable and governed by a different balance of biogeochemical processes. However, any pollutant metal discharged to the aquatic environment enters a complex biogeochemical cycle, and predicting its fate requires multidisciplinary modelling of that biogeochemical cycle and its constituent physical, chemical, and biological processes. Such models are now under development: for example Harris *et al.*[1] have described a model for the dispersal of pollutants in the Tamar estuary. On a much larger scale, the requirement for detailed safety cases for radioactive waste disposal proposals has stimulated the development of radiological models predicting the potential dose to man over periods of millions of years for the longer lived nuclides.[2] As the requirement for predictive modelling of the fate and effects of pollutant discharges grows, it will be increasingly important for the models used to be based on a detailed knowledge of the individual processes involved.

3 CHEMICAL SPECIATION OF METAL POLLUTANTS

Natural waters are extremely complex chemical environments. Not only do they contain almost every element in the periodic table (Figure 1), but also dissolved organic matter of largely unknown composition (the terms fulvic acid and humic acid are used to give a veneer of chemical respectability), and colloidal and particulate material, both inanimate and living. The range of chemical processes affecting pollutant metals can be illustrated by the case of copper in natural waters. Copper dissolved at trace levels in distilled water is present at the cation Cu^{2+} at thermodynamic equilibrium. In a natural water this cation is subject to the range of interactions shown in Figure 2. The distribution of copper between the different chemical species and forms shown is termed its chemical speciation, and it is chemical speciation which holds the key to understanding the geochemical and biological reactivity of trace metals. The problem of chemical speciation can be tackled in two ways: by chemical modelling and by direct measurement on natural waters.

[1] J. R. W. Harris, A. J. Bale, B. L. Bayne, R. F. C. Mantoura, A. W. Morris, L. A. Nelson, P. J. Radford, R. J. Uncles, S. A. Weston, and J. Widdows, *Ecol. Modell.*, 1984, **22**, 253–284.
[2] A. A. Krol and D. Read, 'Investigation of the Requirements for a Unified Approach to Modelling Chemistry in the PRA of Radioactive Waste Disposal,' Report DoE/RW/87.003, Atkins Research and Development, Epsom, Surrey, 1986.

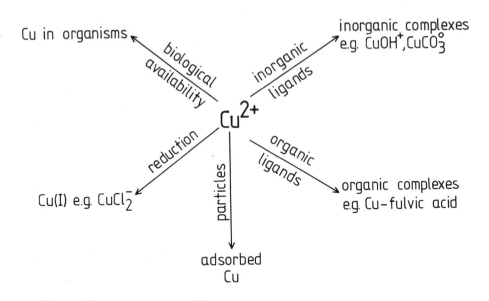

Figure 2   *Major processes involving trace metals in water, taking copper in seawater as an example*

Table 1   *Metal complexation and adsorption models for natural waters*

	Ligand Description	*Competition by other cations*	*State-of-the-Art*
Inorganic Complexation	small number of well-defined ligands	can be readily treated as side reactions of the ligand	accuracy of models is limited only by the thermodynamic data available
Organic complexation	polydisperse range of binding sites	empirical treatments only	modelling is based on empirical constants which are functions of pH, *etc.*
Adsorption (pure phases)	single type of binding sites in immobilized array	treated as side reactions	surface complexation modelling provides a successful approach
Adsorption (natural particles)	No working models are yet available; material comprises both inorganic particles and natural organic matter		empirical K_D values provide the best available parameterization

Equilibrium thermodynamic modelling provides a powerful means of predicting chemical speciation within a complex mixture, and the development of quantitative chemical models appropriate to multicomponent, multiphase systems has been a major preoccupation of aquatic chemists in recent decades. The accuracy of these models is dependent upon (i) identification and correct parameterization of all relevant processes, (ii) the accuracy of the parameters and input data used, and (iii) the correctness of the equilibrium assumption. Direct measurements on natural waters and related model systems therefore provide a complementary approach to modelling which (i) allows the thermodynamic models to be further developed and refined, (ii) allows non-equilibrium behaviour to be identified, and (iii) provides data which can be used for empirical state-of-the-art modelling in areas where detailed chemical models are not yet available. The discussion which follows will consider our understanding of the processes shown in Figure 2 largely in terms of modelling, and will also identify the problems which remain to be solved. We will consider first complexation and adsorption processes, which share a number of common features and problems as outlined in Table 1.

3.1 Inorganic Complexation

Descriptions of inorganic complexation are based on equilibrium thermodynamic models. Each possible complexation reaction is described by a thermodynamic equilibrium constant K°, *e.g.*, for the complex $Cu\,CO_3^0$

$$K^\circ(CuCO_3^0) = a(CuCO_3^0)/a(Cu^{2+})\,a(CO_3^{2-}) \qquad 1$$

where $a(X)$ represents the activity of the component X. For practical purposes we are more interested in the concentration $[X]$ than the activity $a(X)$, so we define a concentration product or conditional constant K^*

$$K^*(CuCO_3^0) = [CuCO_3^0]/[Cu^{2+}]\,[CO_3^{2-}]$$
$$= K^\circ(CuCO_3^0)\,\gamma(Cu^{2+})\,\gamma(CO_3^{2-})\,/\,\gamma(CuCO_3^0) \qquad 2$$

where $\gamma(X)$ is the activity coefficient of the component X. The activity coefficient effects are not trivial: for example $K^*(CuCO_3^0)$ in seawater is about 10 times smaller than K°. It follows that to characterize an inorganic complexation reaction accurately, we need to know either the conditional constant in the medium of interest, or the thermodynamic constant K° and the appropriate activity coefficients. Good activity coefficient models based on the comprehensive Pitzer equations are now available for most of the major and minor elements of seawater (concentrations $> 1\,\mu M$, Figure 1),[3] and the modelling approaches used are equally applicable to the trace elements. However, the detailed data necessary are not forthcoming in most cases, although Millero and Byrne[4] have successfully applied the Pitzer approach to lead chloride complexation. Trace

[3] F. J. Millero, *Thalassia Jugosl.*, 1982, **18**, 253–291.
[4] F. J. Millero and R. H. Byrne, *Geochim. Cosmochim. Acta*, 1984, **48**, 1145–1150.

metal speciation models are generally based on simpler activity coefficient models and relatively sparse data, but nevertheless show an encouraging degree of agreement: a comparison of models for copper in seawater is shown in Table 2. Such models are gradually improving as more objective methods are used to assess the available thermodynamic data, and as more data become available. Unfortunately, the complexes of most interest in natural waters are often those of the hydroxide and carbonate ion which present considerable problems in thermodynamic measurements, and consequently have not been studied in as much detail as 'easier' complexes such as those of the halides.

Table 2 *Inorganic chemical speciation models for copper in seawater*[a]

Reference	Cu^{2+}	OH^-	Cl^-	Ligand F^-	SO_4^{2-}	CO_3^{2-}
b	4	16	2	c	d	79
b	1	73	1	c	d	25
e	9	8	3	d	1	79
f	3	6	1	c	d	90[g]
h	4	6	i	i	i	86[j]

a figures show the percentage of copper bound with each ligand according to the calculations of the different authors
b D. Dyrssen and M. Wedborg, in 'Chemistry and Biogeochemistry of Estuaries,' ed. E. Olausson and I. Cato, John Wiley, Chichester, 1980, pp. 71–119: two different models proposed depending on the value chosen for $K^*(Cu(OH)_2)$
c not considered in this model
d <1% of total copper calculated to be bound to this ligand
e D. R. Turner, M. Whitfield, and A. G. Dickson, *Geochim. Cosmochim. Acta*, 1981, **45**, 855–881.
f J. L. Symes and D. R. Kester, *Mar. Chem.*, 1985, **16**, 189–211
g includes 6% $CuCO_3OH^-$
h R. H. Byrne and W. L. Miller, *Geochim. Cosmochim. Acta*, 1985, **49**, 1837–1844
i <4% of total copper calculated to be bound to this ligand
j includes 8% $CuCO_3OH^-$

These problems become particularly acute for the actinides such as plutonium which need to be modelled in connection with radioactive waste disposal assessments. Many of the thermodynamic data available in the literature have been measured in strong acid solution (usually in the range 2 M to 4 M), conditions far removed from those of fresh, estuarine, or sea waters, and often show large differences between different workers. This has stimulated a major international thermodynamic database project coordinated by the Nuclear Energy Agency of the OECD.[5] This project aims to collect all available thermodynamic data for selected actinide elements and other radionuclides, to identify the most reliable data in cases of conflict, and to recommend the best available conditional constants for use in modelling the elements concerned in natural waters. It is expected that this task will take several years for each element, and that the results will need updating as new information becomes available.

[5] A. B. Muller, *Rad. Waste Man. Nucl. Fuel Cycle*, 1985, **6**, 131–141.

3.2 Organic Complexation

For the small fraction of organic matter which has been identified as low molecular weight compounds such as amino acids, complexation of trace metals can be modelled in exactly the same way as inorganic complexation discussed above. However, the bulk of dissolved organic matter in natural waters consists of polydisperse material of unknown structure which is operationally divided into humic and fulvic acids. A widely used method for investigating the metal-binding capability of natural waters is complexing capacity titration.[6] This consists of titrating a trace metal (usually copper) into a natural water sample, and measuring the concentration of uncomplexed metal. From the resulting titration curve an estimate of the organic ligand concentration and its stability constant can be obtained. These estimates should be considered as operational parameters rather than true concentrations or stability constants for two reasons. Firstly, the titrations cover only a limited range of metal concentrations and thus provide information on a similarly limited range of the total binding capacity. In addition, problems arising from the selectivity of anodic stripping voltammetry, the most popular technique for detection of the uncomplexed metal, have led to some controversy about the interpretation of complexing capacity titrations.[7,8]

Laboratory measurements have shown that humic and fulvic acids can form strong complexes with a number of trace metals,[9,10] in some cases dominating the dissolved trace metal speciation. Modelling of this complexation necessarily proceeds on a more pragmatic basis than that of inorganic complexation where the identity and concentration of each ligand can be defined. The two major problems to be addressed are the polydispersity of the binding sites, and competition by major ions for the binding sites (Table 1). Laboratory studies of complexation by fulvic acid have concentrated mainly on copper, and have largely been confined to fixed experimental conditions (pH, ionic strength). Under these conditions, competition by the hydrogen ion and the major salt cations for binding sites is constant and is thus implicitly included in the empirical binding constants measured. The polydispersity of the binding sites nevertheless remains and is manifested as an apparent dependence of the binding constant on the degree of binding. Polydispersity can also be viewed as a probability distribution of binding strengths, natural organics being characterized by a distribution function of finite width in contrast to the single value exhibited by discrete ligands (Figure 3). Although a number of workers have attempted to map such distribution functions from titration data, empirical models which postulate a small number of discrete binding sites are both simpler to use and provide a better statistical fit to the experimental data.[11,12]

[6] T. A. Neubecker and H. E. Allen, *Water Res.*, 1983, **17**, 1–14.

[7] G. A. Bhat, J. H. Weber, J. R. Tuschall, and P. L. Brezonik, *Anal. Chem.*, 1982, **54**, 2116–2117.

[8] M. S. Shuman, P. L. Brezonik, and J. R. Tuschall, *Anal. Chem.*, 1982, **54**, 998–1001.

[9] R. F. C. Mantoura, A. G. Dickson, and J. P. Riley, *Est. Cstl. Mar. Sci.*, 1978, **6**, 387–408.

[10] G. R. Choppin and B. Allard, in 'Handbook on the Physics and Chemistry of the Actinides,' ed. A. J. Freeman and C. Keller, Elsevier, Amsterdam, 1985, pp. 407–429.

[11] S. E. Cabaniss, M. S. Shuman, and B. J. Collins, in 'Complexation of Trace Metals in Natural Waters,' ed. C. J. M. Kramer and J. C. Duinker, Nijhoff/Junk, The Hague, 1984, pp. 165–179.

[12] D. R. Turner, M. S. Varney, M. Whitfield, R. F. C. Mantoura, and J. P. Riley, *Geochim. Cosmochim. Acta*, 1986, **50**, 289–297.

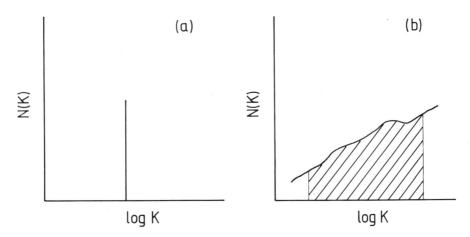

Figure 3   *Concentration of binding sites of a given strength $N(K)$ as a function of binding strength
$\log K$. (a) small monodentate ligand (e.g., carbonate) with concentration given by the
height of the vertical line, (b) polydisperse ligands (e.g., natural organic matter) with
concentration of ligand in a given $\log K$ range related to the area under the curve*

Organic binding models valid for a single pH and ionic medium are, however,
of limited practical application, and more general models are required which
take account of competition by the proton and major cations. Cabaniss and
Shuman[13] have achieved some success with a purely empirical approach to pH
dependence, and it has been proposed that ionic medium and pH dependence be
treated by modelling the organic matter as a polyelectrolyte gel.[14] Until these
ideas are developed more fully, we lack an effective means of modelling metal–
organic interactions under a range of chemical conditions.

3.3   Adsorption

This is a key process in the cycling of trace metals in natural waters on a wide
range of timescales. Adsorption onto sinking particles has been identified as a
major control on the composition of seawater over timescales of thousands to
millions of years.[15] The same process also has a major influence on the short term
fate of metal pollutants in dynamic environments such as estuaries.[16]

At the empirical level, adsorption is characterized in terms of a partition
coefficient K_D defined by

$$K_D = \frac{\text{adsorbed metal/mol.kg}^{-1}}{\text{dissolved metal/mol.l}^{-1}} \qquad 3$$

[13] S. E. Cabaniss and M. S. Shuman, *Geochim. Cosmochim. Acta*, 1988, **52**, 185–194.
[14] J. A. Marinsky, in 'Aquatic Surface Chemistry: Chemical Processes at the Particle–Water
Interface,' ed. W. Stumm, John Wiley, New York, 1987, pp. 49–81.
[15] M. Whitfield and D. R. Turner, in 'Aquatic Surface Chemistry: Chemical Processes at the
Particle–Water Interface,' ed. W. Stumm, John Wiley, New York, 1987, pp. 457–493.
[16] A. W. Morris, *Sci. Tot. Environ.*, 1986, **49**, 297–304.

Much effort has been devoted to the measurement of K_D values for incorporation into models of pollutant dispersion, and this remains the most effective means of modelling adsorption at the present time. However, K_D values are by no means constants, being functions of a wide range of variables such as pH, salinity, and particle surface. Although some of these variations can be tackled by an extension of the empirical approach, such as measuring K_D as a function of salinity or pH, an alternative strategy is the development of detailed chemical models analogous to those described above for dissolved phase complexation.

Quantitative chemical models of adsorption processes are still largely limited to adsorption onto pure solid phases. An approach which has proved very effective is that of surface complexation, which assumes the solid surface to consist of an array of immobilized hydroxyl groups. These can gain or lose protons, and the resulting charged groups can form surface complexes with dissolved ions of the opposite charge, *e.g.*,

$$>\text{S—OH} \xrightarrow{-\text{H}^+} >\text{S—O}^- \xrightarrow{\text{Cu}^{2+}} >\text{S—O}^-\text{—Cu}^{2+} \qquad 4$$

$$>\text{S—OH} \xrightarrow{\text{H}^+} >\text{S—OH}_2^+ \xrightarrow{\text{CO}_3^{2-}} >\text{S—OH}_2^+\text{CO}_3^{2-} \qquad 5$$

where $>\text{S}$ represents a surface site. The reactions are therefore expressed in the same form as those occurring in solution, except that in dealing with surfaces it is necessary to take account of the development of electric charge at the solid surface. This is achieved by dividing the stability constant, K defining each surface reaction into two parts:

$$K = K_{int} K_{elect} \qquad 6$$

where K_{int} is the value of K at zero surface charge, and K_{elect} is an electrostatic term which takes account of the effect of the electric potential at the particle surface on the free energy of the aqueous ions: detailed discussions of the theory involved can be found elsewhere.[17,18] This approach has proved to be very successful in modelling the adsorption of metals onto pure solid phases in simple ionic media,[19] and has been extended with some success to pure solid phases in the more complex ionic medium of seawater.[20,21] Real particles in natural waters present two further problems (i) they are rarely pure phases, and (ii) they are often coated with natural organic matter, which draws in all the problems of the previous section. These aspects have not yet been incorporated into detailed chemical models. Although co-adsorption of organic ligands and trace metals onto pure solid phases, and the resulting ternary surface complexes, have been

[17] J. A. Davis, R. O. James, and J. O. Leckie, *J. Coll. Interfac. Sci.*, 1978, **63**, 480–499.
[18] J. C. Westall, in 'Aquatic Surface Chemistry: Chemical Processes at the Particle–Water Interface,' ed. W. Stumm, John Wiley, New York, 1987, pp. 3–32.
[19] J. A. Davis and J. O. Leckie, *J. Coll. Interfac. Sci.*, 1978, **67**, 90–107.
[20] L. S. Balistrieri and J. W. Murray, *Geochim. Cosmochim. Acta*, 1982, **46**, 1253–1265.
[21] A. L. Sanchez, J. W. Murray, and T. H. Sibley, *Geochim. Cosmochim. Acta*, 1985, **49**, 2297–2307.

investigated and modelled in some detail,[22,23] modelling adsorption onto natural particles remains a challenging objective.

3.4 Redox Processes

We have thus far considered only a single oxidation state. However, many trace elements are capable of existing in more than one oxidation state, and a change of oxidation state will usually be accompanied by a significant change of chemical and biological properties. This is an area where non-equilibrium conditions are particularly important, since many redox transformations are extremely slow. Most natural waters contain dissolved oxygen and are therefore strongly oxidizing: the great majority of the elements are present in their highest oxidation states in seawater at thermodynamic equilibrium. However, a significant proportion of many such elements is found in a lower, non-equilibrium oxidation state, which can be formed in a variety of ways. Biological processing can result in metastable species such as arsenic(III) in place of arsenic (V), and a whole range of alkylated metals which can be highly toxic: methylmercury compounds are particularly notorious. Dissolved manganese(II) is produced by the reduction of particulate manganese(III) and (IV) in anoxic interstitial and bottom waters, and when mixed into oxic waters persists for long periods owing to its slow oxidation kinetics. Photochemical processes involving natural organic matter at the water surface can also result in the production of reduced manganese[24] and hydrogen peroxide, which has further consequences for trace metal redox chemistry.[25] It has been shown that the hydrogen peroxide present in seawater is capable of maintaining a significant proportion of dissolved copper in the copper(I) oxidation state, which is stabilized by the high chloride concentration resulting in relatively slow oxidation kinetics. Recent measurements have revealed copper(I) to account for up to 5–10% of total copper in surface sea waters.[26] Thus equilibrium speciation models for copper(II) in sea water do not necessarily tell the whole story.

An extremely topical example of redox effects is provided by plutonium in the Irish Sea. As mentioned above, this element has a complicated redox chemistry. At equilibrium, plutonium is present in seawater in an oxidized form, most probably Pu(V) rather than Pu(VI),[27] which associates with particles considerably less strongly than the reduced forms Pu(III)/Pu(IV). Investigations of the fate of plutonium discharged into the Irish Sea from Sellafield on the North West coast of England have revealed that much of the plutonium has been immobilized in adsorbed form on a large mud patch on the seafloor.[28] Redox chemistry is thought to play an important role here with the adsorbed plutonium being in the reduced Pu(III)/Pu(IV) form: it has been shown that natural organic matter is an efficient reducing agent for plutonium.[27]

[22] A. C. M. Bourg and P. W. Schindler, *Chimia*, 1978, **32**, 166–168.
[23] A. C. M. Bourg and P. W. Schindler, *Chimia*, 1979, **33**, 19 21.
[24] W. G. Sunda, S. A. Huntsman, and G. R. Harvey, *Nature*, 1983, **301**, 234–236.
[25] J. W. Moffett and R. G. Zika, *Mar. Chem.*, 1983, **13**, 239–251.
[26] J. W. Moffett and R. G. Zika, *Geochim. Cosmochim. Acta*, 1988, **52**, 1849–1857.
[27] G. R. Choppin, R. A. Roberts, and J. W. Morse, in 'Organic Marine Geochemistry,' ACS Symposium Series 305, ed. M. L. Sohn, American Chemical Society, Washington DC, 1986, pp. 382–388.

3.5 Biological Availability

A great deal of experimental evidence has confirmed that the uptake of trace elements by aquatic organisms depends on the chemical speciation of the trace element, and hence on the whole range of processes discussed above. Copper is perhaps the best studied element in this respect, and many workers have concluded that the free ion Cu^{2+} is the biologically available species, although hydroxide complexes have also been implicated.[29] Similar conclusions have also been reached for other divalent ions, prompting the temptingly simple conclusion that biological availability can be equated with free cation concentrations. Matters are not, however, so simple. It has been shown that copper complexes with a range of low molecular weight organic ligands are also biologically available,[30] presumably through their lipid solubility, and a range of other metals whose free cation concentrations are negligible are taken up as other chemical species. For example, acidification of clay-rich fresh waters brings aluminium into solution to a degree which becomes toxic in the pH range 5.5 to 6. The cation Al^{3+} is certainly not responsible since its concentration is negligible at this pH, and the complex $Al(OH)_2{}^+$ has been proposed as the bioavailable form.[31] Despite these complications, the link between chemical speciation and biological availability is clear and should help guide the assessment of water quality.

4 MONITORING OF METAL POLLUTANTS

The previous section has outlined the range of chemical processes involving pollutant metals. We will now consider how a knowledge of those processes can be used to guide pollution monitoring and water quality assessments. There is increasing concern that measurement of total pollutant metal levels does not lead to a reliable estimate of water quality implications. This is not simply a one-way effect where it is easy to err on the side of caution. For example, total copper measurements, interpreted on the simple assumption that inorganic copper(II) dominates, will often overestimate the likely toxic effects to the biota, since much of the copper will be bound in biologically unavailable organic complexes. The magnitude of this effect will of course depend on the concentration and nature of the natural organic matter present. Conversely, total arsenic or mercury measurements, interpreted on the simple assumption that inorganic arsenic(V) and mercury(II) dominate, may conceal relatively small proportions of highly toxic arsenic(III) or methylmercury compounds and thus underestimate the toxic effect. The clear implication is that monitoring procedures are required which can detect particular chemical species or groups of species in addition to the total burden of pollutant metal; the target chemical species will in most cases

[28] R. J. Pentreath, D. S. Woodhead, P. J. Kershaw, D. F. Jeffries, and M. B. Lovett, *Rapp. P.-v. Reun. Cons. Int. Explor. Mer*, 1986, **186**, 60–69.

[29] D. R. Turner, in 'Metal Ions in Biological Systems,' ed. H. Sigel, Marcel Dekker, New York, 1984, pp. 137–164.

[30] T. M. Florence, B. G. Lumsden, and J. J. Fardy, *Anal. Chim. Acta*, 1983, **151**, 281–295.

[31] S. Helliwell, G. E. Batley, T. M. Florence, and B. G. Lumsden, *Environ. Technol. Letts.*, 1983, **4**, 141–144.

correspond to the biologically available fraction. As discussed above, this will vary from one metal to another, although the free cations will be important for many divalent metals.

4.1 Chemical Measurements

Chemical measurements provide the most direct method of assessing chemical speciation, but are fraught with difficulties.[29] There is as yet no technique sensitive to free cations in the pM or nM concentration ranges: ion-selective electrodes, which have the requisite selectivity, are sensitive only down to a concentration level of about 1 μM at best. Dynamic electrochemical techniques such as anodic stripping voltammetry have been proposed as suitable sensors: these certainly have the required sensitivity, but lack the necessary selectivity since they respond to the free metal ion together with rapidly dissociating complexes (most inorganic complexes together with some organic complexes), and therefore sense a different fraction from that required. Similar restrictions apply to ion-exchange resins. Ligand exchange techniques allied to cathodic stripping analysis or column chromatography may, however, provide useful alternatives after further development. In contrast, good speciation-sensitive measurement techniques are now available for the 'hydride forming elements' of groups IV, V, and VI of the periodic table. Selective reduction of these elements to the hydride or alkyl hydride followed by gas chromatography allows efficient estimation of different oxidation states (*e.g.*, As(III)/As(V) and Te(IV)/Te(VI)) and of a range of metal alkyls.[32]

A further problem which must be faced when a suitable technique is identified is that of accuracy. A parallel can be drawn with the problem of measuring low trace metal concentrations in ocean waters. For many years there was little agreement between different authors, and international intercalibration exercises produced levels of disagreement which can only be described as embarrassing. However, developments in clean handling and analysis techniques led to improved agreement between different authors, and to the measurement of 'oceanographically consistent' trace metal profiles, some of which correlate strongly with nutrient profiles.[33] Any technique which is successful at measuring a trace metal fraction which has been identified as biologically available will be at the limits of our analytical capabilities, as were oceanic trace metal measurements, and will need to go through a comparable sequence of intercalibrations before it can be considered as the basis of water quality management decisions.

4.2 Modelling

In view of the problems involved in direct chemical measurements, modelling must be considered as a possible alternative. In well-defined media, modelling is

[32] M. O. Andreae, in 'Trace Metals in Seawater,' ed. C. S. Wong, E. A. Boyle, K. W. Bruland, J. D. Burton, and E. D. Goldberg, Plenum Press, New York, 1983, pp. 1–20.
[33] C. S. Wong, E. A. Boyle, K. W. Bruland, J. D. Burton, and E. D. Goldberg (ed.), 'Trace Metals in Seawater,' Plenum Press, New York, 1983, 920 pp.

a valid and reasonably accurate method of determining chemical speciation: several studies of biological availability have used chemical modelling to determine chemical speciation in the test medium. However, in natural waters significant problems can arise from interactions such as those with natural organic matter and particles for which adequate models are not yet available, and also from non-equilibrium behaviour. Modelling is therefore restricted at present to situations where these processes play only a minor role, although we can expect that improvements in quantitative process models (Figure 2, Table 1) will widen this scope.

4.3 Bioassay Measurements

Bioassay is arguably the most direct water quality indicator of all, since it assesses directly the ability of the sample to support life. Copper again provides a convenient example. It has been shown that the ^{14}C-glucose turnover rate of a marine bacterial isolate is a direct function of the free copper concentration in its growth medium,[34] and techniques based on this principle have been used successfully to measure free copper concentrations at the pM level in complexing capacity titrations both in laboratory experiments and in natural samples using natural bacterial populations.[35,36] The drawback is the lack of selectivity: metabolic rates reflect a range of environmental factors including pollutants, so that it is not possible to infer the concentration or even the presence of a particular pollutant from bioassay measurements in a monitoring application.

Bioassay can, however, play an important role where specific biological responses to individual pollutants can be observed. The response of the dog whelk *Nucella lapillus* to tributyltin (TBT) is noteworthy in this respect. It has been shown[37,38] that TBT concentrations as low as 3 ng l^{-1} Sn are capable of inducing imposex (development of male characteristics) and sterility in females. Other alkyltins tested induced either minor effects or none at all, and dog whelk imposex provides an effective bioassay for TBT. The incidence of dog whelk imposex and sterility was found to be so high that TBT has now been banned as an antifoulant for small boats in the UK.

5 CONCLUDING REMARKS

Our knowledge of the chemistry of metal pollutants in natural waters is growing steadily. The goal is an ability to predict the fate and effects of pollutant discharges, complemented by an ability to monitor the dispersal of discharges in terms of water quality standards which take account of the chemistry and biology of individual elements. In terms of modelling, which must form the basis of our ability to predict the fate of pollutant discharges into natural waters, two

[34] P. A. Gillespie and R. F. Vaccaro, *Limnol. Oceanogr.*, 1978, **23**, 543–548.
[35] W. G. Sunda and P. A. Gillespie, *J. Mar. Res.*, 1979, **37**, 761–777.
[36] W. G. Sunda and R. L. Ferguson, in 'Trace Metals in Seawater,' ed. C. S. Wong, E. A. Boyle, K. W. Bruland, J. D. Burton, and E. D. Goldberg, Plenum Press, New York, 1983, pp. 871–891.
[37] P. E. Gibbs, P. L. Pascoe, and G. R. Burt, *J. Mar. Biol. Assoc. UK*, 1988, **68**, 715–731.
[38] G. W. Bryan, P. E. Gibbs, and G. R. Burt, *J. Mar. Biol. Assoc. UK*, 1988, **68**, 733–744.

complementary approaches are required. Empirical parameterization of the key processes involved, which may necessarily be somewhat rough and ready in places, allows the development of practical models, although the limitations of these models must be clearly understood. At the same time, it is vital to continue research into each of the chemical processes involved in order to develop improved and more accurate parameterization to feed into the next generation of models. In terms of monitoring techniques and water quality standards, considerable effort will be required in improving our knowledge of which chemical species are bioavailable, and in developing techniques with the appropriate sensitivity and selectivity to measure them.

CHAPTER 3

Sewage and Sewage Sludge Treatment

J. N. LESTER

1 INTRODUCTION

It is estimated that the volume of water used daily in England and Wales (exclusive of water abstracted for cooling purposes) amounts to 5000×10^6 gal (23×10^6 m^3) or approximately 95 gal (430 l) per capita per day. Domestic use accounts for nearly 1800×10^6 gal (8×10^6 m^3) of this average daily total. Nearly all the water used domestically and approximately 1500×10^6 gal (6.8×10^6 m^3) of the water used by industry each day is discharged to the sewers, yielding a total sewage flow of 3100×10^6 gal (14.1×10^6 m^3) or about 60 gal (275 l) per capita per day.

The sewage from approximately 44 million people in England and Wales is treated by conventional wastewater treatment processes, that from about a further 6 million people is discharged without treatment to the sea and some 1 to 2 million people are not connected to the sewerage system. To achieve this degree of wastewater treatment requires some 5 000 sewage treatment works serving populations in excess of 10 000; these are distributed throughout the ten Water Authorities, soon to be privatized, in England and Wales. The sewerage systems which carry the sewage to the site of treatment, or point of discharge, are of two types. Foul sewers carry only domestic and industrial effluent. In areas serviced in this way there are entirely separate systems for the collection of stormwater which is discharged directly to natural water courses. However, in older towns and cities considerable use has been made of combined foul and stormwater systems. The use of combined sewage systems leads to very significant changes in the flow of sewage during storms. However, even in foul sewers significant changes in the flow occur due to variations in the pattern of domestic and industrial water usage which is essentially diurnal, and at its greatest during the day. Infiltration will also influence the flow in the sewage system. Although a properly laid sewer is watertight when constructed, ground movement and aging may allow water to enter the sewer if it is below the water table. The combined total of average daily flows to a sewage treatment works is called the dry weather flow (DWF). The DWF is an important value in the design and operation of the sewage treatment works and other flows are expressed in terms of it. DWF is defined as the daily rate of flow of sewage (including both domestic and trade waste), together with infiltration, if any, in a sewer in dry weather. This may be measured after a period of 7 consecutive days during which the rainfall has not exceeded 0.25 mm.

The DWF may be calculated from the following formula:

$$DWF = PQ + I + E$$

where, P = population served
Q = average domestic water consumption $(1d^{-1})$
I = rate of infiltration $(1d^{-1})$
E = volume (in litres) of industrial effluent discharged to sewers in 24 hours

1.1 Objectives of Sewage Treatment

Water pollution in the United Kingdom was already a serious problem by 1850. It is probable that the early endeavours to control water pollution were considerably stimulated by the state of the lower reaches of the River Thames which at the point where it passed the Houses of Parliament was grossly polluted. An early solution to these problems was sought through the construction of interceptor sewers. These collected all the sewage draining to the River Thames and carried it several miles down the river before discharging it to the estuary on the ebb tide. From there it moved towards the sea and in so doing received greater dilution. Despite these measures and the passing of the first Act of Parliament to control water pollution in 1876 the situation continued to deteriorate. The requirement for, and the objectives of, sewage treatment were first outlined by the Royal Commission on Sewage Disposal (1898–1915). The objectives of sewage treatment have developed significantly since this report; however, the standards described then are still applicable in many areas and this report provided the framework around which the United Kingdom wastewater industry has developed.

Originally the objective of sewage treatment was to avoid pestilence and nuisance (disease and odour) and to protect the sources of potable supply.

During sewage treatment disease-causing organisms may be destroyed or concentrated in the sludges produced; similarly offensive materials may be concentrated in the sludges or biodegraded. As a consequence the quantities of these agents present in the sewage effluent is much less than in the untreated sewage and their dilution in the receiving water far greater. The benefits of sewage treatment are not limited to greater dilution however, since each receiving water has a certain capacity for 'self purification.' Providing sewage treatment reduces the burden of polluting material to a value less than this capacity then the ecosystem of the receiving water will complete the treatment of the residual materials present in the sewage effluent. Thus sewage treatment in conjunction with the selection of appropriate points for sewage effluent discharge has resulted in the elimination of water-borne disease in the UK and many other advanced countries. However, as the population has expanded and become urbanized with a concomitant development of water-consuming industries an additional requirement has been placed upon sewage treatment.

It is now the objective of sewage treatment in many parts of the UK to produce a sewage effluent which after varying degrees of dilution and self purification is suitable for abstraction for treatment to produce a potable supply. This indirect re-use affects some 30% of all water supplies in the UK.

1.2 The Importance of Water Re-use

That the United Kingdom practises indirect re-use to a greater extent than most other countries may appear surprising given the annual rainfall. Indeed that re-use should be important in global terms given the abundance of water on the earth's surface may also be considered improbable in all but the most arid regions. However, two important factors readily explain this situation; a vast amount of the available water is too saline to be used as a potable supply (the salinity is too costly to remove in all but the most extreme cases) and secondly the non-uniform distribution of the population and the available water supply. The available water supply is determined by the rainfall, the ability of the environment to store water (essentially the size of lakes and rivers which are small in the United Kingdom) and their location, *i.e.*, Wales has an abundance of suitable water supplies, but limited population, whilst South East England has a large population with limited water resources.

It has been estimated that of the water falling on the United Kingdom 50% is not available for use as a result of run-off to the sea. Of the remainder approximately 17% is utilized. Current predictions suggest that by 2000 AD the amount of water used will have doubled. Thus the potential reserves are very limited. However, because demand and supply are not geographically proximate re-use is already essential. As a consequence the traditional concept of water supply employing single-purpose reservoirs impounding unused river water has been abandoned in favour of multi-purpose schemes designed to permit repeated use of the water before it reaches the sea. In these schemes sewage treatment plays a vital role in addition to being an integral part of the hydrological cycle.

1.3 Criteria for Sewage Treatment

Sewage is a complex mixture of suspended and dissolved materials, both categories constitute organic pollution. The strength of sewage and the quality of sewage effluent are described in terms of their suspended solids (SS) and biochemical oxygen demand (BOD); these two measures were either proposed or devised by the Royal Commission (1898–1915).

The SS are determined by weighing after the filtration of a known volume of sample through a standard glassfibre filter paper, the results are expressed in $mg\,l^{-1}$.

Dissolved pollutants are determined by the BOD they exert when incubated for 5 days at 20 °C. Samples require appropriate dilution with oxygen saturated water and suitable replication. The oxygen consumed is determined and the results again expressed in $mg\,l^{-1}$.

The two standards for sewage effluent quality proposed by the Royal Commission were for no more than $30\,mg\,l^{-1}$ of suspended solids and $20\,mg\,l^{-1}$ for BOD, the so called 30 : 20 standard. The Royal Commission envisaged that the effluent of this standard would be diluted 8 : 1 with clean river water having BOD of $2\,mg\,l^{-1}$ or less. This standard was considered to be the normal minimum requirement and was not enforced by statute because the character and use of rivers varied so greatly. It was intended that standards would be introduced locally as required. For example, a river to be used for abstraction of potable supplies would require a higher standard such as the 10 : 10 standard imposed by the Thames Conservancy. Whilst other countries which are members of the European Economic Community have adopted 'uniform emission standards,' that is, the same quality of effluent regardless of the state or use of the river, the United Kingdom has continued with its pragmatic approach whereby effluent standards are set depending on the 'water quality objectives' of the river, which in turn is determined by its function or use. In the 1970's with the reorganization, the water industry's reliance solely on the 30 : 20 standard was abandoned, although this standard is probably still the most commonly applied. Sewage treatment now attempts to consistently produce an effluent with a quality superior to its 'Legal Consent' and attempts to achieve an 'Operating Target,' frequently half the Legal Consent. In addition considerable importance has been placed upon the concentration of ammonia in the effluent. In the case of a works attempting to nitrify the effluent (see Section 2.3.3) the ammonia concentration is frequently limiting. Typical Legal Consent and Operating Targets are outlined in Table 1. It is evident that the Operating Targets included in Table 1 are the same as the Royal Commission 30 : 20 standard.

Table 1   *Legal Consent and Operating Target values for a conventional two stage sewage treatment works*

Parameter	Legal consent Value $(mg\,l^{-1})$	Operating target Value $(mg\,l^{-1})$
SS	50	30
BOD	35	20
ammonia	25	12

1.4 Composition of Sewage

Domestic sewage contains approximately $1000\,mg\,l^{-1}$ of impurities of which about two-thirds are organic. Thus sewage is 99.9% water and 0.1% total solids upon evaporation. When present in sewage approximately 50% of this material is dissolved and 50% suspended (see Figure 1). The main components are: nitrogenous compounds—proteins and urea; carbohydrates—sugars, starches, and cellulose; fats—soap, cooking oil, and greases. Inorganic components include chloride, metallic salts, and road grit where combined sewerage is used. Thus sewage is a dilute, heterogeneous medium which tends to be rich in nitrogen.

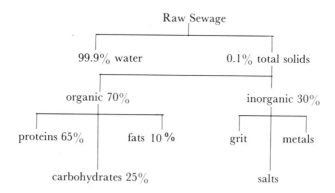

Figure 1 *Composition of a typical raw sewage*

2 SEWAGE TREATMENT PROCESSES

Conventional sewage treatment is a three stage process including preliminary treatment, primary sedimentation, and secondary (biological) treatment; these are presented schematically in Figure 2. In addition some form of sludge treatment facility is frequently employed, typically anaerobic digestion (see Section 3.1).

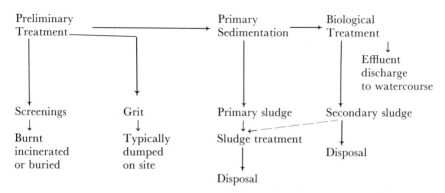

Figure 2 *Flow diagram of a conventional Sewage Treatment works*

2.1 Preliminary Treatment

These treatment processes are intended to remove the larger floating and suspended materials. They do not make a significant contribution to reducing the polluting load, but render the sewage more amenable to treatment by removing large objects which could form blockages or damage equipment.

Floating or very large suspended objects are frequently removed by bar screens, these consist of parallel rods with spaces between them which vary from 40 to 80 mm, through which the influent raw sewage must pass. Material which accumulates on the screen may be removed manually with a rake at small works, but on larger works some form of automatic raking would be used. The material removed from the screens contains a significant amount of putrescible organic matter which is objectionable in nature and may pose a disposal problem. Typically the material is buried or incinerated and less frequently burnt.

If screens have been used to remove the largest suspended and virtually all the floating objects, then it only remains to remove the small stones and grit, which may otherwise damage pumps and valves, to complete the preliminary treatment. This is most frequently achieved by the use of constant velocity grit channels. The channels utilize differential settlement to remove only the heavier grit particles whilst leaving the lighter organic matter in suspension. A velocity of 0.3 ms^{-1} is sufficient to allow the grit to settle whilst maintaining the organic solids in suspension. If the grit channels are to function efficiently the velocity must remain constant regardless of variation of the flow to the works (typically between 0.4 and 9 DWF). This is achieved by using channels with a parabolic cross section controlled by venturi flumes. The grit is removed from the bottom of the channel by a bucket scraper or suction, and organic matter adhering to the grit is removed by washing with the wash water being returned to the sewage. Small sedimentation tanks from which the sewage overflows at such a rate that only grit will settle out may also be used. These are compact and by the introduction of air on one side a rotary motion can be induced in the sewage which washes the grit *in situ*. However, these tanks do not cope with the variation in hydraulic load in such an elegant and effective manner as the grit channel.

To avoid the problems associated with the disposal of screenings comminutors are frequently employed in place of screens. Unlike screens which precede grit removal the comminutors are placed downstream of the grit removal process. The comminutors shred the large solids in the flow without removing them. As a result they are reduced to a suitable size for removal during sedimentation. Comminutors consist of a slotted drum through which the sewage must pass. The drum slowly rotates carrying material which is too large to pass through the drum towards a cutting bar upon which it is shredded before it passes through the drum.

The total flow reaching the sewage treatment works is subjected to both these preliminary treatment processes. However, the works is only able to give full treatment up to a maximum flow of 3 DWF. When the flow to the works exceeds this value the excess flows over a weir to the storm tanks which are normally empty. If the storm is short, no discharge occurs and the contents of the tanks are pumped back into the works when the flow falls below 3 *DWF*. If the storm is prolonged then these tanks will begin to discharge to a nearby watercourse, inevitably causing some pollution. However, this excess flow has been subjected to sedimentation which removes some of the polluting material. Moreover, as a consequence of the storm, flow in the watercourse will be high giving greater dilution.

2.2   Primary Sedimentation

The raw sewage (containing approximately 400 mg l^{-1} SS and 300 mg l^{-1} BOD) at a flowrate of 3 *DWF* or less and with increased homogeneity as a result of the preliminary treatment processes enters the first stage of treatment which reduces its pollutant load, primary sedimentation, or mechanical treatment. Circular (radial flow) or rectangular (horizontal flow) tanks equipped with mechanical sludge scraping devices are normally used (see Figure 3). However, on small works hopper bottom tanks (vertical flow) are preferred; although more expensive to construct these costs are more than offset by savings made as a result of eliminating the requirement for scrapers (see Figure 3).

Removal of particles during sedimentation is controlled by the settling characteristics of the particles (their density, size, and ability to flocculate), the retention time in the tank (h), the surface loading (m^3 m^{-2} d^{-1}) and to a very limited degree the weir overflow rate (m^3 m^{-1} d^{-1}). Retention times are generally between 2 and 6 h; however, the most important design criterion is the surface loading, typical values would be in the range 30 to 45 m^3 m^{-2} d^{-1}. The surface loading rate is obtained by dividing the volume of sewage entering the tank each day (m^3 d^{-1}) by the surface area of the tank (m^2). The retention time may be fixed independently of the surface loading by selection of the tank depth, typically 2 to 4 m, which increases the volume without influencing the surface area. Because they strongly influence the value for surface loading selected, the nature of the particles in the sewage is one of the most important factors in determining the design and efficiency of the sedimentation tank. Of the three factors mentioned before, flocculation is perhaps the most significant.

Four different types of settling can occur:

Class 1 Settling: settlement of discrete particles in accordance with theory (Stokes' Law).
Class 2 Settling: settlement of flocculant particles exhibiting increased velocity during the process.
Zone Settling (Hindered Settlement): at certain concentrations of flocculant particles, the particles are close enough together for the interparticulate forces to hold the particles fixed relative to one another so that the suspension settles as a unit.
Compressive Settling: at high solids concentrations the particles are in contact and the weight of the particles is in part supported by the lower layer of solids.

During primary sedimentation settlement is of the Class 1 or 2 types. However, in secondary sedimentation (see Section 2.4) zone or hindered settlement may occur. Compressive settlement only occurs in special sludge thickening tanks.

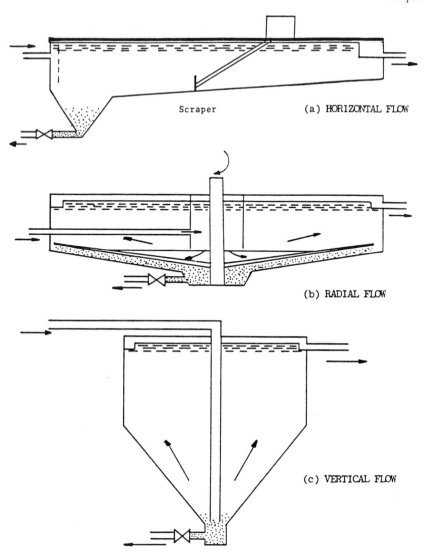

Figure 3 *Types of sedimentation tank*

Primary sedimentation removes approximately 55% of the suspended solids and because some of these solids are biodegradable the BOD is typically reduced by 35%. The floating scum is also removed and combined with the sludge. As a result the effluent from the primary has a SS of approximately 150 mg l^{-1} and a BOD of approximately 200 mg l^{-1}. This may be acceptable for discharge to the sea or some estuaries without further treatment. The solids are concentrated into the primary sludge which is typically removed once a day under the influence of hydrostatic pressure.

2.3 Secondary (Biological) Treatment

There are two principal types of biological sewage treatment:

(i) The percolating filter (also referred to as a trickling or biological filter).
(ii) Activated sludge treatment.

Both types of treatment utilize two vessels, a reactor containing the micro-organisms which oxidize the BOD, and a secondary sedimentation tank, which resembles the circular radial flow primary sedimentation tank, in which the micro-organisms are separated from the final effluent.

The early development of biological sewage treatment is not well documented. However, it is established that the percolating filter was developed to overcome the problems associated with the treatment of sewage by land at 'sewage farms,' where large areas of land were required for each unit volume of sewage treated. It was discovered that approximately 10 times the volume of sewage could be treated in a given area per unit time by passing the sewage through a granular medium supported on underdrains designed to allow the access of air to the microbial film coating the granular bed.

The origins of the percolating filter are present in land treatment and its development was an example of evolution. The second and probably predominant form of biological sewage treatment, the activated sludge process, arose spontaneously and represents an entirely original approach. This process involves the aeration of freely suspended flocculant bacteria, 'the activated sludge floc' in conjunction with settled sewage which together constitute the 'mixed liquor.' Activated sludge treatment continues the trend established by the change from land treatment to the percolating filter in that at the expense of higher operating costs it is possible to treat very much larger volumes of sewage in a smaller area.

The activated sludge process is probably the earliest example of a continuous bacterial (microbial) culture deliberately employed by man, and certainly the largest used to date. Development of the activated sludge process was announced by its originators Fowler, Ardern, and Lockett in 1913, based upon their research at the Davyhulme Sewage Treatment Works, Manchester. These scientists very generously did not patent the process to facilitate its rapid and widescale application.

Development of these two forms of biological sewage treatment has been largely empirical and undertaken without the benefit of information about the fundamental principles of continuous bacterial growth, which began to be developed from the late 1940's when Monod published his work on continuous bacterial growth, although the relevance was not perceived until approximately 10 years later. This lack of microbiological knowledge is highlighted by the fact that the role of micro-organisms in the activated sludge process was not fully accepted until after 1931, prior to this it was accepted by several workers that coagulation of the sewage colloids was the principal mechanism in the activated sludge process, although in the USA the role of bacteria in percolating filters was first recognized in 1889.

(a) SECTION

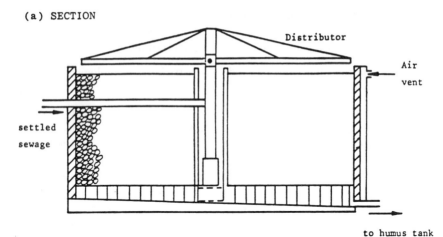

(b) PLAN

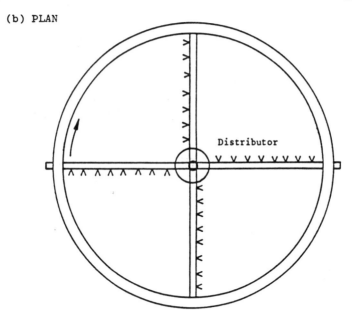

Figure 4   *A percolating filter*

2.3.1   Percolating Filter. These units consist of circular or rectangular beds of broken rock, gravel, clinker, or slag with a typical size in the range of 50–100 mm. The beds are between 1.5 and 2.0 m deep and of very variable diameter or size depending on the population to be served. The proportion of voids (empty spaces) in the assembled bed is normally in the range 45 to 55% (see Figure 4). The settled sewage trickles through interstices of the medium which constitutes a very large surface area on which a microbial film can develop. It is in this gelatinous film containing bacteria, fungi, protozoa, and on the upper surface algae that the oxidation of the BOD in the settled sewage takes place. The percolating filter is in fact a continuous mixed microbial film reactor. Settled sewage is fed onto the surface of the filter by some form of distributor mechanism. On circular filters a rotation system of radial sparge pipes is used which are usually reaction–jet propelled although on larger beds they may be electrically driven. With rectangular beds electrically powered rope hauled arms are used.

The micro-organisms which constitute the gelatinous film appear to be organized, at least near the surface of the filter where algae are present, into three layers (see Figure 5). The upper fungal layer is very thin (0.33 mm), beneath it the main algal layer is approximately 1.2 mm and both are anchored by a basal layer containing algae, fungi, and bacteria of approximately 0.5 mm. However, algae do occur to some extent in all three layers. Beneath the surface where sunlight is excluded and as a consequence the algae are absent, this structure is significantly modified, probably into a form of organization with only two layers. It has been calculated that photosynthesis by algae could provide only 5% or less of the oxygen requirements of the micro-organisms in the filter. Furthermore, photosynthesis would only be an intermittent source of oxygen since it would not occur in the dark and algae are often present only in the summer months. Carbon dioxide generated by other organisms in the filter might however increase the rate of photosynthesis. It has been proposed that algae derive nitrogen and minerals from the sewage and that some may be facultative heterotrophs. The nitrogen fixing so-called 'blue–green algae,' really bacteria, are frequently present in filters.

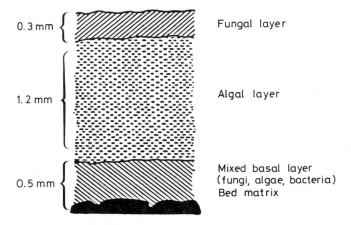

0.3 mm — Fungal layer

1.2 mm — Algal layer

0.5 mm — Mixed basal layer (fungi, algae, bacteria) Bed matrix

Figure 5 *Cross-section of the surface layers of a percolating filter*

Whilst fungi are efficient in the oxidation of the BOD present in the settled sewage they are not desirable as dominant members of the microbial community. They generate more biomass than bacteria, per unit of BOD consumed, thus increasing the sludge disposal problem. Moreover, an accumulation of predominantly fungal film quickly causes blockages of the interstices of the filter bed material, impeding both drainage and aeration. The latter may result in a reduction in the efficiency of treatment which is dependent upon the metabolic activity of aerobic micro-organisms.

Protozoa and certain metazoans (macrofauna) play an important role in the successful performance of the biological filter, although the precise nature of this role is dependent on the extrapolation of observations made in the activated sludge process, which is more amenable to study. However, the similarity in the distribution of organisms within the two processes suggests strongly that their roles are the same in both. The protozoa in particular remove free-swimming bacteria thus preventing turbid effluents, since freely suspended bacteria are not settleable. Certain metazoans may also ingest free-swimming bacteria, but their most important function is to assist in breaking the microbial film which would otherwise block the filter. This film is 'sloughed off' with the treated settled sewage. Protozoa (principally ciliates and flagellates) tend to dominate in the upper layers of the filter, whilst the macrofauna (nematodes, rotifers, annelids, and insect larvae) dominate the lower layers.

If film is not removed satisfactorily, frequently as the result of excessive fungal growth, the condition known as 'ponding' develops. In this condition the surface of the filter is covered in settled sewage, air flow ceases, treatment stops, and the bed becomes anaerobic. Ponding may also be caused by the growth of a sheet or felt of large filamentous algae principally *Phormidian* sp. on the face of the filter.

To minimize film production recirculation of treated effluent is often employed. This reduces film growth by dilution of the settled sewage, improves the flushing action for the removal of loose film, and promotes more uniform distribution of the film with depth.

Treated sewage is subject to secondary sedimentation which is similar to primary sedimentation as a result of which the suspended sloughed off film is consolidated into humus sludge and the final effluent discharged to the receiving water.

2.3.2 Activated Sludge. In the activated sludge process the majority of biological solids removed in the secondary sedimentation tank are recycled (returned sludge) to the aerator. The feedback of most of the cell yield from the sedimentation tank encourages rapid adsorption of the pollutants in the incoming settled sewage and also serves to stabilize the operation over the wide range of dilution rates and substrate concentrations imposed by the diurnal and other fluctuations in the flow and strength of the sewage. Stability is also provided by the continuous inoculation of the reactor with micro-organisms in the sewage and airflows, which are ultimately derived from human and animal excreta, soil run-off, water, and dust. The reactor of the activated sludge plant is usually in the form of long deep channels. Before entering these channels the returned sludge and settled sewage are mixed thereby forming the 'mixed liquor.' The retention time of the 'mixed liquor' in the aerator is typically 3 to 6 hours, during this period it moves down the length of the channel before passing over a weir, prior to secondary sedimentation. The sludge which is not returned to the aerator unit is known as surplus activated sludge and has to be disposed of. In practice the conditions in the aeration unit diverge from the completely mixed conditions commonly used for industrial fermentations and it may be best described as a continuous mixed microbial deep reactor with feed-back.

(a) SCHEMATIC

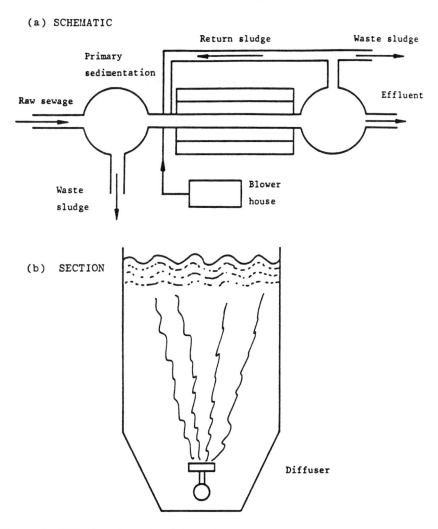

Figure 6 *Diffused aeration activated sludge plant*

The design of the concrete tanks which form the reactor is strongly influenced by the type of aeration to be employed. Two types are available, compressed (diffused air) (see Figure 6) and mechanical (surface aeration) (see Figure 7). In the diffused air system much of the air supplied is required to create turbulence, to avoid sedimentation of the bacteria responsible for oxidation. Surface aeration systems introduce the turbulence mechanically and only provide sufficient air for bacterial oxidation. Both types of system aim to maintain a dissolved oxygen concentration of between 1 and 2 mg l^{-1}.

(a) SCHEMATIC

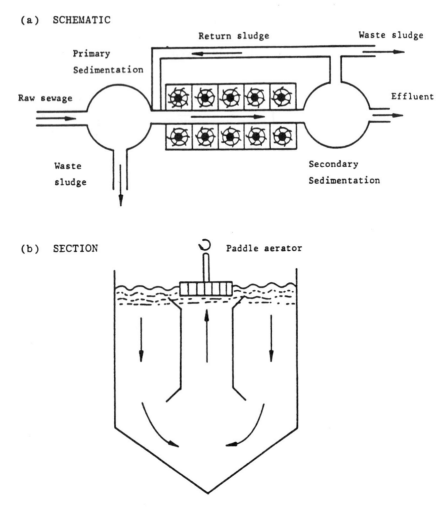

Figure 7 *Mechanically aerated activated sludge plant*

In the diffused air system the air is released through a porous sinter at the base of the tank and this system is characterized by long undivided channels which may be quite narrow (see Figure 6). Mechanical aeration utilizes rotating paddles to agitate the surface thereby incorporating air and creating a rotating current which maintains the bacterial flocs in suspension. Each paddle is located in its own cell which has a hopper shaped bottom, this gives the plant the appearance of a square lattice (see Figure 7). However, beneath the face of the mixed liquor all the cells are connected forming a channel. In both systems the channels are 2–3 m deep and 40–100 m long.

The success of the activated sludge process is dependent on the ability of the micro-organisms to form aggregates (flocs) which are able to settle. It is generally accepted that flocculation can be explained by colloidal phenomena and that bacterial extracellular polymers play an important role, but the precise mechanism is not known. The significance of flocculation to the success of the process is not the only characteristic to distinguish it from other industrial continuous cultures. There are four additional and very significant differences: it utilizes a heterogeneous microbial population, growing in a very dilute multi-substrate medium, many of the bacterial cells are not viable and finally the objectives of the process, which are the complete mineralization of the substrates (principally carbon dioxide, water, ammonia, and/or nitrate) with minimal production of both biomass and metabolites are also unique.

Table 2 *Importance of ciliated protozoa in determining effluent quality*

Effluent property	Ciliates absent	Ciliates present
Chemical Oxygen Demand $(mg\,l^{-1})$	198–254	124–142
Organic Nitrogen $(mg\,l^{-1})$	14–20	7–10
Suspended Solids $(mg\,l^{-1})$	86–118	26–34
Viable bacteria $(10^7\,m\,l^{-1})$	29–42	9–12

The heterogeneous population present in activated sludge includes bacteria, protozoa, rotifers, nematodes, and fungi. The bacteria alone are responsible for the removal of the dissolved organic material, whilst the protozoa and rotifers 'graze,' removing any 'free-swimming' and hence non-settleable bacteria, the protozoans and rotifers being large enough to settle during secondary sedimentation. The role of protozoa in activated sludge has been extensively studied; there are three groups involved: the ciliates, flagellates, and amoebae. It is probably the ciliates (*Ciliophora*) which constitute the greatest number of species with the greatest number present in each species which play the major role in the clarification process. The effect on effluent quality as a consequence of grazing by protozoa is summarized in Table 2. Not only do the protozoa remove free-swimming activated sludge bacteria but they play an important role in the reduction of pathogenic bacteria, including those which cause diphtheria, cholera, typhus, and streptococcal infections. In the absence of protozoa approximately 50% of these types of organisms are removed while in their presence removals rise to 95%. Nematodes have no significant role in the process, whilst the effects of the fungi are generally deleterious and contribute to or cause non-settleable sludge known as 'bulking.' Members of the following bacterial genera have been regularly isolated from activated sludge, *Pseudomonas*, *Acinetobacter*, *Comamonas*, *Lophomonas*, *Nitrosomonas*, *Zoogloea*, *Sphaerotilus*, *Azotobacter*, *Chromobacterium*, *Achromobacter*, *Flavobacterium*, *Alcaligenes*, *Micrococcus*, and *Bacillus*. Attributing the appropriate importance to each genus is a problem which confounds bacteriologists.

Of the principal groups of substrates listed in Section 1.4, only one single substrate (cellulose) was included. Each of the groups includes many substrates for example the 'sugars' identified in sewage include glucose, galactose, mannose, lactose, sucrose, maltose, and arabinose, whilst the nitrogenous compounds include proteins, polypeptides, peptides, amino acids, urea, creatine, and amino-sugars. Since bacteria normally only utilize a single carbon substrate or at the most two, this diversity of substrates in part explains the numerous genera of bacteria isolated from activated sludge, because each substrate under most conditions will sustain one species of bacterium. Moreover as a consequence of the large number of substrates present in the settled sewage the concentration of individual substrates is far less than the 200 mg l^{-1} of BOD present, perhaps 20–40 mg l^{-1} for the most abundant and less than 10 mg l^{-1} for the less common ones. The concentration of each substrate is further reduced in the aeration tank by dilution with the returned activated sludge which is typically mixed 1:1 with settled sewage resulting in a 50% reduction in substrate concentration.

The low substrate concentration means that the bacteria are in a starved condition. As a consequence many of them are 'senescent,' *i.e.*, in that phase between death, as expressed by the loss of viability, and breakdown of the osmotic regulatory system (the moribund state) thus the bacterium is a function-ing biological entity incapable of multiplication. That bacteria could exist in this condition was established at an early stage in a series of inspired experiments by Wooldridge and Standfast who published their results in 1933. They determined the dissolved oxygen concentrations and bacterial numbers (by viable counts) in a series of biochemical oxygen demand bottles containing diluted raw sewage on a daily basis. The viable count reached a maximum on the second day and thereafter fell rapidly. However the consumption of oxygen increased by equal amounts until the fourth day and fell to a negligible value on the fifth day. There was no obvious relationship between viability and oxygen consumption. They tested experimentally the hypothesis that non-viable bac-teria were apparently capable of oxygen uptake by destroying the capacity for division without significantly diminishing enzyme activity. Treatment of *Pseudo-monas fluorescens* with a 0.5% formaldehyde solution prevented division but these bacteria exhibited vigorous oxygen uptake in both sewage and other media. Subsequently they were able to determine the presence of active oxidase and dehydrogenase enzymes in these non-viable bacteria. The effects of low substrate concentration on the viability of the bacteria are compounded by their specific growth rate. It is intended that biological wastewater treatment should result in the production of a final effluent containing negligible BOD. The biochemically oxidizable material in the effluent is composed of compounds originally present in the settled sewage, which have not been completely biodegraded, and bacterial products. Moreover this is to be achieved with the minimal production of biomass. These twin objectives are concomitant with the utilization of a bacterial population with a very low specific growth rate.

Unlike the percolating filter, bacterial growth in the activated sludge process is amenable to the type of description used by bacteriologists for conventional continuous cultures. However although it is amenable to this type of treatment it inevitably appears to be very different from all other continuous cultures. The dilution rates (rate of inflow of settled sewage/aeration tank volume) used are invariably low by the standards of industrial fermentations, typically 0.25 h^{-1}, *i.e.*, one quarter of the aeration tank volume is displaced every hour, therefore the *hydraulic retention time* is four hours. Although in the conventional single pass reactor the dilution rate and the specific growth rate (time required for a doubling of the population) are identical. That is the state in which the rate of production of cells through growth equals the rate of the loss of cells through the overflow. In the activated sludge process because of the recycling of the biomass the specific growth rate is very much lower than the dilution rate, typically in the range $0.002-0.007 \, h^{-1}$. Since under steady-state conditions, the bacteria are only able to grow at the same rate as they are lost from the system, recycling them dramatically lowers their specific growth rate and allows it to be controlled independently of the dilution rate. Under steady-state conditions the specific growth rate is equivalent to the specific rate of sludge wastage (mass of suspended solids lost by sludge wastage and discharged in the effluent in unit time as a proportion of the total mass in the plant) which is the reciprocal of the 'sludge age' or mean cell retention time which is typically 4–9 days. Thus, whilst the retention of the aqueous phase in the system is only 4 h, the retention of the bacterial cells or sludge age is several days. The sludge age (θc) is a value which describes a great deal about the type of activated sludge plant; its purpose, quality of effluent, and bacteriological and biochemical states are all summarized by this term.

The activated sludge process may have up to four phases:

  (i)   clarification, by flocculation of suspended and colloidal matter;
  (ii) oxidation of carbonaceous matter;
  (iii) oxidation of nitrogenous matter (see section 2.3.3);
  (iv) auto-digestion of the activated sludge.

The occurrence of these four phases is directly dependent on increasing sludge age. Those processes which operate at low sludge ages give rapid removal of BOD per unit time, but the effluent is of poor quality. Plants which have high sludge ages give good quality effluents but only a slow rate of removal. Low sludge ages result in actively growing bacteria, and consequently high sludge production, whilst bacteria grown at high sludge ages behave conversely. Figure 8 illustrates the relationships between the growth curve of the bacterial culture and the type of activated sludge plant. By operating continuously the activated sludge process functions only over a small region of the batch growth curve, this region is determined by the specific sludge wastage rate. The region selected determines the type of plant and its performance. These are summarized in Figure 8.

2.3.3 Dispersed Aeration. This type of process is rarely used and is not applicable to the treatment of municipal sewage but may be of use in the preliminary treatment of some industrial wastes. The bacteria are growing rapidly (exponential phase), thus, the process has the ability to remove a large quantity of BOD per unit of biomass and as a consequence a small reactor may be used which is cheap to construct. However, because of their high rate of growth, the bacteria convert much of the BOD into biomass, causing a sludge disposal problem, flocculation is limited so additional treatment is essential to remove solids.

Furthermore, although BOD removal per unit biomass is high, the effluent BOD is also high.

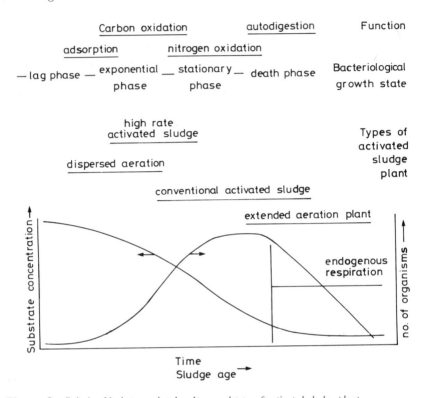

Figure 8 *Relationship between batch culture and type of activated sludge plant*

2.3.4 High-rate Activated Sludge. This shares many features of the previous process; however flocculation proceeds satisfactorily and secondary sedimentation will remove the solids effectively. The growth rate of the bacteria is still high, only carbonaceous material will be oxidized. However, some 60 to 70% of the influent BOD will be removed with a hydraulic retention time of approximately 2 h. This type of process is probably most frequently used for industrial wastes prior to discharge to the sewers, although it is also used for domestic sewage treatment, perhaps most appropriately where effluents are to be discharged to estuarine waters where standards are less stringent.

2.3.5 Conventional Activated Sludge. The two previous processes utilize actively growing bacteria in the exponential phase of growth. They achieve the oxidation of carbon compounds utilizing an exclusively heterotrophic bacterial population. Conventional activated sludge plants operate in the stationary or declining growth phases utilizing senescent bacteria. This very slow growth results in very low residual substrate concentrations and hence low values for effluent BOD. In addition, plants operating at sludge ages towards the upper end of this range contain autotrophic nitrifying bacteria. These organisms convert ammonia to nitrite and nitrate. This further improves the quality of the effluent since ammonia can exert an oxygen demand but nitrate cannot. In addition to maximizing effluent quality conventional activated sludge plants limit the production of new cells. Bacteria which are growing slowly use much of the organic matter available in the maintenance of their cells rather than in the production of new cells. These features have made conventional activated sludge the most widely adopted biological sewage treatment process for medium and large communities. The rate of oxidation is highest at the inlet of the tank and it can be difficult to maintain aerobic conditions. Two solutions to this problem have been adopted. With *tapered aeration* rather than supplying air uniformly along the length of the tank the air is concentrated at the beginning of the tank and progressively reduced along its length. The volume of air supplied remains unchanged but it is distributed according to demand. Alternatively *stepped loading* may be utilized. This aims to make the requirement for air uniform by adding the settled sewage at intervals along the tank, thus distributing the demand.

2.3.6   Extended Aeration. This process operates at very high sludge ages exclusively in the declining phase of growth. The retention time in the aeration tank is between 24 h and 24 days. As a consequence, the available substrate concentration is low and the bacteria undertake endogenous respiration (see Figure 8), that is respiration after the consumption of all available extracellular substrate. The result of utilizing endogenous materials is the breakdown of the sludge, sometimes referred to as auto-digestion. By this means sludge production is minimized and the small amount of material that must be disposed of is highly mineralized and inoffensive. This type of treatment has been extensively used for small communities, whilst capital costs of such plants are high, operating and sludge disposal costs are very low.

2.3.7   Contact Stabilization. The contact stabilization process is a variation of conventional activated sludge used for treating wastes with a high content of biodegradable colloidal and suspended matter. The process utilizes the adsorptive properties of the sludge to remove the polluting material very rapidly (0.5– 1 h) in a small aeration tank. The mixed liquor is then settled and passed into a second aeration tank and aerated for a further 5 to 6 h, during which period the adsorbed material is oxidized. After this the sludge with its adsorptive capacity restored is returned to the contact basin. Although this process requires two aeration tanks, the two are very much smaller than the equivalent single tank, since the mixed liquor suspended solids in the contact basin are typically 2000 mg l^{-1} and in the second tank (digestion unit) they are about 20 000 mg l^{-1}.

2.3.8   Nitrification. The production of a final effluent, with the minimum BOD value, is dependent upon the complete nitrification of the effluent, which involves the conversion of the ammonia present to nitrate. This is a two stage process undertaken by autotrophic bacteria principally from the genera *Nitro-somonas* and *Nitrobacter*. Nitrification occurs in percolating filters and activated sludge plants operated in a suitable manner. The first stage, sometimes referred to as 'nitrosification' involves the oxidation of ammonium ions to nitrite and follows the general formula:

$$NH_4^+ + 1.5O_2 \xrightarrow[\text{\textit{Nitrosomonas}}]{} NO_2^- + 2H^+ + H_2O$$

In the second stage nitrite is oxidized to nitrate:

$$NO_2^- + 0.5O_2 \xrightarrow[\text{\textit{Nitrobacter}}]{} NO_3^-$$

The overall nitrification process is described by the formula:

$$NH_4^+ + 2O_2 \xrightarrow{\hspace{2cm}} NO_3^- + 2H^+ + H_2O$$

Two important points are evident from this last formula. Firstly, nitrification requires a considerable quantity of oxygen. Secondly, hydrogen ions are formed and hence the pH of the wastewater will fall slightly during nitrification.

The settled sewage is effectively self buffering but a fall of 0.2 of a pH unit is frequently observed at the onset of nitrification. In this autotrophic nitrification process, ammonia, or nitrite provide the energy source, oxygen the electron acceptor, ammonia the nitrogen source, and carbon dioxide the carbon source. The carbon dioxide is provided by the heterotrophic oxidation of carbonaceous nutrient, by reaction of the acid produced during nitrification with carbonate or bicarbonate present in the wastewater, or carbon dioxide in the air. Whereas for carbonaceous removal the oxygen requirement is roughly weight for weight with the nutrients oxidized, in the case of ammonia, removal by nitrification requires approximately seven times as much oxygen as is required to achieve the removal of the same quantity of nutrient.

Nitrification significantly increases the cost of sewage treatment since more air is required. Furthermore, because these autotrophic organisms grow only slowly, longer retention periods are also required resulting in higher capital costs. Nor does nitrification result in the production of an entirely acceptable sewage effluent. In areas where water re-use is practised the concentration of nitrate in river waters causes concern. There exists a limit on the concentration of nitrate in drinking water to avoid the occurrence of methaemoglobinaemia (so called 'blue baby' syndrome). As a consequence denitrification is now practised after nitrification in some activated sludge treatment plants. In this anoxic hetero-trophic bacterial process, nitrite and nitrate replace oxygen in the respiratory mechanism and gaseous nitrogen compounds are formed (nitrogen gas, nitrous, and nitric oxides). However, this procedure is not part of conventional sewage treatment practice at present.

2.4 Secondary Sedimentation

Both types of biological treatment require sedimentation to remove suspended matter from the oxidized effluent. Tanks similar to those normally employed for primary sedimentation are generally employed, although, at a higher loading of approximately $40 \, m^3 \, m^{-2} \, d^{-1}$, at $3DWF$, because of the lighter and more homogenous nature of secondary sludge, simpler sludge scrapers are possible and scum removal is not necessary. The association of primary sedimentation tanks and a biological process for secondary treatment, results in a sewage treatment works, as opposed to sewage farms where only land treatment was (is) employed. As an awareness of environmental pollution, in addition to public health, developed in the fifties and sixties, the term water pollution control works was introduced to describe sewage treatment works, although this change of terminology was merely cosmetic. With the recognition of the importance of water re-use the term water reclamation works has found favour in some areas. Such works frequently apply additional tertiary treatment processes.

Sewage treatment results in the production of a final effluent suitable for discharge to the selected receiving water and one or more sludges which may require treatment prior to disposal.

3 SLUDGE TREATMENT AND DISPOSAL

Sludge treatment and disposal is a facet of wastewater treatment which is often given insufficient attention. Sludge treatment and disposal may account for 40% of the operating costs of a wastewater treatment facility. Prior to treatment the sludges contain between 1 and 7% solids (they are therefore nearly all water) which are usually highly putrescible and offensive. A total of 40 million tonnes of sludge, equivalent to 1.3 million tonnes of dry solids, is produced in the UK every year. The sludges are the product of primary sedimentation of raw sewage and the by-product of secondary aerobic treatment of settled sewage. Primary sludge is particularly offensive, with a pronounced faecal odour, and is liable to become putrescent thus causing a nuisance. Secondary sludge consists very largely of bacterial solids. It is less offensive than primary sludge but may still become putrescent. These sludges are sometimes combined during sewage treatment as a consequence of co-settlement of waste activated sludge in the primary sedimentation tanks. The main aims of sludge treatment are to make it easier and cheaper to dispose of the sludge consistent with minimizing any nuisance or adverse effects on the environment generally. A wide range of treatment processes and disposal options have been used, although, recently the cost of energy has reduced the numbers currently employed because of economic considerations.

The most convenient and economical method of disposal at any given site depends on a number of factors. Treatment of sludge is frequently influenced by the final disposal option selected. If sludge is to be disposed to sea from a works where the sludge may be pumped directly to the disposal vessel then little treatment is required. Should the treatment works be close enough to the sea to make that type of disposal feasible, but not close enough to allow direct pumping to the disposal vessel, then economies in transport costs may be achieved by utilizing some type of treatment process to thicken the sludge and reduce its water content prior to transport to the disposal vessel. If sludge is to be disposed of to land it is desirable to reduce transport costs, since the sludge will have to be spread over a wide area, and, in the case of treatment works in urban locations, transported a significant distance to reach suitable land.

At present 67% of the sludge produced in the UK is disposed to land, 29% to sea, and 4% is incinerated. Of the sludge disposed to land approximately two-thirds is applied to agricultural and horticultural land and the remainder is used for land reclamation and land fill.

The processes available for sludge treatment include: thickening by stirring or flotation; digestion, aerobically or anaerobically; heat treatment; composting with domestic refuse; chemical conditioning with either organic or inorganic materials; dewatering, on drying beds, in filter presses, by vacuum filtration, or centrifugation; heat drying; incineration in multiple hearth or fluidized bed furnaces and wet air oxidation. It is not feasible within this presentation to deal with all these processes in depth and the following is confined to the predominant sludge treatment process, anaerobic digestion, mechanical dewatering, and the most frequently utilized disposal option, that to agricultural land.

3.1 Anaerobic Digestion

During anaerobic digestion the organic matter present in the sewage sludge is biologically converted to a gas typically containing 70% methane and 30% carbon dioxide. The process is undertaken in an airtight reactor usually equipped with a floating gas collector. Sludge may be introduced continuously, but more frequently is added intermittently, and the digester operates on a 'fill and draw' process. The methane produced is generally utilized for maintaining the process temperature, heating, and power production by combustion in dual fuel engines which use oil in the absence of methane.

Methane production is only significant at elevated temperatures, when 1 m^3 of methane at STP is produced for every 3 kg of BOD degraded. Digesters are characterized by the temperature at which they operate, those in which gas production is optimum at 35 °C are described as 'mesophilic' whilst those in which gas production is optimum at 55 °C are 'thermophilic,' these terms describe the temperature preferences of the bacteria undertaking the process.

Heat exchangers are used to transfer heat from the treated sludge to the influent sludge. The additional heat is provided by the combustion of methane. Digesters in the UK operate in the mesophilic range, since heat loss from thermophilic digesters would be unacceptable. To minimize heat loss digesters are frequently surrounded by earth banks to provide insulation. For efficient operation the digester requires a mixing system which may be mechanical or utilize the gas produced in the process to provide turbulence. A conventional anaerobic digester is illustrated in Figure 9. The result of anaerobic digestion is to reduce the volatile solids present in the original sludge by 50% and the total solids by 30%. In addition the unpleasant odour associated with the raw sludge is drastically reduced. During the 20 to 40 days required for digestion the sludge is stabilized and emerges with a slightly tarry odour.

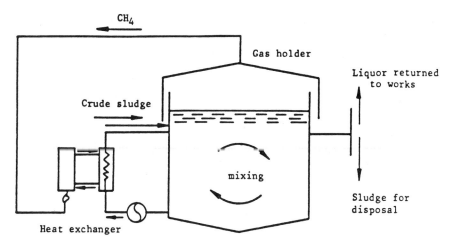

Figure 9 *Schematic diagram of an anaerobic digester*

Traditionally, anaerobic digestion has been considered a two stage process, a non-methanogenic stage followed by a methanogenic stage. The non-methanogenic stage has also been referred to as the acid forming stage since volatile fatty acids are the principal products. However it is now recognized that the first stage may include as many as three steps. The first, involving the hydrolysis of the fats, proteins, and polysaccharides present in the sludge, produces long chain fatty acids, glycerol, short chain peptides, amino acids, monosaccharides, and disaccharides. The second step (acid formation) involves the formation of a range of relatively low molecular weight materials including hydrogen, formic and acetic acids, other fatty acids, ketones, and alcohol. It is now recognized that only hydrogen, formic acid, and acetic acid can be utilized as substrates by the methanogenic bacteria. Thus in the third step compounds other than hydrogen, formic acid, and acetic acid are converted by the obligatory *hydrogen producing* acetogenic (OHPA) bacteria. Some bacteria are able to undertake both steps 1 and 2 and produce hydrogen, formic acid, and acetic acid which therefore do not require step 3. These stages are summarized in Figure 10. Once in operation, with reasonable retention times and volatile solids loadings, routine operation of digesters must include careful monitoring of certain parameters which are used to indicate whether the process is about to fail. The main parameters are volatile acids and hydrogen ion concentration (pH).

Anaeobic digestion is quite sensitive to fairly low concentrations of toxic pollutants, such as heavy metals and chlorinated organics, and to variations in loading rates and other operational aspects. If the balance of the process is upset it is most likely that the methanogenic organisms become inhibited first. This results in a build-up of the intermediate compounds at the stage immediately prior to methane formation. These intermediates collectively are called volatile fatty acids. They include formic, acetic, and butyric acids and can be monitored to determine the state of the process. The volatile acids are important because of their acidic nature. Normally digesters operate in the pH range of neutrality

(6.5–7.5). They also have some resistance to pH change. High concentrations of the volatile acids can cause a reduction in pH sufficient to inhibit bacterial activity to the extent where irreversible failure of the process occurs. Because of their capacity to resist changes in pH, volatile acid concentrations can build up to significant levels before pH change occurs. Therefore they can act as an early warning indicator of impending process failure. Normal levels of volatile acids are 250–1000 mg l^{-1}. If they exceed 2000 mg l^{-1} this could lead very quickly to failure; if they exceed 5000 mg l^{-1} failure is almost inevitable. The adverse effects of volatile acid build-up can be rectified by the addition of lime to restore the balance between acidity and alkalinity.

Anaerobically digested sludge is frequently further dewatered in lagoons prior to disposal. Supernatant liquors are pumped from the surface of the lagoons to the head of the works for treatment.

3.2 Dewatering of Sludge

One of the major objectives of sludge treatment is to reduce the water content. The advantages of this are two-fold. First it reduces the volume of sludge to be handled, which can very often lead to savings in transport costs, and secondly it can improve the physical properties of the sludge making it easier to handle. Sludge can be dewatered in two main ways. Either it can be allowed to dry out naturally or it can be dewatered by forcing the water out mechanically, typically, by either pressure filtration or vacuum filtration.

If sludge is to be dried naturally it is usually spread in layers up to about 2–3 cm thick in special drying beds. These very thin layers permit water loss both by evaporation from the surface and drainage. The sludge lies on a layer of fine ash, over a layer of coarse ash, under which are laid underdrains. The liquor which drains off the sludge goes to a central sump. From there it it pumped back to the main treatment works to undergo aerobic treatment. After a period of about 2 months the solids content increases to about 25% and the sludge can be dug up. This can be done manually although mechanical scrapers which transfer the sludge onto a moving conveyor are sometimes used. Because the sludge is spread in such thin layers, a large land area is required for drying beds. In the UK a drying area of about 0.3 m^2 per head of population is normally required.

Sludges are usually difficult to dewater and their dewaterability can usually be improved by the use of conditioners and this is normally the practice if mechanical dewatering is to be employed. Frequently aluminium and iron compounds are used, such as aluminium chlorohydrate and ferrous sulphate, or alternatively organic polymers called polyelectrolytes may be used. It is not certain how these work but it is probable that they react with the surfaces of the small sludge particles which would appear to be the cause of poor dewatering.

Following conditioning, sludge may be dewatered by pressure filtration. The sludge is pumped at a pressure of 700 kPa into a cloth lined chamber; at this pressure it requires between 2 and 18 hours to form a cake of 25–50% solids. The process can be operated either in batch or continuous manner.

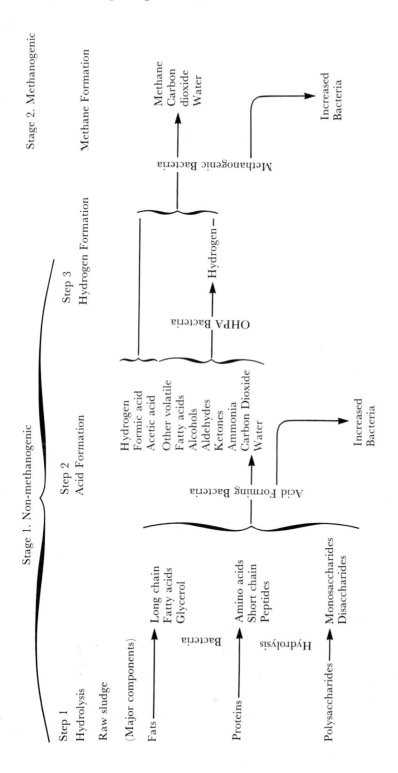

Figure 10 *Biochemical transformations involved in anaerobic digestion*

Alternatively, vacuum filtration may be used, this is invariably a continuous process. A drum, containing several internal segments, revolves on its horizontal axis partially submerged in sludge. A partial vacuum inside the drum causes the sludge to adhere to it in a thin layer and as it rotates out of the wet sludge the water is sucked out. A scraper separates the dried sludge from the outside of the drum. At the point where the scraper is positioned, the vacuum in the corresponding segment inside the drum is released, aiding the release of the sludge from the outside. The pressure difference between inside and outside is 70 kPa, which is only about 10% of the pressure attained in pressure filtration. Hence the dried sludge typically only contains 15–20% solids. A diagram of a vacuum filter is shown in Figure 11.

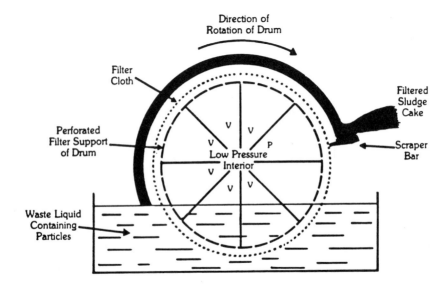

Figure 11   *Vacuum filter*

3.3   Disposal of Sewage Sludge to Land

The practice of disposing sewage sludge to land has several potential benefits. In 1981 a Standing Technical Committee Report on the Disposal of Sewage Sludge to Land had this to say about the objectives of good practice:

'Good practice in sewage sludge disposal, whether to land, to sea, or by incineration, involves striking a balance between economic constraints and the avoidance as far as possible of adverse effects on man, animals, and the environment. Low cost is not always compatible with limited adverse environmental effects and sometimes compromises have to be reached. The best option of treatment within a disposal method may change with time due to changing costs, improved knowledge of treatment processes and environmental effects, experience, and research.

Disposal of sewage sludge involving application to agricultural land has the benefit of resource recovery and the value of the nutrients utilized should be taken into account in assessing the minimum cost to the nation. Disposal of sludge to land can have greater environmental impact than other options and in assessing likely benefits and potential adverse effects consideration should also be given to amenity, formal, and informal recreation and wildlife.'

This report concludes: 'Where ever it is economically justified and environmentally acceptable sewage sludge should be utilized on agricultural land in accordance with . . recommendations . . .'

In coming to this conclusion, the Committee had to weigh up the advantages and disadvantages prior to setting out their recommendations, which included guidelines designed to regulate the quantities of sludge applied to land.

Most of the sludge used in agriculture (about 97%) receives some form of treatment. Nearly half of it is anaerobically digested. Slightly more goes to general arable land than to grazing land. Grass grown for hay or silage is particularly suitable to receive sludge because it reduces the risks of transmission of disease. Fields used for forage crops and cereals are also suitable to receive sludge.

Both dried and liquid sludges are applied to land, the latter from tankers by spraying. Arable land can be ploughed, following the application, which speeds up incorporation of the sludge and can reduce any odour problems. However, when it is applied, even spreading is important to prevent localized 'hotspots.' Liquid sludges may also be injected directly into the soil giving uniform application and almost complete elimination of odour problems.

Ideally, disposal sites should be well away from housing, but have good access. The risk of leachate or run-off contaminating groundwater or surface streams must be carefully considered. If the sludge is to be spread by spraying, care should be taken not to allow drift, especially in windy conditions.

Agricultural land is an important outlet for sewage sludge disposal, with about 40% of the sludge produced at inland wastewater treatment works being disposed of in this way. Nevertheless only a very small fraction of the agricultural land (less than 2%) in the UK receives sludge.

The application of sludge to land may help to slow down the decline in organic matter in soils under modern farming methods, leading to improvements in water holding capacity, porosity, and aggregate stability. The main value of sludge as a fertilizer lies in its nitrogen and phosphorus contents. However, much of the nutrient content may be in organic forms, and thus, be unavailable to plants until mineralization occurs. Although from an economic point of view it may be desirable to apply dried sludge to land significant quantities of available forms of the nutrients may be lost during drying. Liquid digested sludge may contain up to 10% (w/w) of nitrogen but only a fraction of this may be in available forms. For the purpose of calculating the available nitrogen in sludge, it is assumed that 85% of the total nitrogen in liquid digested sludge and 33% of that in dried sludge is available to crops during the growing season.

Since most agricultural soils are deficient in nutrients, fertilizers nearly always have to be added. However, sewage sludge is deficient in potassium, and therefore cannot fulfil complete fertilizer requirements. Furthermore, if all of the sewage sludge produced in the UK were to be applied to agricultural land, it would only provide 4.5% of the country's fertilizer requirement.

Liquid sludges consist mainly of water. Some farmers value sludge solely for its water content which can often help to overcome irrigation problems during dry weather.

Sewage sludge generally represents the non-degradable residue left after the treatment of domestic or mixed industrial/domestic wastewater. This means that it contains may of the materials originally present in the wastewater which could be classified as pollutants.

The potential hazards from the application of sludge to land are protozoal, viral, bacterial, and other pathogens, which are present to the greatest extent in untreated sludges, persistent toxic organic compounds, and toxic heavy metals.

Due to these hazards salad or other crops which may be eaten raw should not be sown until one year after the application of treated sludge. If treated sludge is applied to pasture, animals should not be grazed until 3 weeks after the application. In the case of dairy cattle, whose milk is not to be pasteurized, the period of delay should be 5 weeks.

Heavy metals are of particular concern because they may be detrimental to crop growth or mobilized through the food chain. In many countries there exist guidelines designed to maintain the addition of sludge to land within safe limits. In the UK the permissible concentrations of heavy metals in sludge to be applied to land and the quantities of such sludge which may be spread on the land are based on what is known about normal concentrations of heavy metals in soil and the levels at which adverse effects are likely to occur. Considerable research has been done and continues to be done on this subject. The guidelines in use now represent the present state of knowledge, and will inevitably be revised in the future as more is discovered about the behaviour of heavy metals (and other pollutants) in soil.

There are limits placed on:

(a) the cumulative quantities of metals which may be added to the soil over a period of 30 years. These quantities are expressed in terms of kilograms of metal per hectare of land.

(b) the total concentrations of metals which may be present in the soil. These concentrations are expressed in terms of milligrams of metal per kilogram of soil dry matter. This unit is the same as micrograms of metal per gram of soil dry matter and is sometimes referred to as parts per million.

Table 3 shows the limits placed on each of these quantities. In addition, the annual average quantities of metal which can be added have also been calculated by dividing the cumulative totals by 30, and these are included in the table. The guidelines also specify that the maximum quantity which can be applied in any one year is 6 times the annual average.

The additions of Zn, Cu, and Ni are governed by the Zinc Equivalent Concept which appears to be the limiting factor for up to 75% of the sludges produced in the UK. The Zinc Equivalent Concept assumes that the phytotoxic

Table 3 *Limits on metal additions to arable soils through sewage sludge disposal*

Metal	Cumulative limit kg ha^{-1}	Annual limit kg ha^{-1}	Soil concentration limit mg kg^{-1}
Arsenic	10	0.33	10
Cadmium	5	0.17	3.5
Chromium	1000	33	600
Copper	280	9.3	140
Mercury	2	0.067	1
Molybdenum	4	0.133	4
Nickel	70	2.3	35
Lead	1000	33	550
Selenium	5	0.167	3
Zinc	560	18.6	280
Boron	—	3.5	3.25
Fluorine*	600	—	500

The cumulative period referred to in the Table is 30 years.
The values shown are for arable land.
The values for pasture are the same except for copper, nickel, and zinc which are double the values shown.
Where the soil pH is >7 the values for copper, nickel, and zinc can be doubled.
* Fluorine is not actually a metal but is included because it is desirable to limit its concentration in soil.

effects of Zn, Cu, and Ni are additive in the ratio 1:2:8. The recommended limit of addition of zinc equivalent is 560 kg ha^{-1}. The following is an example of the calculation of the zinc equivalent:

assuming a sludge contains 1000 mg kg^{-1} Zn, 500 mg kg^{-1} Cu, and 60 mg kg^{-1} Ni, then the zinc equivalent is given by:

$$\text{zinc equivalent} = (1 \times 1000) + (2 \times 500) + (8 \times 60)$$
$$= 2480 \text{ mg kg}^{-1}$$

The recommended limit of application of the sludge (in tonnes of dry solids per hectare) over 30 years is given by:

$$\frac{\text{limit of addition of a particular metal (kg ha}^{-1}) \times 1000}{\text{concentration of metal in sludge (mg kg}^{-1})}$$

which in the case of the above example is:

$$\frac{560 \times 1000}{2480} = 225 \text{ t ha}^{-1} \text{ over 30 years}$$

the 560 kg ha^{-1} being the limit of addition for zinc (zinc equivalent).

This calculation is repeated, substituting values for the limits of addition of the other metals and their concentrations in the sludge, and the lowest value obtained is the maximum quantity of the sludge that may be applied.

The behaviour, fate, and significance of organic micropollutants during sludge treatment and disposal is not well understood, and as a consequence, no guidelines for the disposal of sludges contaminated with these materials exist. In the United States the Environmental Protection Agency has indicated 114 compounds of particular concern. Many of these materials fall into the following groups, polynuclear aromatic hydrocarbons, halogenated aliphatic and aromatic hydrocarbons, organochlorine pesticides, polychlorinated biphenyls, and phthalate esters. Most of these materials are hydrophobic and not readily amenable to chemical or biological degradation. During sewage treatment they are intimately associated with the solids and therefore strongly concentrated into the sludges produced. They are therefore in the sludges to be disposed of, however, knowledge of their significance is very limited. It appears that some chlorinated organics present in sewage sludge have been ingested (with soil) by grazing cows and that these materials have been passed to the milk associated with its lipid content.

4 BIBLIOGRAPHY

C. R. Curds and H. A. Hawkes (ed.), 'Ecological Aspects of Used-Water Treatment,' Volume 1 – The Organisms and their Ecology, Academic Press, London, 1975.

Government of Great Britain, 'Water Pollution Control Engineering,' HMSO, London, 1970.

Government of Great Britain, Department of the Environment, 'Taken for Granted,' Report of the Working Party on Sewage Disposal, HMSO, London, 1970.

Government of Great Britain, Department of the Environment/National Water Council, 'Report of the Working Party on the Disposal of Sewage Sludge to Land,' Standing Technical Committee Report No. 5, HMSO, London, 1977.

R. M. Sterritt and J. N. Lester, 'Microbiology for Environmental and Public Health Engineers,' E. and F. N. Spon, London, 1988.

Metcalf and Eddy Inc., 'Wastewater Engineering: Treatment, Disposal, Re-use,' McGraw-Hill Inc., New York, 1979.

The Open University, Environmental Control and Public Health, 'Units 7–8 Water: Distribution, Drainage, Discharge and Disposal,' Administrative Control, The Open University Press, Milton Keynes, 1975.

T. H. Y. Tebbutt, 'Water Science and Technology,' John Murray, London, 1973.

M. Winkler, 'Biological Treatment of Wastewater,' Ellis Horwood Ltd., Chichester, 1981.

J. N. Lester, (ed.), 'Heavy Metals in Wastewater and Sludge Treatment Processes, Volume II: Treatment and Disposal,' CRC Press, Boca Raton, 1987.

J. N. Lester and R. M Sterritt, 'Water Pollution Control, Module 3, Unit 2: The Basic Principles of Biological Wastewater Treatment,' Manpower Services Commission, Sheffield, 1988.

R. M. Sterritt and J. N. Lester, 'Water Pollution Control, Module 3, Unit 4: The Treatment and Disposal of Sludge,' Manpower Services Commission, Sheffield, 1988.

CHAPTER 4

The Treatment of Toxic Wastes

A. JAMES

1 INTRODUCTION

The presence of toxic substances in wastewaters has always been a matter for concern. This concern has become much more pressing with the intentional or unintentional release of an ever larger variety of substances into the environment.

A whole spectrum of difficulties has arisen in attempting to control the toxicity problem. Two major issues may be summarized as follows:

(a) Assessment of environmentally safe concentrations—subtle sub-lethal effects may impair ecological success at concentrations well below those causing death. Chronic effects due to prolonged exposure or bio-accumulation, toxic interactions, and geo-chemical cycles involving toxins all add further complications.
(b) Assessment of biodegradability—persistence of toxins in the environment is clearly undesirable but there are still difficulties in designing suitable tests to assess the biodegradability of newly synthesized compounds.

Because of these doubts and uncertainties the treatment and disposal of toxic wastewaters has remained on an empirical level. This is reflected in the following notes which cover:

(a) Sources of Toxic Wastewaters
(b) Problems in Collection
(c) Pretreatment
(d) Primary and Secondary Treatment
(e) Sludge Treatment and Disposal
(f) Direct Disposal
(g) Case Studies

2 SOURCES AND TYPES OF TOXIC WASTES

Toxic substances are primarily associated with industrial wastes but may be found in all types of wastewaters as shown below:

(a) Domestic Wastewaters—these wastes contain ammoniacal nitrogen in concentrations up to $50 \, \text{mg} \, l^{-1}$ and when septic may contain sulphide at levels up to $50 \, \text{mg} \, l^{-1}$. Both of these can cause damage to aquatic fauna unless diluted and dispersed. The former is particularly damaging to freshwater fish (TL_m for Rainbow Trout is around $2 \, \text{mg} \, l^{-1}$) and the latter acts as an enzyme inhibitor in a wide variety of aquatic organisms at levels of a few mg per litre.

(b) Stormwater—the composition of stormwater is much more varied than domestic wastes and is influenced by the nature of the drainage area and the frequency of storms. But the toxic potential is mainly associated with heavy metals like zinc and lead and is invariably found in the first flush of run-off from a storm.

(c) Agricultural Wastes—these contain a wide variety of materials that are used for fertilizers and pest control. Fertilizers, containing oxidized nitrogen, can cause human toxicological problems. Pesticides and herbicides are the most potent aquatic toxins. Because of their toxicity and persistence in the environment some of these substances like eldrin have had to be prohibited.

Wastes from animal husbandry can also be toxic especially from silage.

(d) Industrial Wastes—the range of toxic substances present in industrial wastes is too wide to catalogue but Table 1 gives an indication of the main types of toxic industrial waste and the toxins they contain.

3 TOXICITY PROBLEMS IN THE COLLECTION SYSTEM

The cost of constructing a wastewater collection system in an urban area is extremely high, often accounting for 70% of the total cost for treatment and disposal. Damage to the fabric of the sewer is therefore to be avoided and strict controls are usually imposed on substances that may be discharged to the sewer. As shown in Table 2 these controls are also intended to control the discharge of substances like cyanides or sulphide that may give rise to poisonous gases which could damage the health of sewer workers. Levels of HCN and H_2S of 0.03% in the atmosphere are toxic and with H_2S there is an additional problem of anaesthesia which makes detection difficult.

Some organic solvents may cause similar difficulties. They tend to be immiscible with water, volatile, and intoxicating, and may also form explosive mixtures.

4 PRE-TREATMENT OF TOXIC WASTES

In general industrial wastewaters are most readily and most economically treated in admixture with domestic wastewaters rather than in isolation. Many

Table 1   *Sources of some common toxins*

Toxin	Sources
Acids—mainly inorganic but some organic causing pH <6	Acid Manufacture Battery Manufacture Chemical Industry Steel Industry
Alkalis—causing pH >9	Brewery wastes Food Industries Chemical Industry Textile Manufacture
Antibiotics	Pharmaceutical Industry
Ammoniacal nitrogen	Coke Manufacture Fertilizer Manufacture Rubber Industry
Chromium—mainly hexavalent but also less toxic trivalent form	Metal processing Tanneries
Cyanide	Coke production Metal Plating
Detergents—mainly anionic but some cationic	Detergent Manufacture Textile Manufacture Laundries Food Industry
Herbicides and Pesticides—mostly chlorinated hydrocarbons	Chemical Industry
Metals—mainly copper, cadmium, cobalt, lead, nickel, mercury, and zinc	Metal processing and plating Chemical Industry
Phenols	Coke Production Oil Refining Wood Preserving
Solvents—mostly benzene, acetone, carbon tetrachloride, and alcohols	Chemical Industry Pharmaceuticals

benefits of scale, balancing nutrient supplementation as well as skilled operation, may be obtained by discharging the industrial wastewaters to a sewer. But there are a number of occasions when this is not possible or not desirable:

(a) Rural areas without convenient sewerage
(b) By-product recovery is economically and technically feasible
(c) Domestic effluent is used ultimately in irrigation
(d) Industrial wastewater does not meet consent conditions for discharge to a sewer

Under these circumstances some form of pretreatment is needed to render the wastewater suitable for discharge, further treatment, or disposal. The main advantages of treatment on site are the possibilities of recovering specific substances in an uncontaminated condition and economies which result from treatment at higher temperatures or concentrations. There may be an important additional advantage; that is the avoidance of contamination of a much larger wastewater stream which would cause difficulties in its disposal. Where the toxic materials are organic in nature there is often a problem in treatment due to inhibition of bacterial growth. It is often easier and cheaper to develop the necessary bacterial flora in an on-site treatment plant.

Table 2   *Typical consent conditions for discharge to sewers*

Parameter	Consent condition
Maximum Temperature	40–45 °C
pH	6–10
Substances producing inflammable vapours	Nil
Cyanide concentration	* 5–10 mg l^{-1}
Sulphide concentration	* 1 mg l^{-1}
Soluble sulphates	* 1250 mg l^{-1}
Synthetic detergents	* 30 mg l^{-1}
Free chlorine	* 100 mg l^{-1}
Mercury	* 0.1 mg l^{-1}
Cadmium	* 2 mg l^{-1}
Chromium	* 5 mg l^{-1}
Lead	* 5 mg l^{-1}
Zinc	* 10 mg l^{-1}
Copper	* 5 mg l^{-1}
Zinc equivalent (Zn + Cd + 2Cu + 8 Ni)	* 35 mg l^{-1}
Total non-ferrous metal	* 30 mg l^{-1}
Total soluble non-ferrous metal	* 10 mg l^{-1}

* *Note*: There are also a large number of specific toxic substances whose discharge to sewers is controlled.

This to some extent depends upon the concentration and toxicity of the substances concerned. In some cases dilution of the wastes by admixture with sewage reduces the toxic inhibition making it preferable to treat the industrial waste and sewage together. Also many industrial wastes are deficient in some nutrient such as nitrogen or phosphorus. The desirable ratio of BOD : N : P is 100 : 5 : 1 and the ratio in domestic wastes is commonly 100 : 18 : 2.5 so that deficiencies in industrial wastes can be balanced.

There are other considerations in deciding for or against pre-treatment such as:

(a) Availability of space—the site may be too restricted or land may be too valuable to be used for a treatment plant.
(b) Availability of expertise—the company may not wish to get involved in effluent treatment.
(c) Sludge and/or odour production may create a nuisance.

Even where it is decided not to carry out pre-treatment of the toxic waste by chemical or biological methods it is often useful to install devices to improve the effluent quality by simple physical means. This includes some form of screening, coarse or fine, to reduce solids. Also some form of balancing to reduce variations in concentration, flow, pH, *etc.*, and some traps to prevent the escape of oil and grease and some grit arrestors.

Every attempt should be made to minimize the quantity of material discharged through good housekeeping. This can take the form of any or all of the following techniques:

(a) Extending the life of process solutions by filtration, topping up, adsorption, *etc.*

(b) Altering the production process to use less toxic compounds, *e.g.*, substitution of copper pyrophosphate for copper cyanide in electroplating solutions.

(c) Dry cleaning prior to wash-down can remove a large proportion of the pollutant in solid form.

(d) Evaporation of strong organic liquors can often produce a burnable product.

(e) Minimizing and segregating any flows which contain toxic materials. In some cases it is necessary to separate wastes for safety reasons, *e.g.*, cyanides or sulphides and acid wastes, trichlorethylene, and alkaline wastes. In other cases it may be desirable to segregate for treatment reasons. However segregation can be very expensive.

Table 3 *Physical methods of pre-treatment*

Process	*Aim*	*Examples*
Screening	Removal of coarse solids	Vegetable canneries, paper mills
Centrifuging	Concentration of solids	Sludge dewatering in chemical industry
Filtration	Concentration of fine solids	Final polishing and sludge dewatering in chemical and metal processing
Sedimentation	Removal of settleable solids	Separation of inorganic solids in ore extraction, coal, and clay production
Flotation	Removal of low specific gravity solids and liquids	Separation of oil, grease, and solids in chemical and food industry
Freezing	Concentration of liquids and sludges	Recovery of pickle liquor and non-ferrous metals
Solvent extraction	Recovery of valuable materials	Coal carbonizing and Plastics manufacture
Ion Exchange	Separation and concentration	Metal processing
Reverse osmosis	Separation of dissolved solids	Desalination of process and wash water
Adsorption	Concentration and removal of trace impurities	Pesticide manufacture, dyestuffs removal

Having minimized so far as possible the types, quantities, and concentrations of any toxic wastes it may still be necessary to treat them prior to discharge either to a sewer or a water course. The processes which are used may be classified as physical, chemical, and biological. The physical processes are summarized in Table 3. Where the toxic wastes contain or are composed of organic materials it may also be necessary to provide some biological treatment especially if the effluent is to be discharged directly into a watercourse. Many different types of process are used but the following are the most popular.

(a) High-rate filtration using plastic media with very high rates of recirculation.
(b) Activated sludge using contact stabilization.

Like all biological processes these can suffer from toxicity problems especially where the concentration of toxin is not constant. In general terms it is easier for bacteria and other micro-organisms to adapt to toxic substances than for organisms like worms, fly larvae, *etc.* For this reason conventional percolating filters have not proved successful—the lack of grazing fauna has led to persistent ponding.

Due to a combination of high organic strength and inhibition from toxic substances it is unusual to obtain complete treatment of toxic industrial wastes by conventional primary and secondary treatment. The effluent from high-rate filters often has a BOD and COD similar to settled sewage and is either suitable for discharge to a sewer or for further biological treatment on site.

Chemical treatment of industrial wastes may be used in addition to and to some extent in place of biological treatment. The aims are somewhat different since biological treatment is mainly a way of oxidizing organic matter or a way of converting it into a settleable form. Chemical treatment is used only for oxidizing particular compounds, like cyanide, since it is expensive and liable to lead to the production of undesirable chlorinated organics. It is mainly used for pH correction and improving the removal of solids. The commonest chemicals in use are shown in Table 4.

Table 4 *Chemicals used in industrial waste treatment*

Chemical	*Purpose*
Calcium hydroxide	pH adjustment, precipitation of metals, and assisting sedimentation
Sodium hydroxide	Used mainly for pH adjustment in place of lime
Sodium carbonate	pH adjustment and precipitation of metals with soluble hydroxide
Carbon dioxide	pH adjustment
Aluminium sulphate	Solids separation
Ferrous sulphate	Solids separation
Chlorine	Oxidation

Table 5 *Amounts of heavy metal ions removed from sewage by sludges*

Heavy metal ion	Primary sedimentation		Percolating filter treatment		Activated-sludge process	
	Metal concentration in crude sewage (mg l^{-1})	Proportion removed by treatment (per cent)	Metal concentration in settled sewage (mg l^{-1})	Proportion removed by treatment (per cent)	Metal concentration in settled sewage (mg l^{-1})	Proportion removed by treatment (per cent)
Copper					0.4	54
Copper	Up to 0.8	45	Up to 0.44	20	Up to 0.44	80
Copper	Up to 5	12			0.4–25	50–79
Copper					28	90–93
Copper						
Dichromate	(as Cr) Up to 1.2	28	(as Cr) Up to 0.86	32	(as Cr) Up to 0.86	67–70
Dichromate					4.0	6.3
Dichromate					0.5–2	ca. 100
Dichromate					5	50
Dichromate					50	10
Iron (Ferric)	3–9	40	1.8–5.4	Nil	1.8–5.4	80
Lead	0.3–0.9	40	0.18–0.54	30	0.18–0.54	90
Nickel	0.1–0.3	20	0.08–10	40	0.08–0.24	30
Nickel					20	31
Nickel					2.5–10	30
Zinc	0.7–1.6	40	0.4–1.0	30	0.4–1.0	60
Zinc					2.5	90
Zinc					2.5	95
Zinc	Up to 5	12			7.5	100
Zinc					15	78
Zinc					20	74

5 PRIMARY AND SECONDARY TREATMENT

Provided that the pre-treatment of toxic industrial wastes is successful then no difficulties should be encountered in subsequent treatment. However no pre-treatment system is perfect and malfunction will occasionally occur mostly due to variations in the manufacturing process. As a result toxic material together with possible overload of organics and solids may be passed on to the subsequent treatment stages.

The effect of toxic materials on primary sedimentation is insignificant since this is a purely physical process of sedimentation and flocculation. However the effect of the primary sedimentation on toxic wastes can be very important. Toxic materials in suspension such as particulate metals are effectively removed. Also the flocculant material has a great capacity for adsorption and removes the majority of dissolved metals, pesticides, and other toxic organics. In one respect this is beneficial since it renders the waste material less inhibitory for biological treatment but it selectively concentrates the toxins in the sludge and may give rise to problems in digestion and in sludge disposal. Some indication of the removal of metals during primary treatment is given in Table 5.

Chemicals may be used to enhance the effectiveness of primary sedimentation, in some cases, removing additional material by precipitation at the same time. Chemical addition can be expensive, often requires pH correction, and may produce large quantities of sludge with a disposal problem. For these reasons chemical enhancement of primary sedimentation is rarely practised at plants treating domestic wastes. Nevertheless for many toxic industrial wastewaters this is an attractive treatment option since it enables industry to avoid secondary biological treatment and enables the waste to be discharged to a sewer, estuary, or the sea.

Where chemically enhanced sedimentation is used the main aim is generally to increase removal of solids, but since many toxins such as metals and chlorinated organics adsorb strongly, their removal is also increased to levels similar to combined primary and secondary treatment. The material employed for enhancement is lime or less frequently aluminium salts sometimes supplemented by polyelectrolytes. Some indications of metal removal achievable by sedimentation, with and without lime, are given in Table 6.

Table 6 *Metal removal by sedimentation*

Metal	Concentration in wastewater	% Removal by sedimentation	% Removal (with lime) by sedimentation
Iron	6.3	48	80
Copper	0.6	28	60
Chromium	0.34	40	58
Lead	0.12	33	55
Mercury	0.028	15	50
Nickel	0.08	15	15
Zinc	0.7	38	70

Whatever the form of primary treatment employed, further treatment is generally brought about by biological processes, either aerobic or anaerobic. The key to successful secondary treatment of wastewaters containing toxins is the adaptation of the micro-organisms to the presence of the toxin. Bacteria, and to a lesser extent protozoa, show a remarkable ability to acclimatize to the presence of toxic substances and a great adaptability in degrading new synthetic organic compounds. Metazoa are less adaptable so forms of treatment that rely on metazoa are best avoided in dealing with toxic wastes.

It is important in biological treatment of a toxic waste that a microbial population is developed which is acclimatized to the presence of the toxin and, in the case of degradable toxins, it is essential that it contains sufficient numbers of organisms which can metabolize the toxins. These twin aspects of acclimatization require great care in the start-up operation and may need a period of several months before successful operation is achieved. Even after start-up is complete particular processes involving sensitive bacterial species like nitrification or methane production can be easily disrupted by shock loads.

Table 7 *Biological processes for treating toxic wastewaters*

Aerobic/ anaerobic	Reactor type	Advantages and disadvantages	Examples of process
Aerobic	Dispersed growth	Tend to be completely mixed therefore dilutes toxin but affects whole biomass. Liable to cause settling problems as well as interfere with oxidation.	Activated sludge and modifications
Aerobic	Fixed film	Tend to be plug flow so no dilution unless recirculation is used. Biomass more robust for shock loads but metazoa more sensitive.	High rate filters (good) Standard rate filters (less suitable)
Anaerobic	Dispersed growth	Tend to be completely mixed and suffer from washout of methanogens. The latter are also more sensitive than acidogens to toxic effects and have a low growth rate.	UASB and contact digester
Anaerobic	Fixed film	Tend to be plug flow but level of attachment not as good as aerobic filters. Need recirculation to dilute toxins.	Anaerobic filters and fluidized bed

Biological processes for treating wastewaters may be divided into aerobic and anaerobic and each division may be subdivided into dispersed growth and fixed film. The resulting four categories of treatment process have advantages and disadvantages for the treatment of toxic wastewaters, as shown in Table 7. The tolerance levels of these processes are difficult to define precisely but some indication is given in Tables 8 and 9.

Table 8 *Toxic levels in aerobic biological treatment*

Toxin	Significant level
Hydrogen ions	pH <6 or >9
Phenols	50–100 mg l^{-1}
Ammoniacal-N	500–1000 mg l^{-1}
Zinc	10–50 mg l^{-1}
Chromium	5–20 mg l^{-1}
Lead	5–30 mg l^{-1}
Alkyl Benzene Sulphonates	3–20 mg l^{-1}
Sulphide	5–50 mg l^{-1}

Note: The wide ranges given in the Table are in part a reflection of the ability of bacteria to acclimatize to toxins but are also caused by physical/chemical interactions which adsorb or precipitate or otherwise remove toxins from solution.

Table 9 *Toxic effects in anaerobic treatment*

Toxin	Inhibitory concentration	
	In Sewage	In Sludge
Chromium	—	2
Cadmium	2	2
Copper	1.5	—
Iron	10	—
Lead	100	—
Nickel	80	—
Zinc	50	—
Detergent	—	2% of Suspended Solids
Benzene	—	50–200
Carbon tetrachloride	—	10
Chloroform	—	0.1
Dichlorophen	1	—
γ-BHC	—	48
Toluene	—	430–860

All concentrations in mg l^{-1}

6 SLUDGE TREATMENT AND DISPOSAL

This topic is discussed in Chapter 3 so the remarks below are limited to the treatment of toxic sludges. These are generated almost exclusively by industrial sources although sludges from silage and other agricultural practices may cause occasional problems. Industrial sludges are generally formed during primary sedimentation and tend to contain high levels of toxins like metals and chlorinated organics which are adsorbed onto the solids. Where domestic wastes are treated along with industrial discharges a similar problem may arise.

The treatment and disposal of toxic wastes may be broken into a series of steps:

(a) Dewatering
(b) Biological treatment
(c) Disposal
(d) Detoxification

with two possible strategies in wind-dispersion or concentration.

Where the industrial wastewater contains substances that are not permitted in the environment the overall strategy is to segregate the waste into a small volume of hazardous material for containment in a long term storage site (at sea or inland). Where the toxic substances in the wastewater are less dangerous, or in dealing with the larger volume of less hazardous waste, the aim is to dispose of these to the environment by ensuring adequate dilution.

Examples of these processes are as follows:

(a) Detoxification of domestic sludges—metal removal using acid hydrolysis in combination with electrolysis.
(b) Containment of radioisotopes by acid extraction, precipitation, followed by vitrification for long term storage.

7 DISPOSAL OF TOXIC WASTES

Ultimately toxic wastes have to be disposed of and this disposal must be in a manner which does not present a short term or a long term hazard to man or the environment. Some hazardous wastes may be rendered innocuous by treatment prior to disposal. Incineration is frequently used for decomposition of organic toxins. However care is required particularly in dealing with halogenated materials since irritant corrosive gases may be produced. There is also a danger that the treatment plant may become too complex. Examples of the successful use of incineration may be found in the treatment of toxic sulphide liquor from the Kraft process for making paper pulp and in the treatment of steel pickle liquor. Both of these have the added advantage of regenerating useful compounds.

The toxic wastewaters that give rise to the most serious problems are metallic wastes and radioactive wastes. These types of wastes have common characteristics in that they contain hazardous elements which cannot be broken down (or not for many decades) and there appear to be no less toxic substances that could be used to replace them.

The two major alternatives for disposing of toxic waste may be described briefly as follows:

(a) To land—where wastewaters contain human toxins great care is required to avoid contamination of groundwater. Where aquifers are at great depth and not directly connected to streams or underground supplies, *e.g.*, some Middle Eastern countries, then land disposal may be preferable to using shallow semi-landlocked seas.

The usual disposal arrangement for less hazardous wastes is a lagoon, which may have a connection with a watercourse, but which also permits infiltration (plus some evaporation and possibly some degradation). In the long term swelling and blinding of the soil may reduce the infiltration capacity.

For more hazardous wastewaters the land disposal policy is segregation followed by long term containment of the hazardous material in impervious disposal sites.

(b) To sea—the sea has an almost unlimited capacity for dilution but the corollary is that it has an almost infinite retention time. Disposal to sea is therefore capable of diluting acute toxins below their toxic threshold but problems may arise with substances that accumulate due to geochemical or biochemical mechanisms.

A further complication with marine disposal is the international aspect. Waste material discharged into the ocean may be transported around the world.

Two means of disposal are commonly practised;

(i) Discharge by pipeline to inshore waters. Dispersion in the buoyant jet can give adequate initial dilution although inshore areas are particularly sensitive to pollution, being used as shellfisheries and recreational zones.

(ii) Deep sea disposal of toxic wastes has in recent years been the subject of several international agreements as the result of which the volume of hazardous disposal has declined and the nature of the waste has changed. Substances like organohalogens, carcinogenic substances, mercury and cadmium compounds, and plastics are all banned. Deep sea disposal of less hazardous material still takes place but in packaged form but only in deep sea.

8   INDUSTRIAL WASTE TREATMENT—CASE STUDIES

The foregoing notes have dealt in general with the problems of treating toxic wastes but it is not possible to explain the complexities in so general a discussion. The following notes therefore describe in more detail two examples of industries that produce toxic wastes and the methods used in treating them.

8.1   Tannery Wastes

The waste is complex arising from the following operations;

(a) Soaking—soil, dung, blood, *etc.*
(b) Dehairing—sodium sulphide and fibre
(c) Liming—calcium hydroxide and protein
(d) Deliming and bating—ammonium chloride and sodium bisulphite or boric, lactic, and sulphuric acids + proteolytic enzymes + protein
(e) Pickling—sulphuric acid, sodium chloride + protein

(f) Tanning—trivalent chromium and other neutral salts
   —catechols and pyrogallols
   —borax
   —alkali

(g) Dyeing—weak acid + dyes

The quality of wastes generated varies from $35–150\,l\,kg^{-1}$ hides. The quality of effluent is also very variable due to batch process.

Parameter	Range
Suspended Solids (SS)	0 8 500
Biochemical Oxygen Demand (BOD)	100–10 000
Sulphide (S^{2-})	0–500
pH	3.0–11.0

These variations can be considerably reduced by balancing. Effluent from tanning may be disposed of in one of four ways:

(a) Discharge to sewer
(b) Discharge to river or estuary
(c) Discharge to sea
(d) Discharge underground

The discharge of tannery waste to a river or estuary is usually subject to severe constraints on solids, BOD, toxins, pH, *etc*. Only a minority of tanneries have access to the sea and though underground disposal is popular in the USA it is not favourably regarded in the UK.

Whether effluents are being discharged to sewers or to rivers the form of the pre-treatment is similar. In cases where space is very limited a mechanically brushed perforated screen is all that is installed but wherever possible a balancing tank should be added. The latter will reduce the need for pH correction, reduce the load of BOD and SS, and enable any treatment units to operate continuously.

If the effluent is to be discharged to sewers some reduction in sulphide and metals may be necessary. Removal of specific contaminants such as chromium sulphide may be desirable but the isolation of drainage lines usually makes this too expensive. It is however usually possible to separate the sulphide-containing liquors (dehairing and liming liquors). The aim is to reduce sulphides from concentrations of up to $2\,500\,mg\,l^{-1}$. Three methods of sulphide removal have been used:

(a) Precipitation by copper as $(FeSO_4)$
(b) Oxidation by chlorine
(c) Oxidation by dissolved air in the presence of a catalyst $(MnSO_4)$

The relative economics of these processes favour the aeration technique. Chromium is less of a problem than sulphide due to the very poor solubility of trivalent chromium at pH values over 6. Also chromium in the spent liquors has been reduced in recent years by improvements in leather technology. Nevertheless at large tanneries recovery of chromium is economically viable.

The aim of the treatment is to reduce the chromium concentration below $20 \, \text{mg} \, l^{-1}$ which is the standard imposed by many Water Authorities for discharge to the sewer. The tanning liquor may be treated with sodium hydroxide, sodium carbonate, or calcium hydroxide. There is a danger with the sodium salts that at high pH values some of the chromium may redissolve. But, where it is intended to recover the chromium for re-use, they are preferred to the use of lime since the presence of calcium salts is undesirable. Recovery of chromium can be achieved by removing the precipitated hydroxide in a filter press, dewatering by centrifuging, and redissolving in sulphuric acid.

Pre-treatment of leather dressing wastes is often carried out prior to discharge to sewer for, although relatively small in quantity, they are a very high polluting load. Substantial reductions in strength may be obtained by precipitation with lime together with iron or aluminium salts. The process requires careful control to minimize the sludge production, which can be up to 80% of the volume of waste treated.

Where tannery effluents have been discharged to sewers there have been some reports of difficulties at the municipal treatment plants. These difficulties seem to have arisen mainly because of the increase in organic concentration rather than any toxic effect. Undiluted tannery wastes may have BOD values around $2\,000 \, \text{mg} \, l^{-1}$ and therefore require a minimum of 3–4 times dilution to avoid problems of oxygen transfer (care is needed in the use of COD data since the ratio of COD : BOD is often twice that found in raw sewage). There have been suggestions that the sulphide present in tannery waste may have an adverse effect on the biological treatment in admixture with sewage, but this is unlikely to be serious at concentrations of $<10 \, \text{mg} \, l^{-1}$. Higher concentrations around $25 \, \text{mg} \, l^{-1}$ are tolerated by the activated sludge process and up to $50 \, \text{mg} \, l^{-1}$ for short periods. Percolating filters are even more resistant and concentrations up to $100 \, \text{mg} \, l^{-1}$ of sulphide can be treated successfully. The severe restrictions imposed on discharges of sulphide are mainly to protect the health of workers in the sewer and the fabric of the sewer and are not aimed primarily at protecting the treatment plant. Metal toxicity from tannery wastes has not been found to be a problem. Most of the metal is in the form of trivalent chromium and around 80% of this is removed during primary sedimentation. The resulting sludges do not give rise to difficulties in digestion provided that the retention time exceeds 21 days.

Experience of treating tannery waste on its own has confirmed the general rule that it is better treated along with domestic waste. Anaerobic treatment has not been too successful with moderate BOD removals and some difficulty in treating the resultant liquor.

Activated sludge appears to have good potential for treating tannery waste although this has not been exploited. Loading rates of 0.5–1.0 kg BOD per kg MLSS per day can be used with BOD removals $\geqslant 90\%$. In particular the oxidation ditch seems an attractive form of the activated sludge process for treating tannery wastes.

Experience with biological filters has been somewhat mixed. It has proved to treat effluent on a conventional stone filter using loadings around 0.12 kg BOD per m^3 per day with removal rates of over 95%. High-rate filtration using plastic media is likewise possible but the economics of this are not attractive when compared with chemical treatment.

8.2 Metal-Processing Wastes

Discharge of metals to the aquatic environment has been a major cause of concern and the treatment of these wastes has consequently attracted considerable attention. Wastes containing metals may arise from a variety of industrial and agricultural operations including tanneries, paint manufacture, battery manufacture, pig wastes, *etc.*, but the main source is from metal processing. The wastes from metal processing may be classified as follows:

(a) Mining—ore production and washing—also contain inert SS
(b) Ore processing—smelting, refining, quenching, gas, scrubbing, *etc.*—also contain sulphides, ammonia, and organics
(c) Machining—metal particles from machining usually mixed with lubricants
(d) Degreasing—metals mostly in solution with cyanides, alkalis, and solvents
(e) Pickling—acids with metals and metallic oxides in solution
(f) Dipping—alkalis with sodium carbonate, dichromate, *etc.*, plus metals
(g) Polishing—particles of metals and abrasives together
(h) Electrochemical or chemical brightening and smoothing—acids, mainly sulphuric, phosphoric, chromic, and nitric with metals in solution
(i) Cleaning—hot alkalis with detergents, cyanides, and dilute acids plus metals in solution
(j) Plating—acids, cyanides, chromium salts, pyrophosphates, sulphamates, and fluoroborates plus metals in solution
(k) Anodizing—chromium, cobalt, nickel, and manganese in solution

The sources of wastes in metal processing are numerous and also extremely variable both in quantity and quality. Metals in the wastes occur in forms ranging from large particles of pure metal in suspension to metallic ions and complexes in solution. The most appropriate method of treatment depends upon the form of the metal, its concentration, pH, other constituents of the waste, and the desired effluent standard. The technique most commonly employed in treating metal processing wastes is precipitation using pH adjustment. The optimum pH for precipitation varies depending on the particular metal and where several metals are involved a compromise pH is used. A typical value is in the range 8.0–9.0. With amphoteric metals, notably zinc, care must be taken to avoid too high a pH to prevent the formation of zincates. It should also be appreciated that other constituents of the waste, *e.g.*, ammonia, can significantly affect the solubility of the metal hydroxides and it is therefore not possible to predict accurately the level of residual metal in the treated effluent.

Whilst the hydroxide precipitation method is satisfactory for most metals encountered in effluents both hexavalent chromium and lead are not precipitated in this way. Hexavalent chromium is present in wastes from metal plating and must first be reduced to the trivalent form before treatment with lime or caustic soda. The reducing agents commonly used are sodium bisulphate, sulphur dioxide, and occasionally ferrous sulphate. The reduction is carried out under acid conditions and subsequent addition of alkali precipitates, trivalent chromium hydroxide.

In the case of lead, the hydrated oxide formed when lime or caustic soda is added to the lead waste has an appreciable solubility and the resulting effluent after removal of solids would normally be unsatisfactory for discharge to sewer or watercourse. However basic lead carbonate has a very low solubility and therefore sodium carbonate can be used in place of lime as the precipitating agent. Like zinc, lead is amphoteric and redissolves as plumbate at high pH, and so careful pH control must be exercised.

A particular type of precipitation system used in the metal plating industry is known as the Integrated Method of Treatment. The principal feature of this system is that the rinsing stage immediately after the metal plating stage is a chemical rinse which precipitates the metal from the liquid around the article being plated. A further water rinse is then required to wash off the treatment chemical. In the case of nickel plating the chemical rinse would contain sodium carbonate to precipitate nickel carbonate, whilst with chromium, a prior stage to effect reduction from hexavalent to trivalent form would be required. The integrated system has the advantages that water re-use can be readily practised and that the metals are not precipitated in a mixture and so can also be recovered. However, it is sometimes difficult to adapt the system to existing plating lines, since it necessitates the placement of an extra tank in the line.

Once the metals have been precipitated from solution, liquid, and solid phases must be separated. The traditional method for this stage of treatment is settlement in either a circular or rectangular tank. In small installations where the effluent flow is less than say $25 \, \text{m}^3/\text{day}$, it is convenient to carry out the effluent treatment on a batch basis and to allow settlement to take place in the same tank as that used for reaction. For larger installations a continuous flow system is required. The size of the tank depends on both the maximum effluent flowrate and on the configuration adopted for the tank. The most common type of settlement vessel is of the vertical upward flow pattern having a central feed well, a peripheral collection launder, and a sludge cone at the bottom. Clarification of the effluent can be enhanced by the use of flocculating agents. Obviously the size and mode of operation of the precipitation system significantly affects quality of the effluent but typical figures for a well-designed, efficiently operated, settlement system for metal hydroxide precipitates would be in the range 10–$30 \, \text{mg} \, \text{l}^{-1}$ suspended solids.

Where space is at a premium, a compact settling system utilizing parallel tilted plates or tubes can be used to perform the separation stage.

There are two factors which make this system efficient in terms of ground area used. These are:

(a) the distance through which a settling particle has to fall to become 'settled' is considerably reduced;

(b) the configuration produces laminar flow conditions which enhance the settling rate and overall efficiency.

Tilted plates can also be used to uprate existing settling tanks.

Flotation may be used as an alternative to settlement. This process, which is gaining in popularity, consists in the carrying of metal hydroxides and other particles in suspension to the surface of the liquid in the flotation vessel by increasing particle buoyancy using the gas bubbles which adhere to the particles. The scum containing the gas bubbles and separated solids is skimmed off. Variations in the process lie mainly in the method of producing the carrier gas bubbles. This may be done by injecting a super-saturated solution of air in water under pressure into the tank—dissolved air flotation—or by injecting air through a diffuser—dispersed air flotation—or by the electrolysis of water to yield fine bubbles of hydrogen and oxygen—electrolytic flotation. The gas bubbles produced in these processes are extremely small, normally in the range 70–150 μm.

The use of direct filtration appears to be a very attractive process for the phase separation but unfortunately is seldom appropriate, mainly because of the tendency for the filter media to blind rapidly. This tendency is largely due to the gelatinous nature of the metal hydroxide precipitates. Occasionally, where a more granular precipitate is obtained, direct filtration can be satisfactory and a high quality effluent can be obtained.

Whilst filtration has only limited application as the main means of solids removal it is frequently used to polish the effluent from a settlement or flotation system to produce a higher quality effluent.

Where the metal is substantially in solution there are various techniques for separation or concentration of the metal so that a high quality treated effluent may be obtained.

8.2.1 Ion Exchange. Ion exchange resins are in general of two types, insoluble organic acids, used for cation exchange, and insoluble organic bases, used for anion exchange. Cation exchangers may be either sulphonic or carboxylic acids while anion exchangers may be either quaternary or tertiary amines.

An ion exchange system consists of a pair of columns or pressure vessels, one containing an anion exchange resin, the other the cation exchange resin. The effluent is continuously pumped through the two columns in series to yield the treated effluent. Where an exceptionally high quality effluent is required a third 'mixed-bed' exchange column can be placed after the cation exchange column.

After a period of running the treated effluent quality deteriorates due to exhaustion of the resin capacity and the resins must be regenerated. It is usual to have two trains of ion exchange columns so that whilst one is being regenerated the other can take the flow.

It is clear that the regeneration liquors will contain all the metals and associated anions which the raw effluent originally contained but they will now be in a concentrated form. These liquors must be treated by the precipitation–solids separation method but the relatively small volume frequently means that the treatment can be done on a batch basis and absolute control over the final effluent quality can be exercised.

Ion exchange systems are particularly suitable for the treatment of metal containing wastes where relatively low concentrations of metal have to be removed. This enables long periods between regeneration of the resins and the production of a good quality water which is usually suitable for re-use. Rinse waters from electroplating operations are frequently treated by this process.

In practice, in addition to metals, the waste may contain other contaminants which it may be necessary to remove to protect the ion exchange resins. Organic contaminants are particularly damaging and should be removed by passage of the effluent through an activated carbon column preceding the ion exchange system.

8.2.2 Evaporation. Evaporation is one of the most common methods used in industry for the concentration of aqueous solutions. Nevertheless, use of this process as a means for effluent treatment is rare and occurs only under special circumstances where the effluent contains a high concentration of a valuable material. The only application of note here is on the concentration of static rinses (drag out) from electroplating operations, especially chromium plating. In this application the rinse liquor is evaporated to a metal concentration which makes the concentrate suitable for direct re-use in the plating bath.

8.2.3 Reverse Osmosis. On a practical scale this process is still in its infancy, and whilst several plants for the treatment of brackish water to yield potable water have been successfully installed, there are certain drawbacks which have limited its development for the treatment of industrial effluents. In particular the process requires high pressures (up to 100 atmospheres) and is thus fairly costly in terms of energy. The delicacy of the membranes is restrictive with regard to solids content and pH of the material being treated. However, the process has been used on effluents from electroplating in the electronic components industry, through the continuous development of the process and improved mechanical strength of membranes its range of applications will almost certainly increase.

8.2.4 Solvent Extraction. The separation of the components of a liquid mixture by treatment with a solvent in which one or more of the desired components is preferentially soluble is known as solvent extraction. It is a process which is widely used in the chemical and petrochemical industries but which has only recently been adopted for the recovery of metals from aqueous solutions. At present it finds only limited application in this sphere largely because of the cost of solvent loss.

8.2.5 Electrodialysis. The cost of this process is very dependent on dissolved solids concentration. It does not find common use in effluent treatment but may be appropriate in certain circumstances where the concentrate is of value. A recent development is a rotary electrode.

The foregoing discussion of the processes available for the treatment of metal-bearing effluents has been reviewed around the removal of metal salts rather than elemental metal. The metals themselves do arise in effluents from metal processing industries and metal fabrication operations. Where the metal is in large particles, such as swarf, simple screening is usually satisfactory.

Smelting operations produce a very fine dust cell 'fume' in the furnace off-gases. Where wet gas scrubbing of these gases is carried out, the resultant liquor, which contains the fume, must be treated in some way. Settlement is usually satisfactory although a polyelectrolyte may be necessary to achieve a satisfactory settling rate.

The conventional precipitation method described above yields a sludge containing 95–99% water. Even with extremely good consolidation, in a deep tank, the sludge will still contain about 90% water. To reduce the bulk for disposal, it is usual to dewater by some mechanical means.

The items of equipment commonly used for this purpose are filter presses, rotary vacuum filters, and centrifuges each of which has its own particular advantages and disadvantages. Both the rotary vacuum filter and centrifuge are continuous processes and their effectiveness is enhanced by the use of a coagulant to condition the sludge. The filter press is a batch process and whilst a coagulant may be added to aid filtration, a handleable cake containing up to 30% dry solids may be obtained on the neat sludge. In general sludge produced from a lime precipitation is more granular and has better dewatering characteristics than sludge produced using sodium hydroxide as the precipitating agent.

Disposal of metal bearing sludges is, without a doubt, a major problem in some areas if satisfactory tips are not available. Where toxic metals are involved the sludge must be disposed of on a tip which is sealed to prevent the pollution of groundwater or it may be incinerated and the residual ash sealed prior to dumping.

If the economics of metal recovery–sludge disposal can demonstrate that separation of waste streams to recover specific metals is attractive then this can ease the sludge disposal system. This can be the case in certain instances particularly with the integrated plating system and with specific ion exchange systems, *e.g.*, chrome recovery. It seems likely that the rising costs for sludge disposal by contractors will probably tip the balance in favour of metal recovery.

9 BIBLIOGRAPHY

R. E. Train, 'Quality Criteria for Water,' Castle House Publishers Ltd., 1979.
J. S. Alabaster and R. Lloyd, 'Water Quality for Freshwater Fish,' Butterworths, London, 1980.
R. C. Curds and H. A. Hawkes, 'Ecological Aspects of Used Water Treatment Processes and their Ecology,' Academic Press, London, 1983, Vol. 2 and 3.
A. Porteous, 'Hazardous Waste Management Handbook,' 1985.
S. J. Arcievala, 'Wastewater Treatment and Disposal Engineering and Ecology in Pollution Control,' Dekker, 1981.
M. J. Hammer, 'Water and Wastewater Technology,' 2nd Edition, Wiley, New York, 1986.

CHAPTER 5

Water Quality and Health

R. F. PACKHAM

1 INTRODUCTION

The main consideration in ensuring the safety of public water supplies is the elimination of the agents of water borne infectious disease. In the last century major epidemics of typhoid and cholera in Britain and elsewhere in Europe were only eliminated when their bacteriological origin was recognized, the contamination of water supplies with sewage was eliminated, and disinfection treatment was introduced. There have been no major epidemics in the UK since the Croydon typhoid epidemic of 1937 following which all water supplies were chlorinated.

Concern about possible health effects of chemicals in drinking water stemmed from the application of sophisticated analytical techniques which revealed the presence of traces of many potentially harmful chemicals. The significance to health of these was uncertain although some would undoubtedly cause concern were they present at much higher concentrations. Any fear about the possibility of health effects was reinforced by evidence that environmental factors which could conceivably include drinking water quality are involved in many chronic diseases such as cardiovascular disease and cancer. Public interest in drinking water quality has been stimulated by some specific pollution problems, *e.g.*, detergents, pesticides, lead, and asbestos.

Acute health effects of chemical constituents of drinking water are very unlikely but effects due to exposure to low concentrations over a lifetime are just feasible. Such evidence as there is, suggests that any such effects are extremely small and difficult to measure, but this is still the subject of much research.

Organic compounds can adversely affect the taste of water at concentrations well below any likely to give rise to toxic effects. This is marked with chlorophenols and certain petroleum products but this natural defence mechanism

83

cannot be depended on. The taste, smell, appearance, 'feel' (hardness/softness) is markedly affected by chemical composition as is its corrosiveness to materials including toxic metals.

2 DRINKING WATER STANDARDS

In the UK water undertakers are required to supply water that is 'wholesome.' This term has never been defined precisely although reference has often been made to various editions of World Health Organization (WHO) Drinking Water Standards. In 1985 a European Commission Directive on the quality of water intended for human consumption[1] came into force bringing, for the first time, a legal requirement to comply with maximum admissible concentrations for over 40 parameters. Unfortunately, the Directive does not include rationale for these limits although about a third of them appear to relate to possible harmful effects on health. The Guidelines for Drinking Water Quality published by the World Health Organization in 1984[2] gives a comprehensive coverage of drinking water quality and provides rationale for the Guidelines Values set. Although the WHO limits have no legal force in themselves they are intended to provide guidance to national and other organizations wishing to set their own standards. Because of rapid developments in this field WHO plan to publish a revised edition of Guidelines for Drinking Water Quality in the early 1990's.

3 LEAD

Lead levels in raw water are normally very low and lead in drinking water is almost entirely due to lead pipes in household plumbing, usually in combination with soft water. Where there is 'plumbosolvency' the concentration of lead increases with the period of contact with the lead pipe, maximum concentrations arising in morning 'first draw' samples. The EC Drinking Water Directive gives a MAC of $50 \, \mu g \, l^{-1}$ but an annotation indicates that this applies to a sample taken after flushing. It adds that if the sample is taken directly or after flushing and the lead content exceeds $100 \, \mu g \, l^{-1}$ frequently or to an appreciable extent 'suitable measures must be taken to reduce the exposure to lead on the part of the consumer.' The WHO Guideline Value for lead is $50 \, \mu g \, l^{-1}$.

It has been shown that exposure to low concentrations of lead in drinking water leads to detectable increases in blood lead levels but there is considerable controversy over the significance of this to health. Concern centres particularly on possible neurophysiological effects influencing learning ability and general behaviour in children. A Department of the Environment circular in 1982[3] advised that action should be taken to reduce lead exposure where blood lead concentration exceeds $25 \, \mu g \, (dl)^{-1}$.

[1] Council of the European Communities, 'Council Directive Relating to the Quality of Water Intended for Human Consumption (80/778/EEC),' 15 July, 1980, *Off. J. EC*, 1980, **L229/11**.
[2] World Health Organization, 'Guidelines for Drinking Water Quality,' WHO, Geneva, 1984.
[3] Department of the Environment, 'Lead in the Environment,' Circular 22/82, HMSO, London, 1982.

Table 1   *Lead in drinking water*[4]

Sample	Percentage of households with lead concentrations		
	$> 50\,\mu g\,l^{-1}$	$> 100\,\mu g\,l^{-1}$	$> 300\,\mu g\,l^{-1}$
Morning first draw	20.4	9	1.6
Random daytime	10.3	4.3	0.9

A survey in Great Britain of lead in drinking water in 1975[4] gave the results shown in Table 1. Following the survey, water authorities have undertaken more detailed surveys to locate areas where there are problems in meeting standards and remedial measures have been introduced including:

1 Changing the water supply to one that is non-plumbosolvent—where this is possible.
2 Raising the pH of the water supply to 8.0–8.5 and sometimes adding orthophosphate to control plumbosolvency. This is the most cost effective approach although the reduction of lead levels below $50\,\mu g\,l^{-1}$ cannot be guaranteed.
3 The removal of all lead plumbing and its replacement by more suitable material. This approach is totally effective but is very costly. Complete lead pipe replacement in UK could cost around £1.5 billion, about one-third of this cost falling on the utilities.
4 Advising householders to flush the tap before taking water for drinking purposes. This is only of value as an interim measure.

4   NITRATE

There is considerable evidence[5] that nitrate levels in water have increased in the United Kingdom since the 1960's and although the increased use of nitrogenous fertilizers has been blamed for this situation, it has been shown that additional factors need to be taken into account. These include changes in land use, in particular the conversion of pasture into arable land, and increased recycling of sewage effluent in lowland rivers.

Limits for nitrate in drinking water are based on its effect on a blood disease, methaemoglobinaemia, in bottle-fed infants. The 1970 WHO European Drinking Water Standards[6] included a desirable limit for nitrate of $50\,mg\,l^{-1}$ (as NO_3), levels between 50 and $100\,mg\,l^{-1}$ (as NO_3) being 'acceptable,' provided that medical authorities are warned of the possible danger of infantile methaemoglobinaemia.

[4] Department of the Environment, 'Lead in Drinking Water—A Survey in Great Britain 1975–76,' Pollution Paper No. 12, HMSO, London, 1977.
[5] Department of the Environment, Central Directorate of Environmental Protection, 'Nitrate in Water,' Pollution Paper No. 26, HMSO, London, 1986.
[6] World Health Organization, 'European Standards for Drinking Water,' WHO, Geneva, 1970.

More recently WHO set a Guideline Value of $10 \, \text{mg} \, l^{-1} \, \text{N}$ ($44.3 \, \text{mg} \, l^{-1} \, \text{NO}_3$) and the EC Drinking Water Directive includes a maximum admissible concentration (MAC) of $50 \, \text{mg} \, l^{-1} \, \text{NO}_3$. Water supplies containing nitrate levels exceeding $50 \, \text{mg} \, l^{-1}$ affect approximately one million people in the UK, but, there is no evidence of a problem with infantile methaemoglobinaemia.

Infantile methaemoglobinaemia is caused by the bacterial reduction of ingested nitrate to nitrite in the stomach. The nitrite combines with haemoglobin in the blood to produce methaemoglobin thereby reducing the oxygen-carrying capacity. Many factors other than water nitrate level are now known to be important in relation to infantile methaemoglobinaemia and it is noteworthy that a high proportion of the cases reported in the world literature relate to rural wells where bacteriological contamination, which can influence methaemoglobinaemia, cannot be ruled out.

There has been considerable interest in the possibility that ingested nitrate can be reduced to nitrite in the adult stomach leading, in the presence of secondary amines, to the endogenous synthesis of *N*-nitroso compounds. The significance of *N*-nitroso compounds is that many are highly carcinogenic in laboratory animals. There has been concern that high nitrate levels in water could be associated with an increased incidence of cancer of the gastro-intestinal and urinary tracts.

The overall picture in the United Kingdom does not support this. Not only have gastric cancer rates been decreasing, while nitrate levels have been increasing, but many of the areas showing the highest rates of gastric cancer have a low level of nitrate in their water supply.

A WHO Working Group[7] reviewed the available evidence for an association between gastric cancer and nitrate levels in water. It concluded: 'There is no convincing evidence of a relationship between gastric cancer and consumption of drinking water containing nitrate at or below the present guideline value: above this level the evidence is inconclusive.'

Although there is no evidence that current levels of nitrate in public water supplies present a health hazard, the EC Maximum Admissible Concentration of $50 \, \text{mg} \, l^{-1}$ (as NO_3) is a mandatory requirement which will necessitate remedial action in UK and other European countries. It is likely that changes in agricultural practice, including controls in the use of nitrogenous fertilizers and the specification of protection zones around wells and boreholes, will be introduced in an attempt to stem the upward trend in nitrate concentrations. Where possible, nitrate levels in drinking water will be reduced by blending, but in some cases it will be necessary to install nitrate removal treatment. The two currently available options are ion exchange or biological denitrification either of which will add significantly to the cost and complexity of water treatment.

5 WATER QUALITY AND CARDIOVASCULAR DISEASE

The hardness of water supplies in Britain tends to follow a north to south-east

---

[7] World Health Organization, 'Health Hazards from Nitrates in Drinking Water,' WHO, Copenhagen, 1985.

gradient with the softest supplies in Scotland, Northern England, and Wales and the hardest in East Anglia and Southern England.

Mortality from cardiovascular disease (CVD) tends to follow the same pattern and several statistical studies have demonstrated a highly significant inverse relationship between cardiovascular disease mortality and water hardness even when other environmental and socio-economic factors are taken into account.

The British Regional Heart Study is a major epidemiological investigation of the roles of the environmental, socio-economic, and personal risk factors in determining the regional variation of CVD mortality in Britain.

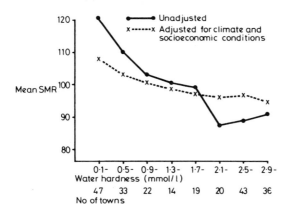

Figure 1   *Water hardness plotted against SMR* for all men and women aged 35–74 with cardiovascular disease for each town. (Water hardness: 1 mmol l^{-1} = calcium carbonate equivalent 100 mg l^{-1})*

* SMR is Standardized Mortality Ratio (actual mortality divided by mortality estimated on basis of age/sex profile of population).

Source: S. J. Pocock, A. G. Shaper, D. G. Cook, R. F. Packham, R. F. Lacey, P. Powell, and P. F. Russell, *Brit. Med. J.*, 1980, **280**, 1243.

The first phase of this research involved a statistical study[8] in which mortality from CVD in 253 large towns (population > 50 000) in England, Scotland, and Wales was examined, in relation to a wide range of climatic and socio-economic factors, as well as a variety of drinking water quality parameters. This work confirmed that, despite considerable scatter in the data (see Figure 1), there is a highly significant inverse statistical relationship between the hardness of drinking water and mortality from CVD. This relation persists when age, sex, socio-economic, and climatic factors are taken into account (see Figure 2) and is not shown for mortality from non-cardiovascular diseases.

On average very soft water towns have about 10% higher cardiovascular mortality than medium-hard or harder water towns after adjustment for socio-economic and climatic factors. Figures 1 and 2 show that most of the variation in mortality takes place at the soft end of the hardness range (below 150 mg l^{-1} (as $CaCO_3$)).

[8] S. J. Pocock *et al.*, *Brit. Med. J.*, 1980, **280**, 1243.

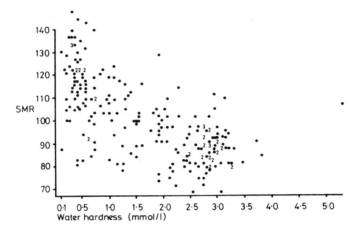

Figure 2   *Geometric means of SMR* (for all men and women aged 35–74 with cardiovascular diseases) for towns grouped according to hardness of water. (Water hardness: 1 mmol l^{-1} = calcium carbonate equivalent 100 mg l^{-1}.)*
* SMR Standardized Mortality Ratio.
Source: S. J. Pocock, A. G. Shaper, D. G. Cook, R. F. Packham, R. F. Lacey, P. Powell, and P. F. Russell, *Brit. Med. J.*, 1980, **280**, 1243.

Out of the large number of water variables examined, water hardness and certain associated water quality parameters, for example, nitrate, calcium, conductivity, carbonate hardness, and silica give the strongest negative correlation with CVD mortality. The proportion of water supply derived from upland sources gives an equally strong but positive correlation. At this stage it can only be concluded that CVD is influenced by water softness or some factor closely associated with it. The relationship could be due either to a harmful factor associated with soft water or a protective factor associated with hard water.

These conclusions have been strengthened by work[9] which has shown that in situations where the hardness of water supplies have changed, there is a corresponding change in the CVD mortality rate, at least for men. The size of the water factor is small in comparison with other CVD risk factors, *e.g.*, heavy smoking doubles the risk of a CVD event while very soft water increases it by 10%.

The principal differences in trace element levels in drinking water in hard and soft water towns are summarized in Table 2.[10] Apart from obvious differences in the major mineral components, the levels of aluminium, iron, manganese, and lead tend to be higher in the soft waters. Where the water is not only soft, but is also acidic, levels of copper as well as lead could be very high. Apart from these extreme situations copper levels tend to be higher in the hard water towns.

While the evidence that drinking water quality affects CVD has been strengthened by recent research, the present situation does not warrant such drastic, unwelcome, costly, and barely practical action as the hardening of soft

[9] R. F. Lacey and A. G. Shaper, *Int. J. Epidemiol.*, 1984, **13**, 18.
[10] R. F. Packham, 'Drinking Water Quality and Health — A Review of DoE Funded Research,' pre-print 5(2) of paper presented 7 April, 1987 to Public Works Congress, 1987.

water supplies. Some restriction is placed on the level of central softening by the EC Drinking Water Directive and the installers of domestic water softening equipment are now offering a hard water tap to provide water for drinking and culinary purposes.

Table 2 *Average trace element concentrations in first-draw and flushed water samples in hard and soft water towns*

Element	Soft water towns		Hard water towns	
	First-draw $\mu g\,l^{-1}$	Flushed $\mu g\,l^{-1}$	First-draw $\mu g\,l^{-1}$	Flushed $\mu g\,l^{-1}$
Silver	3*	3*	3*	3*
Aluminium	190	180	26*	27*
Boron	11	9	59	59
Barium	45	43	88	85
Beryllium	0.1*	0.1*	0.1*	0.1*
Bismuth	41*	48*	42*	51*
Calcium	18 000	17 000	92 000	90 000
Cadmium	2*	2*	2*	2*
Cobalt	5*	5*	5*	5*
Chromium	3*	3*	5	4
Copper	140	24	300	14
Iron	110	100	40	31
Potassium	1 000	970	3 300	3 200
Lithium	3*	3*	7	7
Magnesium	3 300	3 300	12 000	12 000
Manganese	20	22	4	5
Molybdenum	37*	37*	45*	54*
Sodium	8 300	8 100	28 000	28 000
Nickel	20*	18*	18*	16*
Lead	47	11	14	3
Silicon	2 000	2 000	4 600	4 500
Strontium	57	56	350	340
Titanium	4	4	16	14
Vanadium	4	4	12	12
Zinc	63	19	110	9
Zirconium	4*	4*	5*	5*

* Below the analytical limit of detection

6 ORGANIC MICROPOLLUTANTS

The total organic content of drinking water as represented by the concentration of dissolved organic carbon is usually no greater than $5\,mg\,l^{-1}$. Information on the nature of the organic chemicals present in drinking water developed rapidly following the application of GC–MS to water analysis. Worldwide over 2000 different organic compounds have been identified in drinking water mostly at levels below $1\,\mu g\,l^{-1}$. A wide range of substances has been found in almost all types of water.

In a study of fourteen water supplies in the United Kingdom, 343 compounds were identified.[11] The results which are similar to those of other workers[12,13] are summarized in Table 3.

Many of the individual substances found are of anthropogenic origin but a clear distinction between 'natural' and 'man made' is impossible. Irrespective of origin, however, drinking water is found to contain a complex mixture of organic substances at trace levels. The concentrations tend to be higher in water derived from the more polluted sources, *e.g.*, lowland rivers, but some volatile organic compounds, such as trichloroethylene, can occur in groundwater at higher levels than is common in surface waters. Halogenated organic compounds found in drinking water include many that are formed during water chlorination including trihalomethanes (THM) which often occur at higher concentrations $(10–100 \, \mu g \, l^{-1})$ than any other organic components detectable by gas chromatography.

Table 3   *Compound types identified in drinking water*

Compound	Total number identified
Carboxylic acids and esters	
Acids	15
Esters	31
Hydrocarbons	
Aliphatic	68
Alicyclic	5
Aromatic	48
Polynuclear aromatic	20
Halogenated compounds	
Aliphatics	53
Aromatics	8
Ethers	4
Miscellaneous	14
Oxygenated compounds	
Aldehydes	7
Ketones	19
Ethers	11
Miscellaneous	17
Miscellaneous compounds	23

[11] M. Fielding *et al.*, 'Organic Micropollutants in Drinking Water,' Water Research Centre Technical Report TR 159, WRc, Medmenham, Marlow, 1981.
[12] B. L. Carson *et al.*, in 'Advances in Identification and Analysis of Organic Pollutants in Water,' Vol 2, ed. L. H. Keith, Ann Arbor Science, Michigan, 1981, p. 497.
[13] N. D. Bedding *et al.*, *Science of the Total Env.*, 1982, **25**, 143.

There is now a good understanding of the nature of those organic substances that are amenable to GC–MS analysis. Unfortunately such compounds represent only 10–20% of the total organic content of water. Characterization of the substantial non-volatile fraction of organic compounds is a major problem which has been the subject of considerable research.[14] In many waters, humic substances represent a substantial part of the non-volatile fraction, while anthropogenic components include polychlorinated terphenyls, surfactants, pharmaceuticals, and pesticides. Using short-term bioassays for mutagenicity it has been shown that the non-volatile fraction contains mutagenic compounds including a substituted furanone that has become known as MX and related compounds.[15]

The toxicological information available for organic water constituents is defective in many respects and normally relates to levels of exposure that are orders of magnitude greater than those experienced from drinking water. The necessity to extrapolate from data resulting from high exposures with laboratory animals to very low exposures with man leads to considerable difficulty in assigning a limit to many water constituents. These problems are especially great with carcinogenic substances where the best available extrapolation techniques can involve uncertainties of about two orders of magnitude.

Table 4 *Suspected and known carcinogens, mutagens, and promotors identified in drinking water*

Compound	Frequency*	Type	Compound	Frequency*	Type
Acenaphthene	2	M	n-Dodecane	11	P
Benzene	13	C	Eicosane	2	P
1-Bromobutane	1	M	Fluoranthene	1	M
Bromochloromethane	2	M	Hexachloroethane	1	C
Bromodichloromethane	14	M	9-Methyl-fluorene	3	M
Bromoform	14	M	Octadecane	7	P
Carbon tetrachloride	6	C	Phenanthrene	7	M
Chlorodibromomethane	14	M	Phenol	5	P
Chloroform	14	C	Tetrachloroethylene	11	C
n-Decane	14	P	1,1,2,2,-Tetrachloroethane	1	C
Decanoic acid	3	P	Tetradecane	8	P
Decanol	1	P	Trichloroethylene	13	C/M
1,2-Dichloroethane	1	M	1,1,1-Trichloroethane	3	M
1,2-Dichloroethylene	2	M	2,4,6-Trichlorophenol	1	C
Dibromomethane	7	M	2,4,5-Trichlorophenol	1	M
1,4-Dioxane	1	C	Undecane	12	P

* Frequency of occurrence from 14 supplies
M-Mutagen, C-Carcinogen, P-Promotor

[14] B. Crathorne *et al.*, *Env. Sci. Tech.*, 1984, **18**, 797.
[15] M. Fielding and H. Horth, in 'Organic Micropollutants in the Aquatic Environment,' ed. G. Angeletti and A. Bjørseth, Kluwer, London, 1988, p. 284.

Some substances are found in drinking water which undoubtedly would be of concern were they present at much higher concentrations. Table 4 shows some of the carcinogens, mutagens, and promotors that have been identified in drinking water[16]; limits and Guideline Values have been set for some of these substances.

The real significance of the presence of such substances in water at sub-microgram quantities per litre is difficult to assess. In epidemiological studies any small increases in disease incidence due to drinking water are difficult if not impossible to perceive against the normal background incidence due to other more important factors. The possibility that there is some small effect due to trace organics in water cannot however be ruled out. The considerable induction period (20–30 years) for cancer also creates enormous problems in trying to link cause and effect, particularly as exposure information relating to 30 years ago for water constituents of interest today is almost non-existent. Thus the problem of trace organics in water does not easily lend itself to an epidemiological approach nor are long-term toxicological studies feasible at or near to the levels of exposure encountered in water. A wide variety of short-term tests for mutagenicity using bacteria or mammalian cells in tissue culture[17] are proving to be valuable research tools but these cannot be used to improve assessments of safety levels.

7 WASTEWATER RE-USE AND HEALTH

Approximately one-third of water supplies in the UK are derived from lowland rivers used also for conveying treated domestic and industrial wastes to the sea. These rivers are relatively short and often support large populations and considerable industrial activity. The proportion of the total river flow represented by wastewater can therefore often be high particularly during dry periods. The level of this indirect re-use of wastewater for potable purposes is likely to increase in the future.

No precise information is available about the effect on drinking water quality of wastewater recycling. It may be assumed that there will be an increase in the concentration of those waste constituents that are not biochemically degraded during sewage treatment, together with a wide range of metabolic products derived from degradable chemicals.

A DoE funded research programme to investigate the possible significance to health of wastewater re-use[18] had as a major objective the investigation of possible effects of re-use on the incidence of cancer. The analysis of different types of water supply by GC–MS provided no real evidence that drinking water derived from sources containing a high proportion of sewage effluent contains more potentially hazardous substances than those with a low proportion. The health of populations receiving water supplies derived from source waters including differing proportions of sewage effluent was examined. The most

---

[16] J. K. Fawell and H. A. James, in 'Organic Micropollutants in Water,' ed. D. Hammerton, Institute of Biology, London, 1982.

[17] B. J. Kilbey, M. Legator, W. Nichols, and C. Ramel, eds, 'Handbook of Mutagenicity Test Procedures,' 2nd Edn., Elsevier Scientific Publishing Company, Amsterdam, 1984.

[18] R. F. Packham, S. A. A. Beresford, and M. Fielding, *Sci. Total Env.*, 1981, **18**, 167.

detailed study concentrated on the London boroughs.[19] Some of these have groundwater supplies while others have supplies drawn from rivers containing a high proportion of sewage effluent. It was essential in this work to take account of socio-economic factors which are known to have an influence on cancer incidence. When this was done no strong association between cancer mortality and re-use was evident. A national study gave substantially similar results. It has to be recognized, that because of the large number of factors that influence cancer incidence, epidemiological studies of this kind are rather insensitive. The feasibility of large scale case control studies has been considered but it is unlikely that these will proceed in the near future.

8 ORGANIC COMPOUNDS AND DISINFECTION

The discovery in 1974 that those parts of New Orleans served with Mississippi derived water had higher cancer rates than those served with groundwater, and that chloroform, a suspected carcinogen, and other THM were present in the river derived supplies, provoked considerable research in the USA and elsewhere. This has shown that THM are formed when water containing 'natural' organics (*e.g.*, humic substances) is chlorinated in the disinfection process.[20] The formation of THM which can continue in the distribution system can be minimized by removing as much organic material as possible prior to chlorination or by minimizing the level of chlorination consistent with efficient disinfection. The THM have not positively been shown to be hazardous substances at the concentration at which they are present in drinking water.

Table 5 shows the results of a small survey of THM levels in Britain carried out in 1980.[21] Water undertakings in the UK have been advised that the situation does not warrant massive expenditure but that where THM can be minimized, without the commitment of considerable resources, it would be prudent to do so.

Table 5 *Total trihalomethane concentration in water*

Water sample	Total trihalomethane concentration ($\mu g \, l^{-1}$)		
	Minimum	Maximum	Mean
Raw	N.D.*	46	2
Treated	N.D.*	341	34
Stored (7 day) treated	N.D.*	378	52

* N.D. = the concentration of each of the four trihalomethanes in the sample was less than $1 \, \mu g \, l^{-1}$

[19] S. A. Beresford, *Int. J. Epidemiol.*, 1981, **10**, 103.
[20] R. R. Tressell and M. D. Umphres, *J. Am. Water Works Assoc.*, 1978, **70**, 604.
[21] R. A. Hyde, 'Trihalomethanes in Water,' Proceedings of Water Research Seminar, Water Research Centre, Medmenham, Marlow, 16–17 January, 1980.

THM formation will be minimized by using alternative disinfectants such as chlorine dioxide and ozone but there is considerably less information available on the compounds formed when these disinfectants are used.

9 ORGANIC CONTAMINATION OF DRINKING WATER IN DISTRIBUTION SYSTEMS

Potential sources of contamination of water during distribution include leachates from organic linings in pipes or from pipe materials and permeation of plastic pipes by certain organic compounds.

In the UK, about 60% of distribution pipes are made of ductile iron, and, until about 1977, the majority of these were coated internally with coal-tar pitch. Coal-tar pitch can contain up to 50% polycyclic aromatic hydrocarbons (PAH) and there is potential for leaching these substances, which include some known carcinogens, into the water supply. A survey of PAH levels in British waters[22] showed that low levels of these compounds can usually be detected in water that has passed through pipes lined with coal-tar pitch.

Organic compounds added to plastic formulation during pipe manufacture have been shown to leach into drinking water. It has been shown that antioxidants can leach from polyethylene (PE) pipe to give rise to objectionable tastes. Low concentrations of u.v. stabilizers and their degradation products can also be detected in water distributed using PE pipes[23] although the level of extraction usually decreases with time.

The ability of organic compounds to permeate PE has been known for a number of years[24] and it has been shown in the laboratory that low molecular weight, non-polar compounds such as benzene, toluene, trichloroethylene, chlorobenzene, *etc.*, will readily permeate polyethylene. Although this information, together with data on permeation rates, is helpful it cannot be translated easily to a level of soil contamination which might give rise to permeation of a water pipe in practice. This is because compounds adsorbed onto soil particles do not behave in the same was as in aqueous solution or in the vapour phase.

10 OTHER INORGANIC CONSTITUENTS

Drinking water standards and guidelines include limits for many inorganic ions in addition to lead and nitrate. Those set on the basis of toxicity include cadmium, mercury, chromium, arsenic, selenium, and cyanide. None of these represents a general problem in the UK although problems may occur periodically in specific locations. There are a few inorganic water constituents that

[22] R. I. Crane, M. Fielding, T. M. Gibson, and C. P. Steel, 'A Survey of Polycyclic Aromatic Hydrocarbon Levels in British Waters — Water Research Centre Technical Report 158,' WRc, Medmenham, Marlow, 1981.
[23] B. Crathorne *et al.*, 'The effect on water quality from the leaching of chemicals used in formulating plastic pipes,' contribution to discussion of paper 21, 17th International Water Supply Congress, Rio de Janeiro, 12–16 September, 1988.
[24] R. F. Packham and J. K. Fawell, Proceedings of Conference 'The Use of Plastics and Rubber in Water and Effluents,' Plastics and Rubber Institute, London, 1982.

attract media attention from time to time and some notes on the more common of these are included here.

10.1 Aluminium

Aluminium sulphate and other aluminium compounds are widely used as coagulants in the treatment of water for public supply. On addition to water aluminium forms an insoluble 'floc' of hydrolysis product which engulfs colloidal and suspended material in the raw water. The floc is normally removed by filtration often preceded by sedimentation. Inefficient treatment can lead to the breakthrough of floc into the water distribution system where it can give rise to dirty water problems. Limits placed on the maximum level of aluminium in tap water are all set with the objective of limiting such problems; they have no toxicological basis.

Concern about possible toxic effects of aluminium in drinking water stemmed from evidence of its role in dialysis encephalopathy (also called dialysis dementia). High aluminium levels in the brain tissue of patients dying of this disease and an association, between its incidence and high aluminium concentrations in dialysate water, have led to the introduction of measures to minimize aluminium intake from this source. The very low levels of aluminium required (less than $10 \mu g\, l^{-1}$) can only be met by water treatment at the point of use. Water authorities provide information to renal dialysis centres on aluminium levels but accept no responsibility for meeting such a low limit.

High levels of brain tissue aluminium are found in persons dying of a form of premature senile dementia—Alzheimer's disease. One of several mechanisms postulated for this disease ascribes an important role to ingested aluminium. Even in worst cases it is unlikely that water contributes more than 5% of the total dietary intake of aluminium although it has been argued that water aluminium may be more bioavailable than other dietary forms. Concern about water aluminium has been heightened by a recent epidemiological study[25] indicating an association between water aluminium at low levels and the incidence of Alzheimer's disease. Further research is in progress to elucidate these findings.

10.2 Asbestos

Asbestos is a component of many materials used in contact with drinking water including asbestos–cement pipes, tanks, reservoir covers, and certain glands and packing materials. Asbestos is a natural contaminant of water in some parts of the country; its presence in water may also result from pollution from tips containing asbestos waste. It has been shown[26] that drinking water can contain of the order of 1 million chrysotile fibres per litre, usually as short fibres (less than $5 \mu m$ in length). Although there is clear evidence of harmful health effects

[25] C. N. Martyn *et al.*, *Lancet*, 14 January, 1989, **59.**
[26] D. M. Conway and R. F. Lacey, 'Asbestos in Drinking Water — Water Research Centre Technical Report TR 202,' WRc, Medmenham, Marlow, 1984.

associated with inhaled asbestos, evidence of effects associated with ingested asbestos is inconclusive. The World Health Organisation stated in 1984 that there was insufficient evidence to set a Guideline Value for asbestos. More recently the US Environmental Protection Agency have proposed a Maximum Contaminant Level of 7 million fibres l^{-1} (longer than 10 μm). This level has not been encountered in UK water supplies to date but only very limited data is available on asbestos levels.

10.3 Fluoride

Fluoride in water has been the subject of an extended debate but the practical and scientific evidence leaves no room to doubt the benefits to dental health arising from fluoridation to 1 mg l^{-1} and the absence of harmful side effects. The moral issue of mass medication is still argued and fluoridation is still not widely practised although it has official backing.

10.4 Sodium

Sodium is normally present in drinking water at concentrations below 50 μg l^{-1} and the intake from this source will represent no more than 1 or 2% of the total dietary intake. The only group for whom water sodium is of any consequence is bottle fed infants. The infant kidney is less effective than the adult at sodium elimination so there is a danger of hypernatraemia if cow's milk, which has three times the sodium level of human milk, is fed to the infant. The sodium content of milk powders used in baby food is therefore reduced and it is also important not to make the feed up in water containing a high sodium concentration. Normally this will not arise except where a hard water has been softened in a domestic water softener by base exchange. The Department of Health[27] has advised against the use of such water in preparing baby food and it is now good practice to provide a hard water tap for drinking and culinary purposes in domestic water softener installations.

11 CONCLUSIONS

There is only very limited evidence that chemical constituents of drinking water are involved in health problems. In general, water components represent only a very small proportion of the dietary intake of chemicals. Any health effects of such materials are likely to result from long-term exposure and unless this leads to an extremely unusual disease the effect is likely to be very difficult to detect.

Investigations into the basis of the observed statistical relationship between water softness and mortality from cardiovascular disease are of the classical type in which a cause is sought for an observed effect. In much of the other work referred to, it may be said that, possible causes have been identified for which we

[27] Department of Health, 'Present Day Practice in Infant Feeding,' Report on Health and Social Subjects No. 20, HMSO, London, 1980.

are seeking possible effects. This might give rise to the criticism that resources are being wasted in the pursuit of non-existent problems.

The justification for this work is the strong evidence, that exists, that environmental factors are important in several chronic diseases including cancer. The average consumer has little opportunity to exercise choice in relation to the water that he drinks and it is therefore important that we should assess the risks and benefits associated with water constituents which we may be able to control. The quantification of risks is no easy matter and the work described represents only the initial steps in the process. What is urgently needed is a valid technique for comparing these risks with others that are accepted by the public on a day-to-day basis. Thus we need to know how the risk associated with a certain concentration of a toxic chemical in drinking water compares with that involved in crossing a busy street, driving a car, or taking part in recreational activities, *e.g.*, winter sports, swimming, and mountaineering. This approach has been discussed by Pochin.[28]

Risk assessment is necessary to protect public money as well as public health. The enforcement of unnecessarily stringent water quality standards could involve expenditure of millions of pounds which might be used to improve public health more effectively in other ways.

[28] E. E. Pochin, 'The Acceptance of Risk,' *Brit. Med. Bull.*, 1975, **32**, 184–190.

CHAPTER 6

Biological Aspects of Freshwater Pollution

C. F. MASON

1 INTRODUCTION

We can define water pollution as 'the introduction by man into the environment of substances or energy liable to cause hazards to human health, harm to living resources and ecological systems, damage to structure or amenity, or interference with legitimate uses of the environment'.[1,2] Pollutants are therefore chemical or physical in nature and can thus be measured more or less accurately in water. The measured quantities can then be compared with standards of allowable concentrations. Why then do we need to undertake quantitative studies of organisms and biological communities when we know that they can be defined with much less precision than chemical or physical parameters? There are a number of reasons why biological studies are important.

Firstly, the definition of pollution given above includes the adverse effects on living resources and ecological systems, so that such impacts need quantifying. Man, of course, by drinking water, eating fish, and using aquatic ecosystems as recreational resources, is also linked to the freshwater community. We can consider the effects of pollutants we record on the biological community as an early warning system for potential effects on ourselves.

Secondly, animal and plant communities also respond to intermittent pollution which may be missed in a chemical surveillance programme. A chemical survey will indicate to the water manager what is present in the sample at a particular moment in time. However, a plug of pollution could have passed

[1] M. W. Holdgate, 'A Perspective of Environmental Pollution,' Cambridge University Press, Cambridge, 1979.
[2] C. F. Mason, 'Biology of Freshwater Pollution,' Longman Scientific and Technical, Harlow, 1981.

down a river before the water sample was taken or between sampling occasions, while the disreputable manager of a waterside factory may well know which day the river inspector takes his sample so that he can discharge his poisonous waste accordingly. The plug of pollution will kill the most vulnerable members of the aquatic community, the most sensitive species acting as *indicators* of pollution. The amount of change in the aquatic community will be related to the severity of the pollution incident. Because the community can only be restored to its former diversity by reproduction and immigration, its recovery is likely to be slow. If the intermittent pollution occurs with some frequency, the community will remain impoverished. The biologist will be able to detect such damage to the biological community and suggest a more detailed surveillance programme, both biological and chemical, to find the source of pollution.

Thirdly, biological communties may respond to unsuspected or new pollutants in the environment. There are probably more than 1 500 potential pollutants discharged to freshwaters, while the water industry will only routinely test for some 30 determinands, for it is financially prohibitive to look for more compounds. However, if there is a change in the biological community, for example a large fish kill, then a wider screening for pollutants can be initiated. The widespread occurrence of polychlorinated biphenyls (PCBs) in the environment was detected following the large-scale death of seabirds in the Irish Sea and by routine surveillance of biological materials.[1,3]

Finally, some chemicals are accumulated in the bodies of certain organisms, concentrations within them reflecting environmental pollution levels over time. In any particular sample of water, the concentration of a pollutant may be too low to be detectable using routine methods, but nevertheless will be gradually accumulated within the ecosystem to concentrations of considerable concern in some species. Heavy metals, organochlorine pesticides, and PCBs have caused particular problems in aquatic ecosystems and are potential threats to human health.

This review will describe the effects of major types of pollutants on aquatic life, though it must be remembered that any particular source of pollution may contain a range of compounds, such that the effects on ecosystem processes are likely to be complex. The review will conclude with description of some of the methods of assessing the biological effects of pollution.

2 ORGANIC POLLUTION

Organic pollution results when large quantities of organic matter are discharged into a watercourse to be broken down by micro-organisms which utilize oxygen to the detriment of the stream biota. A simple measure of the potential of organic matter for deoxygenating water is given by the biochemical oxygen demand (BOD) which is determined in the laboratory by incubating a sample of water for five days at 20 °C and determining the oxygen utilized.

One of the major sources of organic pollution are effluents from sewage treatment works. In the United Kingdom, such effluents are supposed, as a

[3] S. Jensen, *Ambio*, 1972, **1**, 123.

minimum requirement, to meet the Royal Commission Standard, allowing no more than $30 \, \text{mg} \, l^{-1}$ of suspended solids and $20 \, \text{mg} \, l^{-1}$ BOD (a 30:20 effluent). A dilution with at least eight volumes of river water, having a BOD of no more than $2 \, \text{mg} \, l^{-1}$, is required to achieve this standard. Unfortunately, the design capacities of many sewage treatment works are below the population they are now having to serve (Figure 1) and resources are not available to update the plants, so that organic pollution of the receiving stream occurs periodically. In the majority of the poorer countries of the world there are few, or indeed often no, sewage treatment facilities and the faecal contamination of water results in many parasitic infections and waterborne diseases such as dysentery, cholera, and poliomyelitis.

Figure 1   *A poor quality effluent discharged from a sewage treatment works*

Other sources of organic pollution include industries such as breweries, dairies, and food processing plants. Run-off from the hard surfaces and roads of towns, especially during storm conditions, can be very polluting. Farm effluents have become a particular problem over the last few years, with the intensification of livestock rearing and the overwintering of animals in confined buildings, as well as the increased use of silage to feed them. Silage effluent can be 200 times as strong as settled sewage, measured in terms of BOD. In South West England, a predominantly rural area, some 40% of rivers are failing to comply with their long-term quality objectives, agricultural pollution being the major reason for the decline.

Figure 2 outlines the general effects of an organic effluent on a receiving stream.[1] At the point of entry of the discharge there is a sharp decline in the concentration of oxygen in the water, known as the *oxygen sag curve*. At the same time there is a massive increase in BOD as the micro-organisms added to the stream in the effluent and those already present utilize the oxygen as they break down the organic matter. As the organic matter is depleted, the microbial populations and BOD decline, while the oxygen concentration increases, a process known as *self purification*, assisted by turbulence within the stream and by the photosynthesis of algae and higher plants (macrophytes).

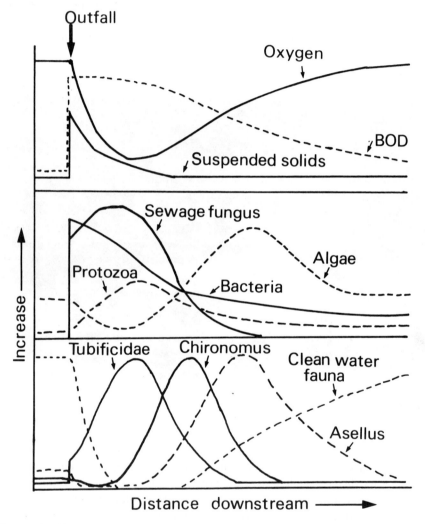

Figure 2 *Changes in water quality and populations of organisms in a river below a discharge of an organic effluent*

The effluent will also contain large amounts of suspended solids which cut out the light immediately below the discharge, thus eliminating photosynthetic organisms. Suspended solids also settle on the stream bed altering the nature of the substratum and smothering many organisms living within it.

Under conditions of fairly heavy pollution sewage fungus develops. This is an attached, macroscopic growth consisting of a whole community of micro-organisms, dominated by *Sphaerotilus natans*, which consists of unbranched filaments of cells enclosed in sheaths of mucilage, and by zoogloeal bacteria. Sewage fungus may form a white or light brown slime over the surface of the substratum, or it may exist as a fluffy, fungoid growth with long streamers trailing into the water.[4] I have traced the development of sewage fungus in a formerly pollution free river for 20 km downstream of a rural milk processing factory whose primitive treatment facility became overloaded one January during the production of the year's supply of chocolate Easter eggs.[5]

Protozoans are chiefly predators of bacteria and respond to changes in bacterial numbers. Attached algae are eliminated immediately below the outfall due to the diminished penetration of light, but they gradually reappear below the zone of gross pollution, *Stigeoclonium tenue* being the initial colonizer. With the decomposition of organic matter large quantities of nitrates and phosphates are released stimulating algal growth and resulting in dense blankets of the filamentous green *Cladophora* smothering the stream bed. Similarly, macrophytes may respond to the increased nutrient concentration, though only *Potamogeton pectinatus* is very tolerant of organic pollution.[6]

Heavy organic pollution affects whole taxonomic groups of macro-invertebrates, rather than individual sensitive species [4,7] and it is only in conditions of mild pollution that the tolerances of individual species within a group assume significance. The groups most affected are those which thrive in waters of high oxygen content and those which live on eroding substrata, the most sensitive being the stoneflies (*Plecoptera*) and mayflies (*Ephemeroptera*). The differences in tolerances of groups of macro-invertebrates form the basis of methods for monitoring, as will be described later in this chapter.

In the most severe pollution the tubificid worms, *Limnodrilus hoffmeisteri* and/or *Tubifex tubifex*, are the only macro-invertebrates to survive. The organic effluent provides an ideal medium for burrowing and feeding, while in the absence of predation and competition, the worms build up dense populations, often approaching one million individuals per square metre of stream bed.[8] These worms contain the pigment haemoglobin, which is involved in oxygen transport, and they can survive anoxic conditions for up to four weeks. As conditions improve slightly downstream, the midge larva *Chironomus riparius*, which also contains haemoglobin, thrives in dense populations and, as the water self-purifies, other species of this large family of flies appear, the proportion of *Chironomus riparius* gradually declining. Below the chironomid zone, the isopod crustacean *Asellus aquaticus* becomes numerous, especially where large growths of *Cladophora* are present. At this point, molluscs, leeches, and the predatory alder

[4] H. A. Hawkes, in 'River Pollution II. Causes and Effects,' ed. L. Klein, Butterworths, London, 1962, p. 311.

[5] C. F. Mason and S. M. Macdonald, 'Otters: Ecology and Conservation,' Cambridge University Press, Cambridge, 1986.

[6] S. M. Haslam, 'River Plants,' Cambridge University Press, Cambridge, 1978.

[7] H. B. N. Hynes, 'The Biology of Polluted Waters,' Liverpool University Press, Liverpool, 1960.

[8] R. O. Brinkhurst, *J. Fish. Res. Board Can.*, 1970, **27**, 1961.

fly (*Sialis lutaria*) may also be present in some numbers. As self-purification progresses downstream the invertebrate community diversifies, though some stonefly and mayfly species, which are sensitive even to the mildest organic pollution, may not recolonize the stream.

Fish are the most mobile members of the aquatic community and they can swim to avoid some pollution incidents. In conditions of chronic organic pollution they are absent below the discharge, reappearing in the *Cladophora/ Asellus* zone, the tolerant 3-spined stickleback (*Gasterosteus aculeatus*) being the first to take advantage of the abundant invertebrate food supply. Organic pollution is usually most severe in the downstream reaches of rivers and may prevent sensitive migratory species, such as salmon (*Salmo salar*) and sea trout (*S. trutta*), from reaching their pollution free breeding grounds in the headwaters.

It must be remembered that many toxic compounds, such as ammonia, may occur in organic effluents, particularly those emanating from sewage treatment works. These make the prediction of the effects on the aquatic community of any particular discharge rather difficult.

3 EUTROPHICATION

It has already been described how the release of nutrients during the breakdown of organic matter stimulates the growth of aquatic plants. This addition of nutrients to a waterbody is known as eutrophication. Other important sources of nutrients include the increasing use of phosphorus-containing detergents (much of this entering the river in sewage effluent), agricultural run-off and leaching of artificial fertilizers, the washing of manure from intensive farming units into water, the burning of fossil fuels which increases the nitrogen content of rain, the felling of forests which causes increased erosion and run-off and bank erosion resulting from recreational boating.

Nitrogen and phosphorus are the two nutrients most implicated in eutrophication and, because growth is normally limited by phosphorus rather than nitrogen, it is the increase in phosphorus which stimulates excessive plant production in freshwaters. Nitrogen is highly soluble and fertilizers form the major source of this element to rivers accounting for some 70% of the annual mass flow of nitrogen in the rivers of East Anglia. Some 50% of the nitrogen applied to crops is lost to water. Phosphorus is largely insoluble, so that it enters water from land mainly by erosion. Some 90% of the phosphorus enters East Anglian rivers in the effluent from sewage works.

The concentrations of nitrate and phosphate in the water of Ardleigh Reservoir, a eutrophic waterbody in East Anglia, are shown in Figure 3.[9] Note that the concentration of nitrate increases during the late winter when fertilizer is applied to growing crops and is washed in large amounts into the streams which feed the reservoir. By contrast, the concentration of phosphate peaks in late summer, when the low flow in the feeder streams consists largely of treated sewage effluent while, at this time of year, much phosphate is released from the reservoir sediments into the water.

[9] M. M. Abdul-Hussein and C. F. Mason, *Hydrobiologia*, 1988, **169**, 265.

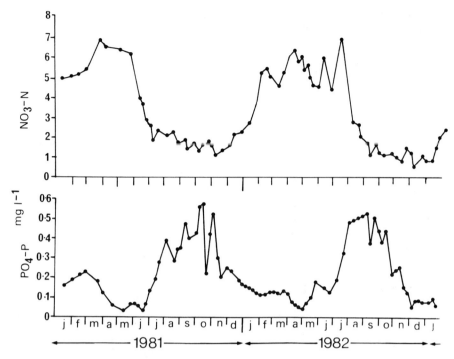

Figure 3 *Concentrations* (mg l^{-1}) *of nitrate and phosphate in Ardleigh Reservoir, eastern England, over two years*

Table 1 lists the guidelines for assessing the trophic status of a waterbody following a survey. Peak phosphorus concentrations in Ardleigh Reservoir were some 250 times the minimum concentration for assigning a waterbody as eutrophic, while, peak concentrations of nitrogen were 10 times the minimum concentration.

Table 1 *Eutrophication survey guidelines for lakes and reservoirs*

	Oligotrophic	Mesotrophic	Eutrophic
Total phosphorus $(\mu\text{g l}^{-1})$	<10	10–20	>20
Total nitrogen $(\mu\text{g l}^{-1})$	<200	200–500	>500
Secchi depth (m)	>3.7	3.7–2.0	<2.0
Hypolimnetic dissolved oxygen (% saturation)	>80	10–80	<10
Chlorophyll-a $(\mu\text{g l}^{-1})$	<4	4–10	>10
Phytoplankton production $(\text{g C m}^{-2}\text{d}^{-1})$	7–25	75–250	350–700

One of the major biological effects of eutrophication, resulting in considerable financial loss, is the stimulation of algal growth especially in water supply reservoirs. As eutrophication proceeds, there is a decline in the species diversity of the phytoplankton and a change in species dominance as overall populations and biomass increase. Figure 4 illustrates the seasonal changes in biomass of the dominant groups of algae in Ardleigh Reservoir.[9] Typical of temperate lakes is the early peak of diatoms (*Bacillariophyta*), followed by a late spring peak of green algae (*Chlorophyta*). Eutrophic lakes are characterized by an enormous summer growth of blue-green algae (*Cyanophyta* or *Cyanobacteria*), in the case of Ardleigh Reservoir mainly *Microcystis aeruginosa*, and these blue-greens are the major cause of water treatment problems.

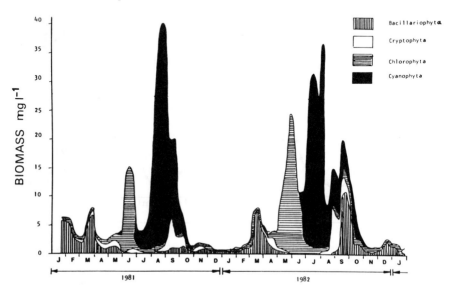

Figure 4 *Seasonal variation in phytoplankton composition and total biomass* (mg l^{-1} *wet weight*) *in Ardleigh Reservoir, eastern England, over two years*

Large populations of algae, and in some cases the zooplankton which they support, may result in the blocking of filters in the treatment works. The reservoir may have to be taken out of service, sometimes for several weeks, because the water becomes untreatable. If the reservoir is the only source of water this can have severe consequences. Some of the smaller algae may pass through the treatment process altogether, to decompose in the water main, giving drinking water an unpleasant taste and odour. The crash of an algal bloom and its subsequent decomposition may also result in a potable water so unpleasant to taste that it is almost undrinkable.

Drinking water which is high in nitrates presents potential health problems. In particular, babies under six months of age who are bottle fed may develop methaemoglobinaemia (blue baby syndrome), the nitrate in their feeds being reduced to nitrite in their acid stomachs and then oxidizing ferrous ions in the haemoglobin of their blood, so lowering its oxygen carrying capacity. There is also some evidence that nitrates may convert to carcinogenic nitrosamines in the adult stomach, though the link between high nitrate levels in drinking water and the incidence of cancer has never been satisfactorily made. The European Health Standard for drinking water recommends that nitrates should not exceed 50 mg

$NO_3 l^{-1}$, but concentrations in some lowland areas of Britian frequently exceed 100 mg $NO_3 l^{-1}$.

The algal blooms associated with excessive amounts of nutrients have other consequences for the aquatic ecosystem. The macrophyte swards of many lakes have been eliminated as the light is reduced on the lake bed, preventing photosynthesis of germinating plants. Growths of epiphytic algae on the leaf surface may also restrict light uptake by aquatic plants so that they become scarce in the lake. Zooplankton use macrophyte canopies as a refuge from fish. Without macrophytes, they are very vulnerable and this itself can accelerate the eutrophication process as the grazing pressure on phytoplankton is reduced in the absence of zooplankton, allowing denser algal blooms to develop.

The Norfolk Broads of eastern England are a series of shallow lakes, formed during medieval times when peat workings became flooded, and famous for their rich flora of *Charophytes* (stoneworts) and aquatic angiosperms, supporting a diverse assemblage of invertebrates. During the 1960's a rapid deterioration set in as nutrient loadings increased and a survey of 28 of the main broads in 1972–73 revealed that 11 were devoid of macrophytes and only 6 had a well developed aquatic flora.[10] The invertebrate fauna was similarly impoverished. Since then the situation has deteriorated further with additional losses of macrophytes and the Norfolk Broads now have some of the highest total phosphorus concentrations of world freshwater lakes.[11]

Fish communities also change as oligotrophic lakes become eutrophicated, those cold water fish with high oxygen requirements, such as salmonids and whitefish, being replaced by less demanding cyprinids. The commercial value of the fishery declines in parallel. Some algae at high densities produce toxins which kill fish. In the Norfolk Broads *Prymnesium parvum* has caused several large fish kills over the last decade. The sudden collapse of an algal bloom may result in rapid deoxygenation of the water and massive kills of fish.

Problems of eutrophication are not restricted to standing waters. Nutrient rich streams support dense growths of aquatic weeds which impede the flow of water, increasing the risk of severe flooding following summer storms. The weeds have to be cut, often twice a year, at great expense and the situation has been made worse by the clearance in the interests of 'river improvement' of bankside trees, which otherwise shade the stream and reduce weed growth.

It is possible to halt the progress of eutrophication.[12] This is most effectively done by controlling inputs of phosphorus because this element is normally limiting to plant growth and much of it in freshwaters is derived from point sources (*i.e.*, sewage treatment works), whereas nitrogen enters the aquatic ecosystem diffusely, via land drainage. Phosphorus can be removed from the effluent at the treatment works by chemical flocculation (tertiary treatment) or before water is pumped into a reservoir. However the reduction in the inflow water may be counterbalanced by the regeneration of phosphorus from the sediments, so-called internal loading, which may necessitate either sediment

[10] C. F. Mason and R. J. Bryant, *Freshwater Biol.*, 1975, **5**, 257.
[11] B. Moss, *Biol. Rev.*, 1983, **58**, 521.
[12] R. C. Loehr, C. S. Martin, and W. Rast, 'Phosphorus Management Strategies for Lakes,' Ann Arbor Science, Michigan, 1980.

removal or chemical flocculation and sealing, for example, with iron or aluminium sulphate.

In the Norfolk Broads the isolation of one small lake (Cockshoot Broad) from the phosphorus rich river water, followed by the removal of lake sediments, resulted in the development of a diverse sward of aquatic macrophytes. By contrast, at another site (Alderfen Broad), inflow water was diverted but the expensive task of sediment removal was not undertaken and any improvement in water quality and macrophyte growth was short-lived due to phosphorus release from the sediments. The reversal of eutrophication is therefore feasible but it is both very complex and very expensive.

4 ACIDIFICATION

Acid rain and acidification of freshwaters has received considerable attention over the past few years, but the problem is not new, for there are observations of lakes losing their fish populations in Scandinavia as early as the 1920's. Studies of the diatom remains in cores of sediment from lochs in South West Scotland have indicated that progressive acidification began around 1850, with an increase in *Tabellaria binalis* and *T. quadriseptata*, species characteristic of acid waters.[13] However, there has certainly been an acceleration of the process in the last three decades. For example, of 87 lakes in southern Norway surveyed in both the periods 1923–49 and 1970–80, 24% had a pH below 5.5 in the earlier period compared to 47% in the second period.[14] The major source of acid is undoubtedly the burning of fossil fuels, releasing oxides of sulphur and nitrogen to the atmosphere. Electricity generating stations are a large source of these pollutants but domestic and industrial sources are also significant, as are the exhausts of vehicles.

The acids either fall directly into waterbodies in precipitation or are washed in from vegetation and soils within the catchment. Three broad categories of water which differ in acidity can be recognized[15]:

1 Those which are permanently acid, with a pH less than 5.6, low electrical conductivity, and an alkalinity close to zero. Such conditions occur in the headwaters of streams and in lakes, where the soils are strongly acid, or in the outflows of peat bogs;

2 Those which are occasionally acid, where pH is normally above 5.6 but because they have low alkalinity (usually less than $5.0 \, \text{mg} \, l^{-1} \, \text{CaCO}_3$) the pH may drop below 5.6 periodically. These include streams and lakes in upstream areas of low conductivity on rocks unable to neutralize acid quickly. Such waters may show episodes of extreme acidity, for instance, during snowmelt or following storms. These may be very damaging to aquatic life but the infrequency of acid events makes the problem difficult to detect;

[13] R. W. Batterbee, R. J. Flower, A. C. Stevenson, and B. Rippey, *Nature (London)*, 1985, **314**, 350.
[14] A. Wellburn, 'Air Pollution and Acid Rain: The Biological Impact,' Longman Scientific and Technical, Harlow, 1988.
[15] United Kingdom Acid Waters Review Group, 'Acidity in United Kingdom Fresh Waters,' Interim Report, Department of the Environment, London, 1986.

3 Those which are never acid, the pH never dropping below 5.6 and the alkalinity always above $5 \, mg \, l^{-1} \, CaCO_3$.

Much of northern and western Britain has a solid geology consisting of granites and acid igneous rocks; there is little or no buffering capacity.[15] The situation is exacerbated in those catchments which have been extensively afforested with conifers. The sulphate ion is very mobile and transfers acidity very efficiently from soils to surface waters. The nitrate ion behaves similarly but is normally taken up by plants first.

In Canada, one experimental lake was artificially acidified, from 1975, by adding sulphuric acid and the pH dropped from 6.8 to around 5.0 over an eight year period. Among the first animals to disappear from the lake were shrimps and minnows. Within a year, at pH 5.8 opussum shrimps (*Mysis relicta*), with an estimated population of almost seven million before the experiment, had all but disappeared. Fathead minnow (*Pimephales promelas*) failed to reproduce and died out a year after the shrimps. Young trout were failing to appear and many of their food items were killed at pH 5.8 and as the pH fell still further cannibalism was noted in trout. At pH 5.6 the exoskeleton of crayfish (*Oronectes virilis*) began to lose its calcium and the animals became infested with a microsporozoan parasite. Some species, such as chironomid midges, did well as they were released from the pressure of predation in the simplified ecosystem. Since no species of fish reproduced at values of pH below 5.4, it was predicted that the lake would become fishless within ten years, based on knowledge of the natural mortalities of long-lived species. Some observations were unexpected in this experiment. There was no decrease in primary production, in the rate of decomposition, or in nutrient concentrations. The changes in the lake were caused solely by changes in hydrogen ion concentration and not by any secondary effects such as aluminium toxicity.[16] Table 2 provides a generalized summary of the sensitivity of aquatic organisms to lowered pH based on studies in Scandinavian lakes. Changes in the community begin at pH 6.5 and most species have disappeared below pH 5.0 leaving just a few species of tolerant insects and some species of phyto- and zooplankton.

Table 2   *Sensitivities of aquatic organisms to lowered pH*

pH	6.0	Crustaceans, molluscs, *etc.*, disappear.
		White moss increases.
pH	5.8	Salmon, char, trout and roach die.
		Sensitive insects, phytoplankton and zooplankton die.
pH	5.5	Whitefish, grayling die.
pH	5.0	Perch, pike die.
pH	4.5	Eels, brook trout die.

[16] D. W. Schindler, K. H. Mills, D. F. Malley, D. L. Finlay, J. A. Shearer, I. J. Davies, M. A. Turner, E. A. Linsey, and D. R. Cruikshank, *Science*, 1985, **228**, 1395.

Considerable research has been directed towards the effects of acidification on fish because of their economic and recreational importance. Figure 5 shows the proportion of lakes in southern Norway which have lost their brown trout (*Salmo trutta*) over the 35 years since 1940. By 1975 over half a total of 2 850 lakes were without trout. The effect of acidity on fish is mediated via the gills. The blood plasma of fish contains high levels of sodium and chloride ions and those ions which are lost in the urine or from the gills must be replaced by active transport, against a large concentration gradient, across the gills. When calcium is present in the water it reduces the egress of sodium and chloride ions and the ingress of hydrogen ions. The main cause of mortality in acid waters is the excessive loss of ions such as sodium which cannot be replaced quickly enough by active transport. When the concentrations of sodium and chloride ions in the blood plasma fall by about a third, the body cells swell and extracellular fluids become more concentrated. To compensate for these changes, potassium may be lost from the cells, but, if this is not eliminated quickly from the body, depolarization of nerve and muscle cells occurs resulting in uncontrolled twitching of the fish prior to death.

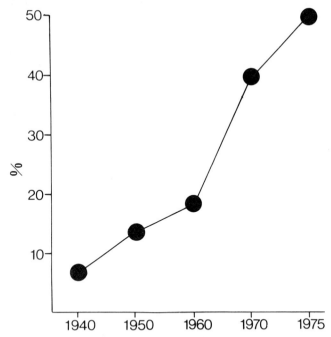

Figure 5   *Percentage of lakes* (n = 2 850) *in Scandinavia which have lost their population of brown trout*

Aluminium has been shown to be toxic to fish in the pH range 5.0–5.5 and during episodes of acidity aluminium ions are frequently present in high concentrations especially where water is draining conifer plantations. Aluminium ions apparently interfere with the regulation by calcium of gill permeability so enhancing the loss of sodium in the critical pH range. They also cause clogging of the gills with mucus and interfere with respiration. The developmental stages of fish are particularly sensitive to acidification and it is thought that aluminium may interfere with the calcification of the skeleton of fish fry resulting in a failure of normal growth. This failure in recruitment results in a gradual decline of the fish population to extinction.

The simplification of the aquatic ecosystem due to acidification may also affect higher levels of the food chain. The dipper (*Cinclus cinclus*) lives along swiftly flowing streams where it feeds mainly on aquatic invertebrates which it collects by 'flying' underwater or foraging on the stream bottom. Dippers in Wales are scarcest along streams with low mean pH (less than 5.5–6.0) and elevated concentrations of aluminium (greater than $0.80–1.0 \, \mathrm{g \, m^{-3}}$). A decline in pH of 1.7 units on one river over the period 1960–1984 resulted in a 70–80% decline in the dipper population. On acidic streams nesting started later and clutch size was smaller than in those dippers breeding elsewhere.[17] If fish populations are eliminated from stretches of water, by acidification, then these will not be used by the piscivorous otter (*Lutra lutra*).[18]

The process of acidification has been reversed, by liming, enabling normal plant and animal communities to be maintained. However, liming is very expensive, and, because of the vast number of fishless lakes, is hardly realistic on a large scale. In West Germany alone it has been estimated that the liming of soils and waters to halt the progress of acidification would cost £15 000 million. Clearly the control of emissions at power stations and from vehicle exhausts is a more sensible and feasible alternative.

It is necessary briefly to mention the problems of acid mine drainage which, for example, affects some 16 000 km of streams and 11 700 $\mathrm{km^2}$ of impoundments in the USA alone. Coal mines are the most important source of pollution. Chemosynthetic bacteria oxidize the mineral iron pyrite with the formation of sulphuric acid, the water flowing from the mine having a pH often below 3.5, resulting in the extermination of much of the stream biota. The problem remains long after the mine has been abandoned. The acidic conditions bring metals, from ores exposed by the mining activity, into solution so that the effluent from the mine is also highly toxic. Many mines are situated in rural localities so aquatic ecosystems in otherwise pristine environments may be destroyed.[2,19]

5 TOXIC CHEMICALS

Some aspects of toxic pollution have already been mentioned but it is now appropriate to describe the modes of action of toxic chemicals. The major types of toxic compounds are:

1 metals, such as, zinc, copper, mercury, cadmium;
2 organic compounds, such as, pesticides, herbicides, polychlorinated biphenyls (PCBs), phenols;

[17] S. J. Ormerod and S. J. Tyler, in 'The Value of Birds,' ed. A. Diamond and F. Fillion, ICBP, Cambridge, 1987, p. 191.
[18] C. F. Mason and S. M. Macdonald, *Mammalia*, 1987, **51**, 81.
[19] C. F. Mason and S. M. Macdonald, *Chemosphere*, 1988, **17**, 1159.

3 gases, such as, chlorine, ammonia;
4 anions, such as, cyanide, sulphide, sulphite;
5 acids and alkalis.

There are a number of terms in regular use in the study of toxic effects:

acute	–	causing an effect (usually death) within a short period;
chronic	–	causing an effect (lethal or sublethal) over a prolonged period of time;
lethal	–	causing death by direct poisoning;
sub-lethal	–	below the level which causes death but which may affect growth, reproduction, or behaviour so that the population may eventually be reduced;
cumulative	–	the effect is increased by successive doses.

A typical example of a toxicity curve is given in Figure 6. Note that both axes are on a logarithmic scale. The median periods for survival are plotted against a range of concentrations. In many cases the relationship between survival time and concentration is curvilinear and there is a concentration below which the organism is likely to survive for long periods ($5 \, \text{mg} \, l^{-1}$ in Figure 3), a concentration known as the incipient LC 50. The lethal concentration (LC) is used where death is the criterion of toxicity. The number indicates the percentage of animals killed at that concentration and it is also usual to indicate the time of exposure. Thus 96 hour LC 50 is the concentration of toxic material which kills fifty per cent of the test organism in ninety six hours. The incipient level is usually taken as the concentration at which fifty per cent of the population can live for an indefinite period of time. Where effects other than death are being sought, for example respiratory stress or behavioural changes, the term used is the effective concentration (EC) which is expressed in a similar way, *e.g.*, 96 hour EC 50.

Effluents are frequently complex mixtures of poisons. If two or more poisons are present together they may exert a combined effect on the organism which is additive, for example mixtures of zinc and copper or ammonia and zinc. In other cases, there may be antagonism, the overall toxicity being less than when compounds are present alone; calcium, for example, antagonizes the toxic effect of lead and aluminium. In some cases the overall effect of mixtures of poisons on an organism may be more than additive (synergistic), for example, mixtures of nickle and chromium.

There has been a large amount of data collected on the acute toxicity of chemicals, especially to fish and invertebrates,[20,21] and this has undoubtedly been of great value in elucidating the mechanisms of toxicity. The value of these data to the river manager is more questionable. Incidents resulting in large mortalities of fish and other organisms are usually accidents over which the river

[20] J. M. Hellawell, 'Biological Indicators of Freshwater Pollution and Environmental Management,' Elsevier Applied Science Publishers, London, 1986.
[21] G. Mance, 'Pollution Threat of Heavy Metals in Aquatic Environments,' Elsevier Applied Science Publishers, London, 1987.

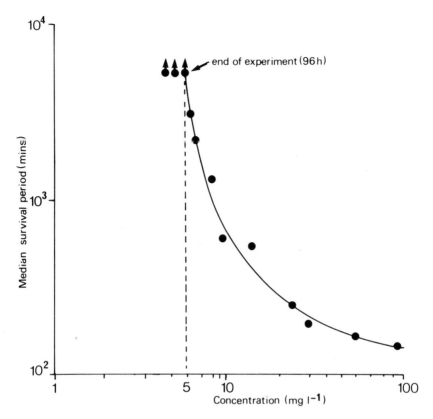

Figure 6 *A typical toxicity curve*

manager has no control. He can merely assess the damage and perhaps restock when conditions improve. In addition, detailed information on toxicity of a range of compounds is available for only a few test organisms, such as, rainbow trout (*Salmo gairdneri*) or *Daphnia pulex*, and it is well known that even closely related species may show very different responses to particular pollutants. It is the sub-lethal effects of pollutants which are of particular concern in many field situations, for low levels of pollutants may result in the gradual loss of populations, without any overt signs of a problem.

Experiments on sub-lethal effects are more difficult to carry out because they invariably take longer and individuals under test may respond very differently to low levels of pollution. Futhermore, the response to pollutants may vary over the lifetime of an organism, developmental stages often being more susceptible. It is necessary to study an organism under experimental conditions over its lifetime to find the weak link in its response to pollution and such long term experiments, possibly over several generations, are essential to discover any carcinogenic, teratogenic, or mutagenic effects of pollutants. Sub-lethal effects may be manifested at the biochemical, physiological, behavioural, or life cycle level.[2] Although it is possible to show small effects, for example, in biochemistry or on growth, at very low levels of pollution, it is essential that they reduce the fitness of an organism in its environment and are not merely within the organism's range of adaptation.

Organisms which are regularly subjected to toxic pollutants may develop tolerance to them. This may be achieved either by functioning normally at high loadings of pollutants or by metabolizing and detoxifying pollutants. Algae living in streams receiving mine drainage are highly tolerant to metals and this adaptation has been shown to be genetically determined. Populations of the isopods, *Asellus aquaticus* and *A. meridianus*, collected from sediments contaminated with copper and lead were found to be less susceptible to these metals than animals from uncontaminated sediments in acute toxicity tests. The metals are stored in the hepato-pancreas. Copper and lead appear to compete for binding sites, lead being more readily bound.[22] There has been more detailed work on acclimation of fish to pollutants. Pre-treatment of young stages of rainbow trout to both arsenic and cadmium reduced the susceptibilty of the fish to later exposure to these metals. It is considered that pre-treatment with low doses of metals stimulates the synthesis of metal-binding proteins which can subsequently bind large doses of metal to produce an inactive complex.[23]

Of particular concern to environmental toxicologists are those compounds which accumulate in tissues, especially the heavy metals and organochlorines (pesticides and PCBs). From often undetectable concentrations in water, organisms may accumulate levels of biological significance. Furthermore, these compounds are passed along the food chain so that top carnivores feeding on contaminated prey may accumulate enormous concentrations of pollutants. For example, it has been estimated that the concentration factor of PCBs from water to carnivorous mammals may be as high as ten million times.[24]

Figure 7 illustrates the frequency distribution of concentrations of total mercury in the muscle of 85 eels (*Anguilla anguilla*) and 79 roach (*Rutilus rutilus*) collected from waters in five regions of Great Britain.[25] It is likely that in the vast majority of rivers and lakes from which the fish originated the water authorities would have found mercury in water below the limit of detection. Eels are probably more contaminated than roach because they spend much of the winter buried within sediments which also accumulate heavy metals. The EEC directive for mercury levels in fish flesh taken for consumption should not exceed $300 \, \mu g \, kg^{-1}$. None of the roach exceeded this standard but 25% of the eels, a species which is commercially exploited, especially for the export market, contained more than $300 \, \mu g \, kg^{-1}$ Hg. Some eels from all regions exceeded the standard but even the mean concentration in eels from North East Scotland and East Anglia was greater than $300 \, \mu g \, kg^{-1}$. Most of the fish contained lead and cadmium also, a good proportion of them above recommended standards for consumption. Eels contained more cadmium than roach but roach were more heavily contaminated with lead. It has recently been cogently argued that we

[22] B. E. Brown, *Nature (London)*, 1978, **276**, 388.
[23] J. H. Beattie and D. Pascoe, *J. Fish Biol.*, 1978, **13**, 631.
[24] S. Tanabe, *Environ. Pollut.*, 1988, **50**, 5.
[25] C. F. Mason, *Chemosphere*, 1987, **16**, 901.

have little knowledge of the long term, low level exposure of populations to toxic metals and over ten billion individuals are being unwittingly exposed to elevated levels of toxic metals and metalloids in the environment.[26]

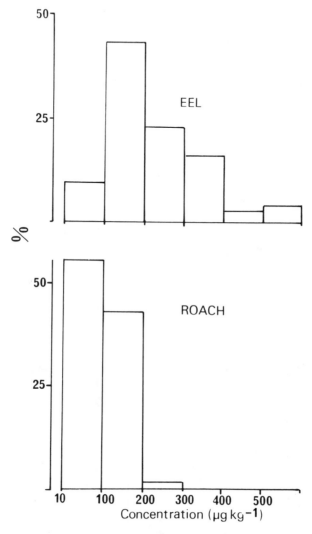

Figure 7 *Concentration* ($\mu g\ kg^{-1}$) *of total mercury in eels and roach from freshwaters in Great Britain*

The otter (*Lutra lutra*) is an amphibious mammal which feeds largely on fish, so it is at the top of the food chain. Populations of otters have declined markedly in Western Europe over the last three decades and the species is now absent from many lowland areas, including most of lowland Britain.[5] The decline in Britain began in the late 1950's, at a time when organochlorine pesticides were being widely introduced into agriculture, so there could be a link. However, at the

[26] J. O. Nriagu, *Environ. Pollut.*, 1988, **50**, 139.

same time there was also an exponential increase in the use of PCBs in many industrial processes and these are widely contaminating the environment, although this was not discovered until the late 1960's. Both pesticides and PCBs are fat soluble and concentrate in animal tissues to be released when the animal mobilizes its fat deposits rapidly, for instance, during food shortage or when pregnant. This sudden release of pollutants may cause rapid death but there are a variety of sub-lethal effects at lower concentrations, including sterility, while the compounds also cross the placenta to the foetus.

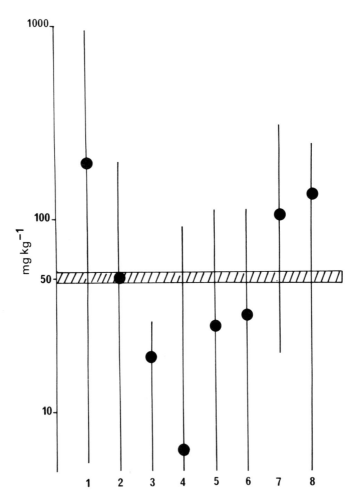

Figure 8   *Mean (and range) of PCBs* (mg kg^{-1} *in lipid) from different European populations of otters (status of population in parentheses). 1, southern Sweden (endangered); 2, northern Sweden (severe decline); 3, coastal Norway (thriving); 4, Scotland (thriving); 5, Wales (expanding, following decline); 6, South West England (expanding, following decline); 7, East Anglia (endangered); 8, The Netherlands (endangered). The hatched line at* 50 mg kg^{-1} *indicates the tissue concentration of PCBs known to cause reproductive failure in mink*

Few otters were analysed during the period of decline but the species has generated considerable concern over the last ten years and otters found dead in several countries have been analysed.[27-30] The means and ranges of PCBs (mg kg^{-1} fat) in otter tissues are shown in Figure 8. A concentration of 50 mg kg^{-1} fat is known to cause reproductive failure in experimentally dosed mink (*Mustela vison*), a species in the same family as the otter. It is quite clear from Figure 8 that PCB concentrations were, on average, greater than 50 mg kg^{-1} in those otter populations considered to be endangered. The steady

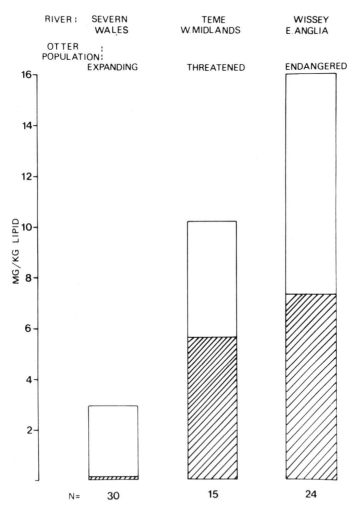

Figure 9   *Mean PCBs (hatched) and mean total organochlorines (mg kg^{-1} lipid) in otter spraints from three rivers*

[27] M. Olsson, L. Reutergårdh, and F. Sandegren, *Sveriges Natur*, 1981, **6**, 234.
[28] C. F. Mason, T. C. Ford, and N. I. Last, *Bull. Environ. Contam. Toxicol.*, 1986, **36**, 656.
[29] S. Broekhuizen and E. M. de Ruiter-Dijkman, *Lutra*, 1988, **31**, 68.
[30] C. F. Mason, *Lutra*, 1988, **31**, 62.

decline noted in a number of otter populations is consistent with suggestions that reproductive failure has resulted from elevated concentrations of PCBs. There is some recent evidence that higher concentrations of PCBs may result in adrenal hyperplasia, one consequence of which is breakdown in the immune system.[31]

Very little information comes from those regions where the species is threatened because otters are rarely found dead. The analysis of droppings (spraints) for organochlorines is one method of overcoming the problem. Otters use their spraints as social signals so they are deposited in conspicuous places. They also contain a scent rich in fatty acids, produced from anal glands. Concentrations of organochlorines in these fatty acids may reflect body burdens. It can be seen in Figure 9 that the average total concentration of organochlorines in otter spraints, and the proportion of PCBs, is inversely proportional to the status of the otter population.[32]

It is certain that the water authorities, from their routine analyses of river water samples, would have considered all of these rivers to be free of organochlorine compounds, which would have been below the level of detection. Nevertheless, by bio-accumulation they have had devastating effects on populations of this top carnivore. It is clear that, in order to protect the ecosystem and indeed man, the routine analysis for both heavy metals and organochlorines in biological samples should be mandatory.

6 THERMAL POLLUTION

Cooling water discharges from electricity generating stations are the main sources of pollution by heat. Such effluents also contain a range of chemical contaminants which, though small in relation to the volume of cooling water, may in fact have a greater impact on the ecology of the receiving stream.[33] An increase in temperature alters the physical environment, in terms of both a reduction in the density of the water and its oxygen concentration, while the metabolism of organisms increases. Cold water species, especially of fish, are very sensitive to changes in temperature and they will disappear if heated effluents are discharged to the headwaters of streams. As the temperature increases, the oxygen consumption and heart rate of a fish will increase to obtain oxygen for increased metabolic processes but, at the same time the oxygen concentration of the water is decreased. For example, at 1 °C a carp (*Cyprinus carpio*) can survive in an oxygen concentration as low as $0.5 \, \text{mg} \, \text{l}^{-1}$, whereas at 35 °C the water must contain $1.5 \, \text{mg} \, \text{l}^{-1}$. The swimming speeds of some species declines at higher temperatures, *e.g.*, trout at 19 °C, making them less efficient predators. Resistance to disease may also change. The bacterium *Chondrococcus columnaris* is innocuous to fish below 10 °C but it invades wounds between 10 °C and 21 °C while it can invade healthy tissues above 21 °C.

[31] I. F. Keymer, G. A. H. Wells, C. F. Mason, and S. M. Macdonald, *Vet. Rec.*, 1988, **122**, 153.
[32] S. M. Macdonald and C. F. Mason, *Acta Theriol.*, 1988, **33**, 415.
[33] T. E. Langford, 'Electricity Generation and the Ecology of Natural Waters,' Liverpool University Press, Liverpool, 1983.

However, it must be remembered that temperature changes are a feature of natural ecosystems so that organisms have the ability to adapt to the altered conditions provided by thermal effluents and although much research has been carried out into thermal pollution, it is now considered to be of little importance compared to other sources of pollution.[33] Indeed, there are some benefits of heated effluents, for growth and productivity may be enhanced.

7 OIL

Compared to the marine situation comparatively little work has been done on the effects of oil in freshwater ecosystems. Nevertheless the chronic pollution of freshwaters with hydrocarbons is widespread. Much of it derives from petrol and oil washed from roads together with the illegal discharge of engine oil. Other sources include boats and irrigation pumps, while accidents involving transporters and spillages from storage tanks are also significant. In 1983, 40% of the pollution incidents which resulted in the closure of water supply intakes involved fuel oils, while 60% of the pollution incidents in the Strathclyde region of Scotland over the period 1983–86 were due to spillages of petroleum products.[34]

The water soluble components of crude oils and refined products may prove toxic to freshwater animals though the prediction of toxic effects is rather difficult owing to the complex chemical nature of discharges. Eggs and young stages of animals are especially vulnerable.

In general terms, the aliphatic compounds of oils are relatively innocuous while the monohydric aromatic compounds are generally toxic, the degree of toxicity increasing with increasing unsaturation. Some components of oils, such as PCBs and lead, will accumulate in tissues. Emulsifiers and dispersants, used to clean up spillages, are themselves often highly toxic. The surface active agents which they contain make membranes more permeable and increase the penetration of toxic compounds into animals. In this way mixtures of oils and dispersants are often more toxic than either applied separately. There are also marked species differences in the susceptibility to particular compounds of oils, for instance the polynuclear aromatic hydrocarbons, further adding to the difficulties of making predictions about toxicity.

The physical properties of floating oil are a special threat to higher vertebrates, especially aquatic birds because contamination reduces buoyancy and insulation, while the ingestion of oil, frequently the result of attempts to clean the plumage, may prove toxic. A further problem is the tainting of flesh, especially of fish, which is detectable to the human palate at very low levels of contamination and renders fish inedible. The major sources of taint are light oils and the middle boiling range of crude oil distillates but there are a number of other sources, such as exhaust from outboard motors, waste from petrochemical factories, refinery wastes, and all crude oils.[34]

---

[34] J. Green and M. Trett, 'Fate and Effects of Oil in Freshwater,' Elsevier Applied Science Publishers, London, 1989.

8 RADIOACTIVITY

Chemically, radionuclides behave in the same way as their non-radioactive isotopes but, if they accumulate up the food chain the radioactive isotopes have much greater significance. Radionuclides come mainly from fall-out from weapons testing and the effluent from nuclear power stations. Because ionizing radiation is highly persistent in the environment, causing cancer and genetic disorders in humans, it has always attracted special concern and the release of radionuclides is strictly monitored and controlled.

Until recently, it was thought that the natural environment was little affected by radionuclides, which are discharged very locally. At only one site in Britain, Trawsfynydd in North Wales, is waste water from a nuclear power station discharged into freshwater, all other nuclear power stations discharging into estuaries or the sea. However, the explosion at the nuclear power station at Chernobyl in the USSR in April, 1986 and the subsequent spread and deposition of radionuclides over large areas of western Europe, emphasized how potentially damaging such pollution can be.

Fish collected immediately from those areas of Britain receiving high deposition of Chernobyl fall-out had only low concentrations of caesium 134 and 137, the main components of the fall-out. Later, higher concentrations were recorded, up to $1000 \, Bq \, kg^{-1}$ in trout (*Salmo trutta*) and $2000 \, Bq \, kg^{-1}$ in perch (*Perca fluviatilis*). The delayed increase in caesium was thought to be due to a number of causes, including, a delay in run-off from the land, the turnover time of water bodies, the build up in the food chain, and the slow turnover time in fish, the biological half-life of caesium in fish being 100 days. The highest deposition of Chernobyl fall-out occurred in areas with soft waters where high concentration factors from water to fish are likely.[35] Concentrations of radionuclides in freshwater fish in 1987 remained broadly similar to those in 1986.[36]

Because wild caught freshwater fish feature little in the diet of the vast majority of the human population it was considered that the Chernobyl accident presented no cause for concern to human health.[35] However, other species within the aquatic ecosystem, are more exclusively piscivorous. Figure 10 shows the radioactivity in otter spraints collected from river banks in central Wales and Galloway. Central Wales was outwith the main deposition area for Chernobyl fallout. In the September following the accident, radioactivity in otter spraints was more than double that in a sample collected some fifteen months prior to the accident and stored in a deep freeze. By the following January levels had returned to normal. By contrast, Galloway, which is an area of soft waters, received large quantities of Chernobyl fall-out and average radioactivity was more than six times that in Wales ($13\,000 \, Bq \, kg^{-1}$, with a maximum of $79\,500 \, Bq \, kg^{-1}$ in one sample).[37] Unfortunately, there was no pre-Chernobyl

[35] W. C. Camplin, N. T. Mitchell, D. R. P. Leonard, and D. F. Jefferies, 'Radioactivity in surface and coastal waters of the British Isles, Monitoring of fallout from the Chernobyl Reactor Accident,' Report No. 15, Ministry of Agriculture, Fisheries, and Food, Lowestoft, 1986.

[36] G. J. Hunt, 'Radioactivity in surface and coastal waters of the British Isles, 1987,' Report No. 19, Ministry of Agriculture, Fisheries, and Food, Lowestoft, 1988.

[37] C. F. Mason and S. M. Macdonald, *Water, Air, Soil Pollut.*, 1988, **37**, 131.

sample but levels were still high in the following January, as has been found for other biological materials from this area. Whether or not such high levels of radioactivity could adversely affect otters is unknown and it would be impossible to disentangle any mortality caused by radiation from other mortality factors, a major problem with many ecotoxicological studies.

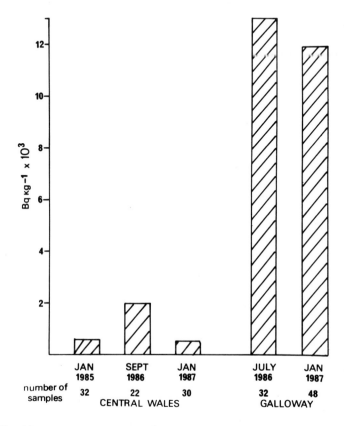

Figure 10   *Mean concentrations* $(Bq\ kg^{-1})$ *of radioactivity in otter spraints*

9 BIOLOGICAL MONITORING OF POLLUTION IN FRESHWATERS

The biological assessment of pollutants includes both laboratory and field techniques.[2,20] Two widely used laboratory methods are bio-stimulation and toxicity testing. Bio-stimulation tests are used mostly for evaluating the nutrient status of water bodies and are normally carried out with algae, the most frequently used species including *Selenastrum capricornutum*, *Asterionella formosa*, and *Microcystis aeruginosa*. The Standard Bottle Test measures either the maximum specific growth rate or the maximum standing crop. The specific growth rate (μ) is :

$$\mu = \frac{\ln\ (X_2/X_1)}{t_2 - t_1}$$

where X_1 is the initial biomass concentration at time t_1 and X_2 is the final biomass concentration at time t_2. The maximum standing crop is the maximum algal biomass achieved during incubation and may be obtained by direct determination of dry weight, by counting cells, by absorbance, by chlorophyll measurement, or by total cell carbon. The algae initially present in the test water must be removed by filtration or autoclaving, which may alter the water quality, while care must be taken to avoid problems of nutrient depletion and the build up of metabolic wastes.

The chief uses of toxicity tests are for a preliminary screening of chemicals, for monitoring effluents to determine the extent of risk to aquatic organisms, and, for those effluents which are toxic, to determine which component is causing death so that it can receive especial treatment. The simplest type of test is the static test in which an organism is placed in a standard tank in the water under investigation for 48–96 h. There are normally a series of tanks with test water of different dilutions. More sophisticated techniques involve the periodic replacement of test water or indeed continuous flow systems. Fish are normally used as test organisms. In the United States, the main test species have been fathead minnows (*Pimephales promelas*) and bluegill sunfish (*Lepomis macrochirus*), though the tendency now is to use a range of test species. Much toxicity work in Britain has been with rainbow trout but the tropical harlequin (*Rasbora heteromorpha*) has become increasingly popular because it is smaller and has a similar sensitivity to pollutants. The planktonic crustacean *Daphnia magna* is also sensitive to pollutants, can be easily cultured, and has a high reproductive rate. Toxicity tests have also been carried out in the field using caged fish.

Fishes show distinct physiological and behavioural responses to low levels of pollutants and recent studies have attempted to use these to devise automatic alarm systems. Automatic fish monitors should provide rapid indications of a deteriorating water quality and have potential uses for monitoring river waters and raw waters which are abstracted for potable supply and for monitoring effluents from treatment plants. Fish alarm systems have monitored movement and respiration. Movements are monitored with photosensitive cells and respiration by changes of potential between two electrodes.[38]

It has already been shown that pollutant levels in water are frequently below the limits of detection but that some organisms can accumulate large quantities of some contaminants which reflect the pollution level over periods of time. Because the degree to which organisms concentrate pollutants varies, for large-scale surveys a single, widespread species is needed. Care must be taken in the interpretation of results because many factors may influence the total pollutant

[38] G. P. Evans, D. Johnson, and C. Withell, 'Development of the WRc MIII Fish Monitor: description of the system and its response to some commonly encountered pollutants,' WRc Tech. Rep. 233, Medmenham, 1986.

content and concentration. These include the age of the organism, its sex, size, and weight as well as the time of year, sampling position, and the relative levels of other pollutants in the tissues. In addition to sampling organisms from the field to measure pollutant levels, they can also be placed in the field in cages so that uptake rates can be measured over defined periods of time.

Fish have frequently been used as biomonitors because they can be considered as a *critical material*, forming part of the diet of man. Eels (*Anguilla anguilla*) occur in most freshwater habitats and are hence particularly useful for large-scale surveys. Molluscs and macrophytes have also been used locally while mosses have proved especially valuable for monitoring pollution by metals.[39,40,41]

The biological assessment of water quality in the field may involve a number of levels of effort; survey, surveillance, monitoring, or research.[2] It is usually impossible to investigate the entire fauna and flora in a pollution study because of constraints of time and the wide range of sampling methods required for different groups of organisms. A survey or monitoring programme should therefore be based on those oganisms which are most likely to provide the right information to answer the particular question being posed. To be suitable, the chosen group must meet a number of criteria.[20] For example, the presence or absence of organisms must be a function of water quality rather than other ecological factors and the group must be relatively easy to sample and identify. In Great Britain the most favoured group for routine biological surveillance are the macro-invertebrates living in the stream bed.

To sample sites for macro-invertebrates a pond net (mesh size 900 µm) is used. The net is placed on the stream bed, facing downstream, and the feet are shuffled on the substratum, animals being dislodged and floating into the net. As much of the stream bed as possible is sampled and additional habitats, such as aquatic macrophytes or marginal vegetation, are also sampled by sweeping the net through them. The sampling period is three minutes and the objective is to obtain a comprehensive list of taxa with the minimum of sampling effort.[42] These samples are normally collected from eroding substrata (rocks, gravels, *etc.*). If there are long stretches of stream consisting of depositing substrata (silts and muds) or which are too deep to wade then sampling may have to be restricted to the margins. Alternatively artificial substrata, which are colonized by invertebrates, can be left *in situ* for later collection and analysis.[2,20]

Once the invertebrates have been identified and counted, there are two commonly used methods for data presentation and interpretation, diversity indices and biotic indices. Diversity indices take into account the number of species within the collection (species richness) and the relative abundance of species within the collection (evenness). It is argued that a community from an unstressed, *i.e.*, pollution free, environment will contain a large number of species (high richness), many at fairly low densities (high evenness), so that the calculated diversity index will be high. As pollution stress increases, species will

[39] P. J. Say, J. P. C. Harding, and B. A. Whitton, *Environ. Pollut. B.*, 1981, **2**, 295.
[40] C. Mouvet, *Verh. Int. Verein. Limnol.*, 1985, **22**, 2420.
[41] M. G. Kelly, C. Girton, and B. A. Whitton, *Water Res.*, 1987, **21**, 1429.
[42] M. T. Furse, J. F. Wright, P. D. Armitage, and D. Moss, *Water Res.*, 1981, **15**, 679.

gradually decline in number and disappear (low richness), while a few tolerant species will build up big populations in the absence of predation and competition (low evenness) resulting in a low diversity index. Diversity indices take no account of the tolerances of individual species to pollution. Of the many diversity indices which have been used,[43] the most popular is the Shannon–Wiener index:

$$H' = - \sum_{i=1}^{s} \frac{n_i}{n} \ln \frac{n_i}{n}$$

where n_i is the number of individuals of species i in the sample, n is the total number of individuals in the sample, and s is the number of species in the sample.

Biotic indices take account of the sensitivities of different species to pollution, that is, species are used as indicators of pollution. Those species which are sensitive to pollution (such as stoneflies) are given a high score, tolerant species (such as tubificid worms) are given a low score. In Britain, the biotic index at present most in favour is the Biological Monitoring Working Party (BMWP) Score, devised to be used in a national river pollution survey in 1980 (Table 3).[44]

Table 3   *The BMWP score system*

Families	Score
Siphlonuridae, Heptageniidae, Leptophlebidae, Ephemerellidae, Potamanthidae, Ephemeridae, Taeniopterygidae, Leuctridae, Capniidae, Perlodidae, Perlidae, Chloroperlidae, Aphelocheiridae, Phryganeidae, Molannidae, Beraeidae, Odontoceridae, Leptoceridae, Goeridae, Lepidostomatidae, Brachycentridae, Sericostomatidae	10
Astacidae, Lestidae, Agriidae, Gomphidae, Cordulegasteridae, Aeshnidae, Corduliidae, Libellulidae, Psychomyiidae, Philopotamidae	8
Caenidae, Nemouridae, Rhyacophilidae, Polycentropidae, Limnephilidae	7
Neritidae, Viviparidae, Ancylidae, Hydroptilidae, Unionidae, Corophiidae, Gammaridae, Platycnemididae, Coenagriidae	6
Mesoveliidae, Hydrometridae, Gerridae, Nepidae, Naucoridae, Notonectidae, Pleidae, Corixidae, Haliplidae, Hygrobiidae, Dytiscidae, Gyrinidae, Hydrophilidae, Clambidae, Helodidae, Dryopidae, Elminthidae, Chrysomelidae, Curculionidae, Hydropsychidae, Tipulidae, Simuliidae, Planariidae, Dendrocoelidae	5
Baetidae, Sialidae, Piscicolidae	4
Valvatidae, Hydrobiidae, Lymnaeidae, Physidae, Planorbidae, Sphaeriidae, Glossiphoniidae, Hirudidae, Erpobdellidae, Asellidae	3
Chironomidae	2
Oligochaeta (whole class)	1

Identification is necessary only to the family level. Score values for individual families reflect their pollution tolerance based on current knowledge of distri-

[43] H. G. Washington, *Water Res.*, 1984, **18**, 653.
[44] National Water Council, 'River quality: the 1980 survey and future outlook,' NWC, London, 1981.

bution and abundance. Each family present is given a score depending on its position in Table 3 and a site score is calculated by summing the individual family scores. This total score can then be divided by the number of families recorded in the sample to derive the average score per taxon (ASPT), which is less sensitive to sample size and sampling effort, and hence gives more information for less effort.[45] High ASPT scores characterize clean, upland sites which contain large numbers of high scoring taxa such as stoneflies, mayflies, and caddisflies. Lower ASPT values are obtained from slow-flowing, lowland sites which are dominated by molluscs, chironomids, and tubificid worms.

With neither the diversity index nor the BMWP score can the value obtained be directly related to a particular pollutional state. At present they have to be used in relative terms only.

10 CONCLUSIONS

Successive surveys of river quality in England and Wales have claimed that conditions have been steadily improving. The percentage of rivers in Class 3 and 4 (poor and grossly polluted) was 13% in 1958, 9% in 1970, 8% in 1975, and 7% in 1980,[44] this latter value amounting to 2 810 km out of a total length of 38 740 km. Tidal rivers have shown a similar improvement, though 16% of 2 800 km were in class 3 and 4 in 1980. These improvements were the result of increased investment, both in terms of capital works on sewage treatment plants and in staff skilled at pollution monitoring. Unfortunately, 1980 represented a high point, for expenditure on both capital works and staff have been slashed since then. At the same time the government abolished the routine collection and publishing of national trends in water quality as an expensive waste of time. However, regional surveys have given every indication that water quality has declined sharply since 1980, major causes being a deterioration in the quality of sewage effluents and increased farm pollution, this latter severely affecting many previously unpolluted rivers in rural areas. There is now a tendency to solve water quality problems by revising consent conditions for discharges downwards rather than by upgrading treatment facilities and prosecuting offenders. Despite government assurances, most organizations concerned with environmental protection and conservation are deeply concerned about the impending privatization of the water industry, where profit is certain to take precedence over environmental protection.

The decline in water quality we have witnessed in England and Wales over the past few years demonstrates the importance of a continuing programme of surveillance and monitoring. We must also remember that newly synthesized materials are constantly being added to our waterways as traces or in effluents and the long-term effects of these are largely unknown. Complacency over pollution is potentially dangerous. Constant vigilance is required to protect our water resources and the biologist should have a central role in the management team.

[45] P. D. Armitage, D. Moss, J. F. Wright, and M. T. Furse, *Water Res.*, 1983, **17**, 333.

CHAPTER 7

Important Air Pollutants and Their Chemical Analysis

R. M. HARRISON

1 INTRODUCTION

Before any detailed consideration of this topic, some description of basic terminology is worthwhile. Air pollutants may exist in *gaseous* or *particulate* form. The former include substances such as sulphur dioxide and ozone. Their concentrations are expressed most commonly either in mass per unit volume (μg^{-3} of air) or as a volume mixing ratio (1 ppm $= 10^{-6}$; 1 ppb $= 10^{-9}$). Particulate air pollutants are very diverse in character, including both organic and inorganic substances with diameters ranging from <0.01 to $>100\,\mu m$. Since very fine aerosol particles grow rapidly by coagulation, and large particles sediment rapidly under gravitational influence, the major part (by mass) exists in the $0.1 - 10\,\mu m$ range. Particles $< ca.$ $2\,\mu m$ are generally formed by growth of smaller particles, generated by condensation processes, whilst larger particles arise from mechanical disintegration processes. Thus the major part of the aerosol of $<2\,\mu m$ comprises man made components (*e.g.*, lead from motor exhausts, ammonium sulphate from atmospheric oxidation of sulphur dioxide), whilst the $>2\,\mu m$ material is frequently natural in origin (*e.g.*, wind-blown soil, marine aerosol). This is obviously not a rigid division. The size distribution typical of ambient aerosols is shown in Figure 1.

1.1 Pollutant Cycles

Pollutants are emitted from *sources* and are removed from the atmosphere by *sinks*. A typical cycle appears in Figure 2. Most pollutants have both natural and man made sources; although the natural source is often of sizeable magnitude in

127

global terms, on a local scale in populated areas pollutant sources are usually predominant.

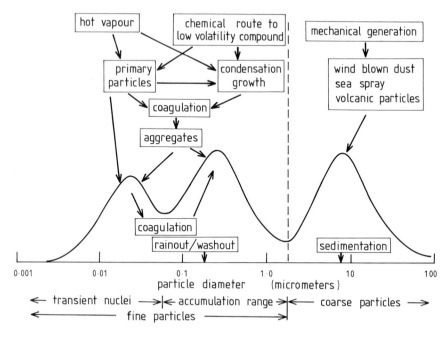

Figure 1 *Schematic diagram of the size distribution (expressed as surface area per increment in log particle diameter) and formation mechanisms for atmospheric aerosols (adapted from reference 1)*

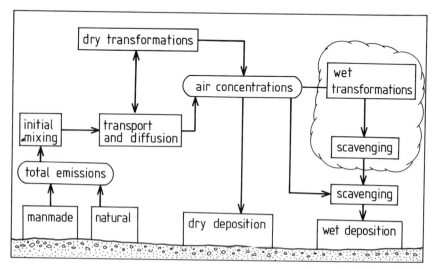

Figure 2 *Typical atmospheric cycle of a pollutant*
Reprinted with permission from *Environ. Sci. Technol.*, 1988, **22**, 241.

[1] K. T. Whitby, *Atmos. Environ.*, 1978, **12**, 135.

Sink processes include both dry and wet mechanisms. Dry deposition involves the transfer and removal of gases and particles at land and sea surfaces without the intervention of rain or snow. For gases removed at the surface, dry deposition is driven by a concentration gradient caused by surface depletion; for particles this mechanism operates in parallel with gravitational settling of the large particles. The efficiency of dry deposition is described by the *deposition velocity*, V_g, defined as

$$V_g(\text{m s}^{-1}) = \frac{\text{Flux to surface } (\mu\text{g m}^{-2}\text{ s}^{-1})}{\text{Atmospheric concentration } (\mu\text{g m}^{-3})}$$

Some typical values of deposition velocity are given in Table 1. For gases, such as sulphur dioxide, which have a fairly high V_g, dry deposition has little influence upon near-surface concentrations, but may appreciably influence ambient levels at large downwind distances.

Table 1   *Some typical values of deposition velocity*

Pollutant	Surface	Deposition Velocity (cm s^{-1})
SO_2	grass	1.0
SO_2	ocean	0.5
SO_2	soil	0.7
SO_2	forest	2.0
O_3	dry grass	0.5
O_3	wet grass	0.2
O_3	snow	0.1
HNO_3	grass	2.0
CO	soil	0.05
Aerosol ($<2.5\,\mu$m)	grass	0.15

Wet deposition describes scavenging by precipitation (rain, snow, hail, *etc.*) and is made up of two components, *rainout* which describes incorporation within the cloud layer, and *washout* describing scavenging by falling raindrops. The overall efficiency is described by the *scavenging ratio*, W, often rather misleadingly referred to as the Washout Factor.

$$W = \frac{\text{Concentration in rainwater } (\text{mg kg}^{-1})}{\text{Concentration in air } (\text{mg kg}^{-1})}$$

Typical values of scavenging ratio are given in Table 2. A large value implies efficient scavenging, perhaps resulting from extensive vertical mixing into the cloud layer, where scavenging is most efficient. A related deposition process termed 'occult' deposition occurs when pollutants are deposited by fogwater deposition on surfaces. Pollutant concentrations in fogwater are typically much greater than in rainwater hence the process may be significant despite the modest volumes of water deposited.

Table 2 *Typical scavenging ratios*

Species	W
Cl^-	600
SO_4^{2-}	700
Na	560
K	620
Mg	850
Ca	1890
Cd	390
Pb	320
Zn	870

Table 3 *Sulphur dioxide: estimated UK emissions from fuel combustion by type of consumer and fuel* (million tonnes)*

(a) *By type of consumer*

	1976	1980	1983	1986	Percentage of total in 1986
Domestic	0.28	0.23	0.19	0.20	5
Commercial/ public service†	0.19	0.19	0.14	0.13	4
Power stations	2.69	2.87	2.54	2.60	70
Refineries	0.28	0.28	0.16	0.17	4
Agriculture	0.03	0.02	0.01	0.01	—
Other industry‡	1.45	1.04	0.62	0.57	15
Rail transport	0.02	0.01	—	—	—
Road transport	0.05	0.04	0.04	0.05	1
All consumers⁺ (ground-based)	4.99	4.68	3.71	3.74	100

(b) *By type of fuel*

	1976	1980	1983	1986	
Coal	2.75	3.02	2.84	2.82	75
Solid smokeless fuel	0.13	0.11	0.10	0.09	2
Petroleum:					
Motor spirit	0.02	0.01	0.02	0.02	—
Derv	0.04	0.03	0.03	0.03	1
Gas	0.16	0.10	0.05	0.05	1
Fuel oil	1.62	1.13	0.53	0.56	15
Refinery fuel	0.28	0.28	0.16	0.17	4
All fuels⁺	4.99	4.68	3.71	3.74	100

* Excludes emissions from chemical and other processes which probably amount to a few per cent of emissions from fuel combustion.
† Includes miscellaneous consumers.
‡ Excludes power stations, refineries, and agriculture.
⁺ Excludes emissions from fuels used for water transport.
Source: Warren Spring Laboratory, Department of Trade and Industry

Another sink process involves chemical conversion of one pollutant to another (termed dry transformations in Figure 2). Thus atmospheric oxidation to sulphuric acid is a sink for sulphur dioxide. For many pollutants, a major sink is atmospheric reaction with the hydroxyl radical (OH). Such reactions are described in Chapter 8 on Tropospheric Chemistry. Since pollutants are continually emitted into, and removed from the atmosphere, they have an associated atmospheric lifetime or residence time defined in Chapter 8.

2 SPECIFIC AIR POLLUTANTS

2.1 Sulphur Dioxide

The major source of sulphur dioxide is the combustion of fossil fuels containing sulphur. These are predominantly coal and fuel oil, since natural gas, petrol, and diesel fuels have a relatively low sulphur content. Table 3 shows an emission inventory for sulphur dioxide, both by type of consumer and type of fuel. The marked decline in domestic emissions during the 1970's arose primarily from the switch by many consumers to natural gas during this period. Trends in urban concentrations of sulphur dioxide and smoke are shown in Table 4.

Table 4 *Trends in average UK urban concentrations of smoke* and sulphur dioxide*

	Sulphur dioxide	Smoke
	Concentration index 1981/82 = 100 ($50\,\mu g\,m^{-3}$)	*Concentration index* 1981/82 = 100 ($23\,\mu g\,m^{-3}$)
1975/76	135	140
1976/77	135	145
1977/78	118	113
1978/79	126	113
1979/80	114	109
1980/81	96	83
1981/82	100	100
1982/83	81	74
1983/84	79	78
1984/85	76	67
1985/86	73	67

Source: 1975–82: National Survey of Smoke and Sulphur Dioxide
    1982–86: UK Smoke and Sulphur Dioxide Monitoring Networks, Warren Spring Laboratory, Department of Trade and Industry.
* Smoke as given in this table is taken to be suspended matter collected on filter paper in accordance with British Standard (BS) 1747: Part 2.

Whilst emissions of this pollutant have fallen by 25% between 1976 and 1986, average urban concentrations have reduced by 46% over the same period. The improvement in urban air quality since 1960 is even greater (see Figure 3). This arises from a shift from low level to high level sources. The low level sources (*e.g.*, domestic chimneys) affect the local urban area, whilst a high level source, such as a power station, disperses its pollutants over a far greater area. Whilst this trend in emission heights has benefited British urban areas, it has contributed to another problem, that of pollutant transport. Sulphur dioxide, and other pollutants, emitted at a high level may be transported over very large distances by the atmosphere. During such transport processes, oxidation of sulphur and nitrogen oxides to sulphuric and nitric acids proceeds, hence giving rise to an 'acid rain' problem at great downwind distances. This problem is most acute in Scandinavia which itself emits little sulphur dioxide. Several environmental problems are associated with acid rain, including the killing of fish in acidified lake waters and the leaching of nutrients from soils (see Chapter 6).

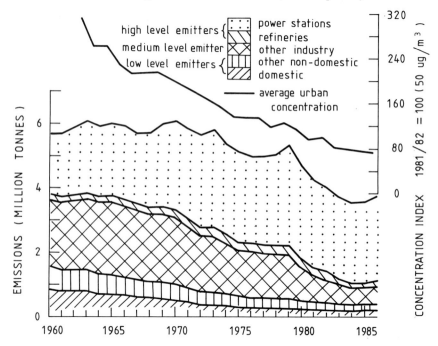

Figure 3   *Trends in UK sulphur dioxide emissions from fuel combustion and average UK urban concentrations*[2]

The adverse effects of sulphur dioxide itself include damage to human respiratory function, especially when exposure is in conjunction with particulates, as in the London smogs of the 1950's. Sulphur dioxide is also damaging to plants at modestly elevated concentrations. Whilst Britain does not set ambient air quality standards for pollutants, the US Environmental Protection Agency standards for this pollutant are $80 \, \mu g \, m^{-3}$ (annual average) and $365 \, \mu g \, m^{-3}$

---

[2] Department of the Environment, 'Digest of Environmental Protection and Water Statistics,' HMSO, 1987.

(24 h average) and the World Health Organization long-term goal is 40–60 µg m^{-3} (annual mean) with 98% of measurements below 100–150 µg m^{-3}. The UK is subject to an EEC air quality standard for sulphur dioxide, which is dependent upon the level of smoke and is described in a later section.

2.1.1 Chemical Analysis of Sulphur Dioxide. Sulphur dioxide may be determined by many procedures of widely differing sensitivity and specificity. The method used in the UK 'National Survey of Air Pollution' involves absorption of the sulphur dioxide in hydrogen peroxide. The resultant acid is determined by acid base titration or by conductivity measurement. Although acidic or basic compounds interfere, these techniques can yield useful results in urban areas where sulphur dioxide concentrations are high and levels of interfering substances are low. In rural areas this is not the case, due to natural production of ammonia, and misleading results are obtained. The West–Gaeke technique, in which sulphur dioxide is collected by reaction with potassium or sodium tetrachloro-mercurate(II), forming the disulphitomercurate(II) and then determined, colorimetrically, after addition of acidic pararosaniline methyl sulphonic acid, has been refined to the point where the effects of known interfering substances have been minimized or eliminated.

$$[HgCl_4]^{2-} + 2SO_2 + 2H_2O \rightarrow [Hg(SO_3)_2]^{2-} + 4Cl^- + 4H^+$$

By using flow systems, continuous monitors have been built using this reaction but the response time is relatively long, of the order of several minutes, and short-lived concentration peaks are not accurately measured.

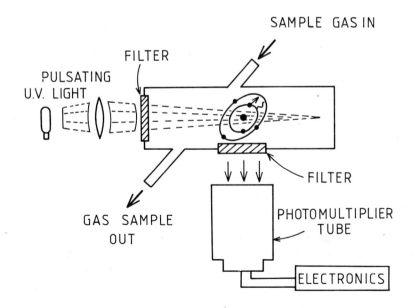

SAMPLE GAS IN

FILTER

PULSATING
U.V. LIGHT

GAS SAMPLE
OUT

FILTER

PHOTOMULTIPLIER
TUBE

ELECTRONICS

Figure 4 *Pulsed fluorescent analyser for sulphur dioxide*

Table 5 *Summary of commonly employed methods for measurement of air pollutants*

Pollutant	Measurement Technique	Sample Collection Period	Response time (continuous technique)[a]	Minimum detectable concentrations
Sulphur Dioxide	Absorption in H_2O_2 and titration	24 h		2 ppb
	Absorption in tetrachloromercurate/ spectrophotometry	15 min		10 ppb
	Flame photometric analyser		25 s	0.5 ppb
	Gas phase fluorescence		2 min	0.5 ppb
Oxides of Nitrogen	Chemiluminescent reaction with ozone		1 s	0.5 ppb
Total hydrocarbons	Flame ionization analyser		0.5 s	10 ppb C
Specific hydrocarbons	Gas chromatography/ flame ionization detector	b		< 1 ppb
Carbon monoxide	Catalytic methanation FID	c		10 ppb
	Electrochemical cell		25 s	1 ppm
	Non-dispersive infrared		5 s	0.5 ppm
Ozone	Chemiluminescent reaction with ethene		3 s	1 ppb
	U.v. absorption		30 s	1 ppb
Peroxyacetyl nitrate	Gas chromatography/ electron capture detection	c		1 ppb
Particulates	High volume sampler	24 h		$5 \mu g\ m^{-3}$
	β-gauge	min/hours		$5 \mu g\ m^{-3}$

(a) Time taken for a 90 per cent response to an instantaneous concentration change
(b) Grab samples of air collected in an inert container and concentrated prior to analysis
(c) Instantaneous concentrations measured on a cyclic basis by flushing the contents of a sample loop into the instrument

In the flame photometric sulphur analyser, gaseous sulphur compounds are burned in a reducing hydrogen–air flame and the emission of the S_2 species at 394 nm is measured. The technique has a very fast response as polluted air may be passed directly to the flame. The sensitivity is similar for all volatile sulphur compounds and in urban air the level of total volatile sulphur, as given by a flame photometric analyser, approximates closely to the sulphur dioxide level.

When several sulphur compounds are present in air at comparable concentrations, gas chromatographic separation is possible from a small air sample at ppb levels using the flame photometric analyser as detector.

An alternative technique is specific for measurement of sulphur dioxide at levels down to 1 ppb. Fluorescence, excited by radiation in the region of 214 nm, is measured (Figure 4) and a very wide linear range of response is found. Instruments covering the range of 1–5000 ppm and 0–0.5 ppm of SO_2 are available. The commonly used techniques for analysis of SO_2 and other pollutants are summarized in Table 5, and are described in more detail elsewhere.[3,4]

2.2 Suspended Particulate Matter

The chemical nature and composition of suspended particles is discussed in some detail in Chapter 8. Much of the total comprises secondary materials, such as ammonium sulphate and ammonium nitrate, formed from atmospheric reactions of other air pollutants. No emission inventory for the secondary particles can be readily given as this depends upon emission of the primary gases and their rates of chemical conversion. The main source of primary man made particulate pollutants is combustion of fossil fuels, especially coal. Trends in emissions of 'smoke' from coal combustion are given in Table 6, and data going back to 1960 are shown graphically in Figure 5.

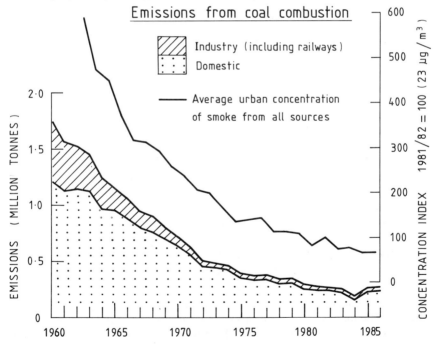

Figure 5 *Smoke emissions from coal combustion and average urban concentrations*[2]

[3] R. M. Harrison and R. Perry (ed.), 'Handbook of Air Pollution Analysis,' Chapman and Hall, 1986.
[4] R. M. Harrison, *CRC Crit. Rev. Anal. Chem.*, 1984, **15**, 1.

Table 6 *Trends in estimated UK emissions of smoke from coal combustion: by source*
(million tonnes)

	Domestic	Industry, etc*	Railways	All sources
1960	1.21	0.35	0.19	1.75
1965	0.95	0.14	0.06	1.15
1970	0.64	0.08	—	0.72
1975	0.35	0.04	—	0.39
1976	0.33	0.04	—	0.37
1977	0.34	0.04	—	0.38
1978	0.30	0.03	—	0.34
1979	0.31	0.04	—	0.35
1980	0.25	0.03	—	0.29
1981	0.24	0.03	—	0.27
1982	0.24	0.03	—	0.26
1983	0.22	0.03	—	0.25
1984	0.16	0.02	—	0.19
1985	0.23	0.03	—	0.26
1986	0.24	0.03	—	0.27

Source: Warren Spring Laboratory, Department of Trade and Industry
* Final energy users (*i.e.*, energy other than fuel conversion industries); includes collieries, public
services, agriculture, and miscellaneous.
Note: The emission factors in calculating these figures are given below.
Domestic: Smoke is taken as 3.5 per cent of the weight of coal burnt in domestic open fires.
Industry, *etc*: For 1958, smoke is taken as 0.97 per cent of the weight of coal burnt; for 1962 it is taken
as 0.5 per cent and 1971 and subsequent years, 0.3 per cent. For the intervening years, intermediate
proportionate values are used.
Railways: An estimate of 2 per cent of the weight of coal burnt is used.

The massive decrease in urban concentrations of 'smoke' shown in Figure 5
(about 90% since 1962–3) has a number of causes. Firstly, the measurement
method is important. In the UK National Survey of Air Pollution (and in many
other countries), smoke is determined by collection of a sample of airborne
particles on a filter, which is then examined for light reflection to estimate its
blackness. By use of calibration tables, the concentration of 'standard smoke' is
determined from the degree of soiling (darkness) of the filter. The calibration
graphs were constructed many years ago when urban particles were dominated by
coal smoke. Nowadays, there are many other important sources of particles and
the intrinsic darkness of urban particulate is much lower. Research has shown
that the method measures elemental carbon and thus the dramatic reduction in
inefficient combustion of bituminous coal in domestic grates over recent years
has led to a massive drop in the emission of particles containing black elemental
carbon. The main source of black smoke in urban air in the UK is now believed
to be diesel-engined road vehicles. There is no universally accepted definition of
'smoke,' although this is generally taken to mean fine suspended particulate
(< 15 μm) arising from the incomplete combustion of fuels. Approximately two-
thirds of urban properties in the UK are subject to smoke control orders which
compel the use of smokeless fuels or combustion of coal smokelessly. The

imposition of such control orders, together with a switch to more convenient fuels, especially natural gas, has led to greatly reduced urban emissions of coal smoke. Power stations use very large tonnages of coal, but are generally operated in such a way as to minimize emission of particles in general, and smoke in particular.

Rather than measuring concentrations of 'smoke,' many countries use measurements of total suspended particulate or TSP. In this measurement, particles are again collected from the air by filtration, but are then quantified gravimetrically. Thus both dark- and light-coloured particles are determined according only to their relative mass. There are problems in efficient sampling of the larger particles and, for this reason, the United States has moved towards the sampling of only particles of less than 10 μm aerodynamic diameter (termed PM 10) by use of a size-selective sampling inlet.

One of the main facets of particulate air pollution is the diminution in visibility caused by light scattering due to the particles. Scattering efficiency per unit mass of aerosol is critically dependent upon the size distribution of the aerosol, but aerosol mass loading above is a good predictor of visibility impairment, as indicated by Figure 6.

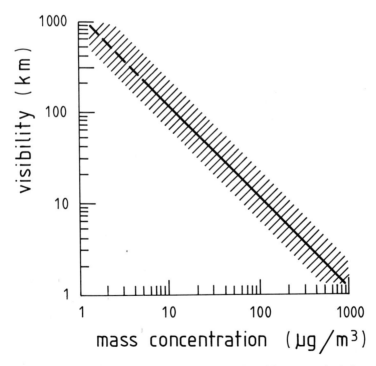

Figure 6 *Atmospheric visibility as a function of suspended particle concentration (after ref 5). The relationship is approximate only, but most data fit within the shaded area*

[5] R. J. Charlson, *J. Air Pollut. Contr. Assoc.*, 1968, **18**, 652.

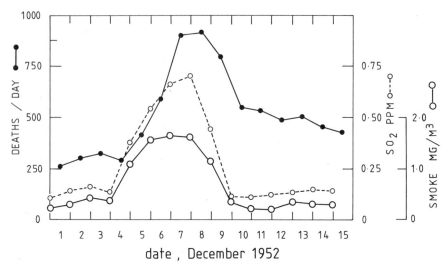

Figure 7 *Daily concentrations of smoke and sulphur dioxide together with mortality rates in the London smog of December, 1952 (adapted from ref 6)*
Adapted from *Fundamentals of Air Pollution, Second Edition* by Arthur Stern, copyright © 1984 by Harcourt Brace Jovanovich, reprinted by permission of the publishers.

Table 7 *EEC air quality standards for smoke and sulphur dioxide*

(a) *Smoke*

Yearly	Median of daily values throughout the year	$80\,\mu g\,m^{-3}$ $(68\,\mu g\,m^{-3})$
Winter	Median of daily values from 1 October to 31 March	$130\,\mu g\,m^{-3}$ $(111\,\mu g\,m^{-3})$
Peak	98 percentile of daily values throughout the year	$250\,\mu g\,m^{-3}$ $(213\,\mu g\,m^{-3})$

(b) *Sulphur dioxide*

	Smoke concentration $\leqslant 40\,\mu g\,m^{-3}$ $(34\,\mu g\,m^{-3})$	$120\,\mu g\,m^{-3}$
Yearly		
	Smoke concentration $> 40\,\mu g\,m^{-3}$ $(34\,\mu g\,m^{-3})$	$80\,\mu g\,m^{-3}$
	Smoke concentration $\leqslant 60\,\mu g\,m^{-3}$ $(51\,\mu g\,m^{-3})$	$180\,\mu g\,m^{-3}$
Winter		
	Smoke concentration $> 60\,\mu g\,m^{-3}$ $(51\,\mu g\,m^{-3})$	$130\,\mu g\,m^{-3}$
	Smoke concentration $\leqslant 150\,\mu g\,m^{-3}$ $(128\,\mu g\,m^{-3})$	$350\,\mu g\,m^{-3}$
Peak		
	Smoke concentration $> 150\,\mu g\,m^{-3}$ $(128\,\mu g\,m^{-3})$	$250\,\mu g\,m^{-3}$

Figures in parentheses relate to the UK method of smoke measurement

[6] A. C. Stern, R. W. Boubel, D. B. Turner, and D. L. Fox, 'Fundamentals of Air Pollution,' Second Edition, Academic Press, 1984.

There has been great emphasis in the UK upon the control of smoke and sulphur dioxide. This arose from the London smogs which culminated in the premature deaths of an estimated 4 000 individuals over a 4 day period in December, 1952. The smogs were caused primarily by low-level emissions from coal combustion during periods of meteorology unsuitable for effective pollutant dispersal. The combination with fog (smog = smoke + fog) led to dramatic losses in visibility. Smoke and sulphur dioxide act synergistically in adversely affecting humans, *i.e.*, the effect of the two together is more severe than the sum of the two acting separately. It is believed that the smoke particles act as a 'carrier' taking SO_2 to the lower parts of the respiratory system which it would not reach alone because of absorption on the walls of the upper respiratory tract. This synergism is reflected in the mortality data exhibited in Figure 7 for the December, 1952 smog; neither smoke nor SO_2 alone at these levels would be expected to cause significant mortality. The synergistic effect is reflected in the EEC ambient air quality standards for these pollutants shown in Table 7. The first part of the table shows the limit values for smoke, although by a method not directly comparable with that used in the UK, and in parentheses equivalent values by the UK method are listed. The second part of the table gives the limits for sulphur dioxide, which are dependent upon the smoke concentration, given in the same manner as the above table. The US Environmental Protection Agency standards for total suspended particulates and other pollutants are given in Table 8.

Table 8 *United States federal primary and secondary ambient air quality standards*

Pollutant	Type of standard	Averaging time	Frequency parameter	Concentration μg m^{-3}	ppm
Sulphur oxides (as sulphur dioxide)	Primary	24 h	Annual maximum*	365	0.14
		1 year	Arithmetic mean	80	0.03
	Secondary	3 hr	Annual maximum*	1 300	0.5
Particulate matter	Primary	24 hr	Annual maximum*	260	—
		24 hr	Annual geometric mean	70	—
	Secondary	24 hr	Annual maximum*	150	—
		24 hr	Annual geometric mean	60	—
Carbon monoxide	Primary and secondary	1 hr	Annual maximum*	40 000	35
		8 hr	Annual maximum*	10 000	9
Ozone	Primary and secondary	1 hr	Annual maximum*	235	0.12
Nitrogen dioxide	Primary and secondary	1 year	Arithmetic mean	100	0.05
Lead	Primary and secondary	3 months	Arithmetic mean	1.5	—

* Not to be exceeded more than once a year.

National primary ambient air quality standards define levels of air quality designed with an adequate margin of safety to protect the public health.

National secondary ambient air quality standards define levels of air quality judged necessary to protect the public welfare from any known or anticipated adverse effects of a pollutant.

Table 9 *Typical concentrations of some metals in airborne particles*

Metal	Urban air	Rural air
	(ng m^{-3})	
As	5–300	1–20
Cd	0.5–200	0.5–10
Ni	1–500	1–50
Pb	10–10 000	5–500
V	10–100	3–50
Zn	200–2 000	5–100
Co	0.2–20	0.1–5
Cr	2–200	1–20
Cu	10–1 000	2–100
Fe	100–10 000	100–10 000

2.2.1 Metals in the Atmosphere. Typical atmospheric concentrations of trace metals are given in Table 9. These are selected from data collected in many parts of the world. The interest in trace metals has arisen primarily as a result of the potential toxicity of several; most specifically lead, cadmium, mercury, and beryllium. As far as exposure of the general population is concerned, lead has been the cause of most concern and many countries have acted to limit emissions of lead to the atmosphere. This is feasible as airborne lead has a predominant source in combustion of leaded petrol (gasoline). The lead is added as an octane improver in the form of tetraalkyl lead compounds, notably tetramethyllead $(CH_3)_4Pb$ and tetraethyllead $(C_2H_5)_4Pb$. Upon combustion in the engine, these are emitted predominantly as an aerosol of fine particles of inorganic lead.[7] In the UK there has been a steady reduction in the maximum permitted level of lead in gasoline, which stood at 0.84 g l^{-1} in 1972 (although such high concentrations were not generally used). By 1981 this had been reduced to 0.40 g l^{-1} and in January, 1986 fell sharply to 0.15 g l^{-1} in line with most other western European countries. This has led to a steep reduction in vehicular lead emissions (Table 10) which has been almost paralleled by the fall of lead in air concentrations. Population blood lead concentrations fell slightly between 1985 and 1986 but little more, if at all, than between 1984 and 1985 (when petrol lead remained constant) and only slightly more for groups heavily exposed to airborne lead than for less exposed groups.[2] As blood lead is a good indicator of recent lead exposure, the data are indicative of the importance of other non-atmospheric routes of exposure to lead, such as the diet, drinking water, and beverages.

The current ambient air quality standards for lead are: USEPA, 1.5 μg m^{-3} (quarterly average) and EEC, 2 μg m^{-3} (annual average).

---

[7] R. M. Harrison and D. P. H. Laxen, 'Lead Pollution: Causes and Control, Chapman and Hall, 1981.

Table 10 *Consumption of petrol and estimated emissions of lead from petrol-engined road vehicles in the UK*

	Consumption of petrol		Estimated emissions of lead from petrol-engined road vehicles*	
	Million tonnes	Index 1975 = 100	Thousand tonnes	Index 1975 = 100
1975	16.12	100	7.4	100
1976	16.88	105	7.6	103
1977	17.34	108	7.4	100
1978	18.35	114	7.3	99
1979	18.68	116	7.3	99
1980	19.15	119	7.5	101
1981	18.72	116	6.7	91
1982	19.25	119	6.8	92
1983	19.57	121	6.9	93
1984	20.22	125	7.2	97
1985	20.40	127	6.5	88
1986	21.47	133	2.9	39

Sources: Department of Energy, Warren Spring Laboratory, Department of Trade and Industry
* These figures are based on lead contents for petrol published by the Institute of Petroleum. It has been assumed that only 70 per cent of this lead in petrol is emitted from vehicle exhausts, the remainder being retained in lubricating oil and exhaust systems.

2.2.2 Chemical Analysis of Particulate Pollutants. There are many possible analyses which can be carried out. A few of the more common ones are exemplified below. *Lead and other metals:* Particles are typically collected by air filtration and may be analysed by a number of techniques. Non-destructive methods include X-ray fluorescence (both wavelength dispersive and energy dispersive) and instrumental neutron activation analysis. Both methods involve costly instrumentation but can give data on a range of elements from a single analysis. Alternatively, the sample may be dissolved in oxidizing acids and the solution analysed by methods such as atomic absorption spectrometry, inductively coupled plasma (ICP) emission spectrometry, ICP–mass spectrometry, or anodic stripping voltammetry.

Polynuclear aromatic hydrocarbons: These compounds have long been of interest as some, most notably benzo(a)pyrene, are carcinogenic in experimental animals. The structures of some of the more commonly encountered compounds appear in Figure 8. These compounds may be separated by gas–liquid chromatography, with detection by flame ionization detection (after clean-up) or electron capture. Nowadays, the most popular method is high performance liquid chromatography; if both u.v. absorption and fluorescence detection are used a considerable number of compounds may be determined with high specificity.

Sulphate and nitrate: These ions are often analysed in the context of acid rain related studies. Until recently, analysis was by rather laborious procedures involving

Structural Formula	Name	Carcinogenic Activity
	Phenanthrene	0
	Anthracene	0
	Pyrene	0
	Fluoranthene	+
	Cyclopenta(cd)pyrene	+
	Benz(a)anthracene	+
	Chrysene	0/+
	Benzo(a)pyrene	+ +

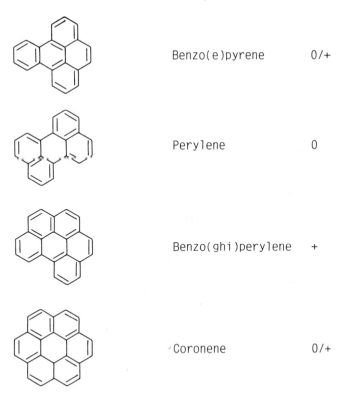

	Benzo(e)pyrene	0/+
	Perylene	0
	Benzo(ghi)perylene	+
	Coronene	0/+

Figure 8 *Structures of some commonly encountered polynuclear aromatic hydrocarbons. Relative carcinogenic activity in experimental animals is indicated on a scale from 0 (inactive) to ++ (appreciable activity)*

reduction of nitrate to nitrite which was determined spectrophotometrically after diazotization and azo coupling reactions. A number of colorimetric reagents were available for sulphate and a barium chloride turbidimetric procedure was also in use. These have in the past few years become largely supplanted in most laboratories by ion chromatography, in which the anions are separated by ion exchange and detected most commonly by conductivity after suppression of the conductivity of the eluent.

2.3 Oxides of Nitrogen

The most abundant nitrogen oxide in the atmosphere is nitrous oxide, N_2O. This is chemically rather unreactive and is formed by natural microbiological processes in the soil. It is not normally considered as a pollutant, although it does have an effect upon stratospheric ozone concentrations and there is concern that use of nitrogenous fertilizers may be increasing atmospheric levels of nitrous oxide.

The pollutant nitrogen oxides of concern are nitric oxide, NO, and nitrogen dioxide, NO_2. By far the major proportion of emitted NO_x (as the sum of the two compounds is known) is in the form of NO, although most of the atmospheric burden is usually in the form of NO_2. The major conversion mechanism is the very rapid reaction of NO with ambient ozone, the alternative third order reaction with molecular oxygen being relatively very slow at ambient air concentrations.

The major source of NO_x is the high temperature combination of atmospheric nitrogen and oxygen, in combustion processes, there being also a lesser contribution from combustion of nitrogen contained in the fuel. An emission inventory for the UK appears in Table 11, with trends shown graphically in Figure 9.

Table 11 *UK emissions of* NO_x *(as* NO_2*) by source* (thousand tonnes)

	1976	1980	1983	1986	Per cent of total 1986
Domestic	50	52	51	57	3
Commercial/public service	41	44	43	45	2
Power stations	774	851	763	783	40
Refineries	44	42	36	38	2
Agriculture (fuel use)	5	4	3	3	—
Other industry	292	216	186	179	9
Rail transport	42	41	37	37	2
Road transport	603	670	688	784	40
Incineration and agricultural burning	12	12	12	12	1
All sources (ground-based)	1 863	1 932	1 820	1 937	100

Typical ambient air concentrations of NO_x are normally within the range 5–100 ppb (roughly 10–200 µg m^{-3}) in urban areas and < 20 ppb at rural sites. The US Environmental Protection Agency (USEPA) ambient air quality standard for NO_2 is 100 µg m^{-3} (annual average). The EEC directive limit is a 98 percentile value of 105 ppb hourly average, not to be exceeded in a year, and guide values are 71 ppb and 26 ppb as 98 and 50 percentiles hourly averages, not to be exceeded in a year. The direct effects of exposure to oxides of nitrogen include human respiratory tract irritation and damage to plants. Indirect effects arise from the essential role of NO_2 in photochemical smog reactions, and its oxidation to nitric acid contributing to acid rain problems.

Analysis of oxides of nitrogen: Manual wet chemical procedures for NO_x are based upon conversion of NO_2 to nitrite, which is then determined typically by diazotization and azo dye formation. This, however, was found to be very problematic when reliable sources of NO_2 calibration gas became available, since the collection efficiency for NO_2 was found to be variable, and the stoichiometry of the NO_2 to nitrite conversion uncertain. This finding led to further method development and introduction of improved procedures such as that involving the use of arsenite in the collection solution to enhance the formation of nitrite. Another problem has been associated with the non-quantitative conversion of NO to NO_2 in order to determine NO_x.

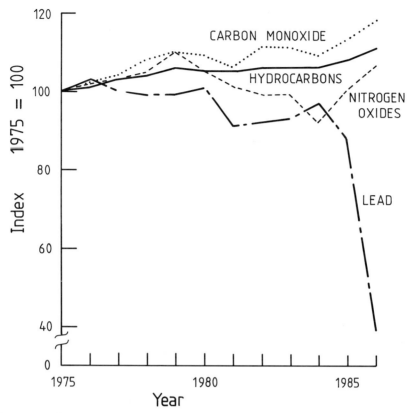

Figure 9 *Trends in UK emissions of hydrocarbons, carbon monoxide, nitrogen oxides, and lead (from petrol-engined road vehicles only)*[2]

Instrumental analysers not subject to these problems have now been available for a good number of years. The currently favoured technique for determination of oxides of nitrogen is based upon the chemiluminescent reaction of nitrogen oxide and ozone to give an electronically excited nitrogen dioxide which emits light in the 600–3 000 nm region with a maximum intensity near 1 200 nm:

$$NO + O_3 \rightarrow NO_2{}^* + O_2$$
$$NO_2{}^* \rightarrow NO_2 + h\nu$$

In the presence of excess ozone generated within the instrument, the light emission varies linearly with the concentration of nitrogen oxide from 1 ppb to 10^4 ppm. The apparatus is shown in Figure 10.

The method is believed free of interference for measurement of NO and may be used to measure NO_x by prior conversion of NO_2 to NO in a tube heated to about 650 °C. Dependent upon which converter is used, some interference in the NO_x mode is likely from compounds such as peroxyacetyl nitrate and nitric acid. Some instruments incorporate two reaction chambers, one running permanently in the NO mode, the other analysing NO_x after NO_2 to NO conversion. Thus pseudo-real-time NO_2 concentrations may be measured.

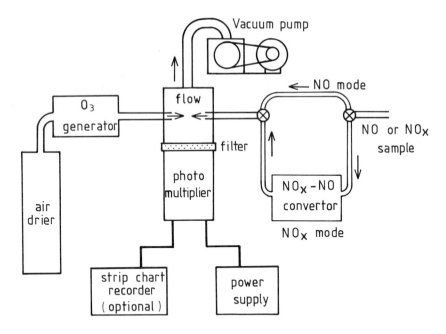

Figure 10   *Chemiluminescent analyser for oxides of nitrogen*

2.4   Carbon Monoxide

This pollutant is very much associated with petrol-engined vehicles (see Table 12). Car exhaust gases contain several per cent carbon monoxide, under normal running conditions, and greater concentrations when cold and choked. Most other combustion processes are relatively efficient and cause rather little CO emission.

Table 12   *UK emissions of* CO, *by source* (thousand tonnes)

	1976	1980	1983	1986	Per cent of total 1986
Domestic	641	532	463	483	9
Commercial/public service	13	11	12	11	—
Power stations	52	53	46	48	1
Refineries	4	4	4	4	—
Agriculture (fuel use)	1	1	1	1	—
Other industry	111	77	74	75	1
Rail transport	16	15	14	14	—
Road transport	3 804	4 295	4 437	4 748	85
Incineration and agricultural burning	220	220	220	220	4
All sources (ground-based)	4 862	5 208	5 269	5 602	100

The major sink process is conversion to CO_2 by reaction with the hydroxyl radical (see Chapter 8). This process is however rather slow and the reduction in CO level away from the source areas is almost entirely a function of atmospheric dilution processes. Carbon monoxide can be a problem in heavily trafficked areas, especially, in confined 'street canyons' where concentrations may reach 50 ppm or more. The adverse effect of CO is due to reaction with haemoglobin to form carboxyhaemoglobin which is relatively stable, causing a reduction in the oxygen-carrying capacity of the blood. The USEPA ambient air quality standard for CO is 9 ppm (8 hour average) and 35 ppm (1 hour average).

2.4.1 Chemical Analysis of Carbon Monoxide. Non-dispersive infra-red is used to measure carbon monoxide in street air where levels encountered normally lie within the range 1–50 ppm. Because of partial overlap of absorption bands, carbon dioxide and water vapour interfere. The latter may be removed by passing the air sample through a drying agent, and the former interference by interposing a cell of carbon dioxide between the sample and reference cells and the detectors. Using a long-path cell, the i.r. absorbance of polluted air at the wavelength corresponding to the C—O stretching vibration is continuously determined relative to that of reference air containing no CO. This is achieved without dispersion of the i.r. radiation by using cells containing CO at a reduced pressure as detectors for two beams which are chopped at a frequency of about 10 Hz and passed respectively through sample and reference cells. Absorption of radiation by the CO causes a differential pressure between the two detector cells which is sensed by a flexible diaphragm between them and used to generate an electrical signal.

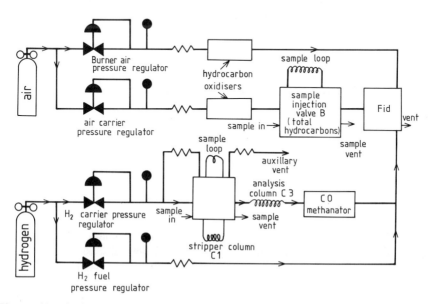

Figure 11 *Analyser giving sequential measurements of methane, carbon monoxide, and total hydrocarbons*

A gas chromatographic technique, although not allowing continuous measurement of CO, has a substantially higher sensitivity. The basis of the method is catalytic reduction of the carbon monoxide to methane and detection by flame ionization detector (FID) (Figure 11). This principle is used in commercial instruments which measure methane, carbon monoxide, and total hydrocarbons. Hydrogen carrier gas flushes a small air sample from a sample loop through a stripper column, (C1) packed with an adsorbent porous polymer, for sufficient time to allow passage of methane and carbon monoxide but not the heavier compounds which are subsequently removed by back-flushing. Passage of the methane and carbon monoxide through a chromatographic column containing molecular sieves (C3) causes separation of these compounds which are passed to the FID via a catalytic methanator. Sub-ppm levels are readily determined. Modification of the instrument allows passage of C_2 hydrocarbons through the stripper column, and these are separated on a second column of porous polymer and passed directly to the detector. Total hydrocarbons are determined by passage of a 10 cm^3 air sample directly into the FID (valve B). The carrier gas is air, purified by catalytic oxidation of impurities and this, rather than more conventional carrier gases, eliminates problems resulting from radical changes in flame characteristics upon introduction of the polluted air sample into the detector.

An analyser for continuous determination of carbon monoxide at levels down to 1 ppm uses an electrochemical cell. Gas diffuses through a semi-permeable membrane into the cell and at an electrode CO is oxidized to CO_2 at a rate proportional to the concentration of CO in the air. The response time is fairly short and interferences from other air pollutants at normally encountered levels are minimal. Analysers based upon electrochemical cells are also available for measurement of sulphur dioxide and oxides of nitrogen.

2.5 Hydrocarbons

Major sources of hydrocarbons in air (Table 13) are the evaporation of solvents and fuels, and the partial combustion of fuels. Obviously such processes give rise to an enormous range of individual compounds, and careful analytical work has shown measureable levels of in excess of 200 hydrocarbon compounds in some ambient air samples. An example of analytical data for C_2–C_6 compounds sampled from urban air in Lancaster, North West England[8] appears in Table 14. Methane is not included as this has a global background concentration of around 1.6 ppm and is hence considerably in excess of other hydrocarbons. Its concentration, as well as that of ethane, propane, and the butanes may be elevated by leakage of natural gas from piped supplies or from fossil reservoirs. Many higher molecular weight compounds, including aromatics, arise also from vehicular emissions. Polynuclear aromatic hydrocarbons, discussed earlier in the context of particulate pollutants, are partitioned between particulate and gas phases, the lower molecular weight species existing predominantly in the vapour phase.

[8] I. Colbeck and R. M. Harrison, *Atmos. Environ.*, 1985, **19**, 1899.

Table 13 *UK emissions of hydrocarbons, by source* (thousand tonnes)

	1976	1980	1983	1986	Per cent of total 1986
Domestic	102	81	70	78	4
Commercial/public service	1	1	1	1	—
Power stations	13	14	13	13	1
Refineries	1	1	1	1	—
Other industry (fuel use)	4	4	3	3	
Rail transport	11	10	10	10	—
Road transport*	456	509	531	573	28
Incineration and agricultural burning	38	38	38	38	2
Gas leakage[+]	142	173	180	201	10
Industrial processes and solvent evaporation[†]	1 043	1 055	1 058	1 068	52
Forests[‡]	80	80	80	80	4
All sources (ground-based)	1 890	1 966	1 984	2 065	100

Source: Warren Spring Laboratory, Department of Trade and Industry
* Includes evaporative emissions from the petrol tank and carburettor of petrol-engined vehicles
[+] Gas leakage is an estimate of losses during transmissions along the distribution system
[†] Including evaporation of motor spirit during production, storage and distribution
[‡] An order of magnitude estimate of natural emissions from forests

Concentrations of total hydrocarbons are commonly reported excluding methane, in units of ppb C (see later). Concentrations of non-methane hydrocarbons in polluted urban air are typically ∼500–5 000 ppb C, the major part comprising alkanes. Both Canada and the US have at one time had an ambient air quality standard for non-methane hydrocarbons of 240 ppb C between 6.00 to 9.00 a.m., designed to limit formation of photochemical air pollution, for which hydrocarbons are an essential precursor. Because of the great diversity of hydrocarbon compounds in polluted air, it is not possible to make a clear statement on toxic levels. Some individual compounds such as benzene are of concern due to human toxicity, whilst others, particularly ethene which is a plant growth hormone may affect the development of crops.

2.5.1 Chemical Analysis of Hydrocarbons. Analysis of total hydrocarbons or non-methane hydrocarbons is complicated by the great complexity of the mixture present and as a rule no attempt is made to analyse every compound individually. At one time, use of non-dispersive infra-red (see section on CO) was common. This utilized hexane vapour in the detector and thus responded to compounds whose infra-red spectrum overlapped with that of hexane, most notable in the regions of the C—H stretching vibration. Since not all hydrocarbons absorb at the same wavelength, and absorption intensities differ, the sensitivity per carbon atom in the molecule is similar for alkanes other than methane but is substantially less for alkenes and aromatics and almost nil for ethyne.

Table 14   *Concentrations of specific* C$_2$–C$_6$ *hydrocarbons measured in Lancaster, NW England* (ppb C)

Compound	Arithmetic mean	Standard deviation
Ethane	54.2	28.1
Ethene	42.3	9.3
Acetylene	10.3	6.0
Propane	12.2	5.8
Propene	9.9	7.1
Isobutane	40.7	10.2
n-Butane	16.6	12.8
1-Butene	11.0	8.0
2-Methylpropene	1.2	1.6
trans But-2-ene	18.9	17.0
cis But-2-ene	8.7	7.2
n-Pentane	11.2	7.1
2-Methylbutane	81.2	65.7
2,2-Dimethylpropane	58.8	16.6
1-Pentene	10.3	4.9
trans Pent-2-ene	19.7	12.4
cis Pent-2-ene	13.6	17.7
2-Methylbut-1-ene	0.4	0.3
3-Methylbutene	38.3	17.7
n-Hexane	10.1	4.5
2,2-Dimethylbutane	47.9	16.9
2-Methylpentane	21.2	17.5
3-Methylpentane	40.2	20.8
1-Hexane	10.4	8.3
trans Hex-3-ene	5.1	5.4
cis Hex-3-ene	6.3	7.1

The universal choice nowadays for total hydrocarbon analysis is the flame ionization analyser, based upon use of ambient air as oxidant in an air/hydrogen flame ionization detector in which the presence of hydrocarbons leads to enhanced conductivity of the flame. Sensitivity per carbon atom of individual hydrocarbons varies little, whilst oxygenated and halogenated compounds cause a lower response. Water vapour, carbon dioxide, and carbon monoxide do not significantly affect the reading. The instrument is calibrated with methane and results are reported as ppb C, or parts per billion carbon since 25 ppb butane (100 ppb C), 50 ppb ethane (100 ppb C), and 100 ppb methane (100 ppb C) all give the same response. Methane is determined in such instruments by selectively scrubbing all other hydrocarbons from the air stream prior to analysis and the non-methane hydrocarbon concentration is derived by difference.

Determination of specific hydrocarbons in ambient air normally requires a pre-concentration stage in which air is drawn through an adsorbent such as a porous polymer or activated carbon, or a tube where freeze-out of the compounds by reduced temperature occurs, followed by injection into a gas liquid chromatograph. Excellent separations of many compounds have been achieved, the best results coming from use of capillary columns. Detection may be by flame

ionization, or by passage into a mass spectrometer (GC–MS), the l
techniques allowing a more positive identification of individual compounds.

2.6 Carbon Dioxide

Increasingly, CO_2 is viewed as an air pollutant because of its importance as a
'greenhouse gas.' Its sources lie in animal and plant respiration, and anoxic
decomposition processes and the main sink is in photosynthesis. It may be both
absorbed or released by the oceans, which contain an appreciable concentration
of bicarbonate. The source drawing most attention, fossil fuel combustion, is a
relatively minor one, but a small imbalance in the CO_2 cycle is leading to a
steady increase in atmospheric concentration, exemplified in Figure 12.

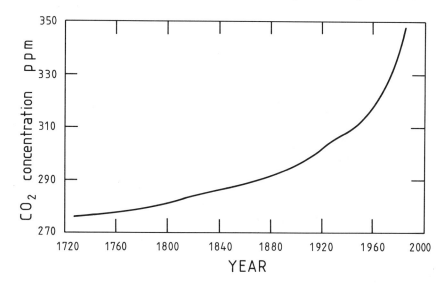

Figure 12 *Trends in concentrations of atmospheric carbon dioxide*

2.7 Secondary Pollutants

2.7.1 Ozone. Atmospheric reactions involving oxides of nitrogen and hydrocarbons cause the formation of a wide range of secondary products. The most
important of these is ozone. In severe photochemical smogs, such as occur in
Southern California, levels of ozone may exceed 400 ppb.

In Britain, we do not experience the classic Los Angeles type of smog.
Nonetheless, the same chemical processes give rise to elevated levels of ozone,
often in a regional phenomenon extending over hundreds of miles simultaneously.
Thus hydrocarbon and NO_x emissions over wide areas of Europe react in the
presence of sunlight causing large scale pollution, which is further extended by
atmospheric transport of the ozone. The phenomenon is crucially dependent
upon meteorological conditions, and hence, in Britain is observed on only

perhaps 10–30 days in each year on average. Concentrations of ozone measured at ground-level commonly exceed 100 ppb during such 'episodes' and have on one severe occasion been observed to exceed 250 ppb in Southern England. These levels may be compared with a background of ozone at ground-level arising from downward diffusion of stratospheric ozone and general tropospheric production of 20–50 ppb. This is seen in Figure 13 showing measurement data from North West England.

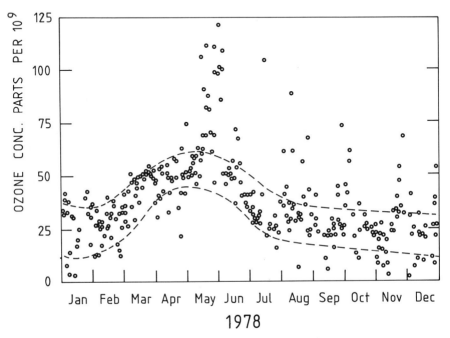

Figure 13   *Daily maximum hourly ozone concentrations at a rural site near Heysham, Lancs during 1978. The dashed lines represent an estimated 'background'*[9]

There is evidence of adverse health effects arising from human exposure to ozone. The USEPA ambient air quality standard is 120 ppb, not to be exceeded as an hourly average more than once per year. Damage to crop plants may occur at levels below this, with some varieties showing adverse effects at levels as low as 50 ppb. Economic losses to crops in the United States due to ozone damage are considerable.

Chemical Analysis of Ozone: No specific wet chemical method is available for ozone analysis and concentrations of 'oxidant' in the air have been determined by measurement of the capacity of an air sample to oxidize neutral buffered potassium iodide. Thus the method is sensitive not only to ozone but to other oxidizing substances in the air including peroxyacyl nitrates and nitrogen

[9] R. M. Harrison and C. D. Holman, *Chem. Brit.*, 1982, **18**, 563.

dioxide, with a lesser sensitivity. The overall reaction with ozone was assumed to be as below with a consequent stoichiometry of I_2/O_3 equal to 1.0.

$$O_3 + 2H^+ + 2I^- \rightarrow I_2 + H_2O + O_2$$
$$I_2 + KI \rightarrow KI_3$$

Some workers, however, have found a stoichiometry for I_2/O_3 of 1.5 and observed an enhanced formation of iodine and a reduced oxygen formation with increasing pH and postulated a more complex mechanism. The feasibility of iodide oxidation to iodate has been shown and the possibility of catalysis of this reaction by a glass frit. Hence, in addition to the lack of specificity of the reagent the stoichiometry of the reaction with ozone is in doubt. The use of a borate, rather than a phosphate buffer, has been shown to overcome many of these problems.

Appropriate chemical filters improve the selectivity of methods based upon the oxidation of neutral buffered halide. Sulphur dioxide produces a negative interference equal to the equivalent molar concentration of oxidant and may be eliminated by incorporating a chromic acid paper absorber in the sampling line. The chromic acid does, however, oxidize nitrogen oxide to nitrogen dioxide causing further interference in the presence of NO.

Ozone may be continuously and specifically determined by measurement of the light emitted by the chemiluminescent reaction of ozone and ethene:

$$C_2H_4 + O_3 \rightarrow \text{Ozonide} \rightarrow CH_2O^*$$
$$CH_2O^* \rightarrow CH_2O + h\nu$$

Emission is centred on 435 nm and hence no interference from the reaction of ozone and nitrogen oxide occurs. The method is sensitive to as little as 1 ppb of ozone.

The u.v. absorption of ozone at 254 nm may be used for its determination at levels down to 1 ppb. Interferences from other u.v.-absorbing air pollutants such as mercury and hydrocarbons may be minimized by taking two readings. The first reading is of the absorbance of an air sample after catalytic conversion of ozone to oxygen and the second is of an unchanged air sample, the difference in absorbance being due to the ozone content of the air. Available instruments perform this procedure automatically and give read-out in digital form. Although truly continuous measurement of ozone levels is not possible, response is fast and readings may be taken at intervals of less than one minute. The instrument is shown diagrammatically in Figure 14.

2.7.2 Peroxyacetyl Nitrate (PAN). PAN is a product of atmospheric photochemical reactions and is characteristic of photochemical smog (see Chapter 8)

$$CH_3-\underset{\underset{O}{\|}}{C}-O-O-NO_2 \quad \text{PAN}$$

Levels in Southern California lie typically within the range 5–50 ppb on smoggy days. In Europe, the formation is far less favoured and concentrations are more usually <10 ppb.

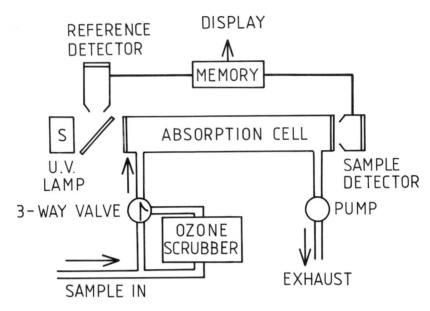

Figure 14 *A u.v. photometric analyser for ozone*

Chemical analysis of peroxyacetyl nitrate: PAN may be determined by long-path i.r. measurements, the greatest sensitivity being achieved when Fourier transform methods are used. The most sensitive and specific routine technique involves gas chromatographic separation and detection of specific peroxyacyl nitrates by electron capture. The detection limit of below 1 ppb permits it to be used as a direct atmospheric monitor under circumstances of high pollution.

3 INDOOR AIR QUALITY

Until rather recently, the emphasis in air quality evaluation has centred upon the outdoor environment. Recently, however, it has become clear that exposures to pollutants indoors may be very important.[10]

Pollutants with indoor sources may build up to appreciable levels because of the slowness of air exchange. An example is oxides of nitrogen from gas cookers and flueless gas and kerosene heaters which can readily exceed outdoor concentrations. Kerosene heaters can also be an important source of carbon monoxide and sulphur dioxide. Building materials and furnishings can also release a wide range of pollutants, such as formaldehyde from chipboard and hydrocarbons from paints, cleaners, adhesives, timber and furnishings. The tendency towards lower ventilation levels (energy efficient houses) has tended to exacerbate this problem.

Pollutants with a predominantly outdoor source may be reduced to rather low levels indoors due to the high surface area/volume ratios indoors leading to

[10] R. M. Harrison, I. Colbeck, and A. Simmons, *Environ. Technol. Lett.*, 1988, **9**, 521.

extremely efficient dry deposition of pollutants such as ozone and sulphur dioxide.

4. APPENDIX

Air Pollutant Concentration Units.

Probably the most logical unit of air pollutant concentration is mass per unit mass, *i.e.*, $\mu g\,kg^{-1}$ or $mg\,kg^{-1}$. This is, however, very rarely used. The commonest units are mass per unit volume (usually $\mu g\,m^{-3}$) or volume per unit volume, otherwise known as a volume mixing ratio (ppm or ppb). For particulate pollutants the volume mixing ratio is inapplicable.

Much confusion arises in the interconversion of $\mu g\,m^{-3}$ and ppm. Whilst the volume mixing ratio is independent of temperature and pressure for an ideal gas (and air pollutant behaviour is close to ideal), the mass per unit volume unit is dependent on T and P conditions, and hence these will be taken into account.

Example 1 Convert 0.1 ppm nitrogen dioxide to $\mu g\,m^{-3}$ at 20 °C and 750 torr.

46 g NO_2 occupy 22.41 l at STP

$$46\text{ g }NO_2\text{ occupy }22.41 \times \frac{293}{273} \times \frac{760}{750}\text{ l}$$

$$= 24.37\text{ l at 20 °C and 750 torr}$$

0.1 ppm NO_2 is 10^{-7} l NO_2 in 1 litre, or

$$\dotsb 10^{-4}\text{ l }NO_2\text{ in 1 m}^3$$

$$10^{-4}\text{ l }NO_2\text{ at 20 °C and 750 torr contain }46 \times \frac{10^{-4}}{24.37}\text{ g}$$

$$= 189\,\mu g\,NO_2$$

$$\therefore NO_2\text{ concentration } = 189\,\mu g\,m^{-3}$$

Example 2 Convert 100 $\mu g\,m^{-3}$ ozone at 25 °C and 765 torr to ppb.

48 g ozone occupy 22.41 l at STP

$$\dotsb \text{ occupy }22.41 \times \frac{298}{273} \times \frac{760}{765}\text{ l}$$

$$= 24.30\text{ l at 25 °C and 765 torr}$$

$$100\,\mu g\text{ ozone occupy }24.30 \times \frac{100 \times 10^{-6}}{48}\text{ l}$$

$$= 50.6 \times 10^{-6}\text{ l at 25 °C and 765 torr}$$

$$\therefore \text{ Volume mixing ratio } = 50.6 \times 10^{-6}\text{ (l) } \div 1000\text{ (l)}$$

$$= 50.6 \times 10^{-9}$$

$$= 51\text{ ppb}$$

CHAPTER 8

Chemistry of the Troposphere

R. M. HARRISON

1 INTRODUCTION

The atmosphere may conveniently be divided into a number of vertical bands reflective of its temperature structure. These are illustrated by Figure 1. The lowest part, typically about 12 km in depth, is termed the troposphere and is characterized by a general diminution of temperature with height. The rate of temperature decrease, termed the lapse rate, is typically around $9.8\ \mathrm{K\ km^{-1}}$ close to ground level but may vary appreciably on a short-term basis. The troposphere may be considered in two smaller components: the part in contact with the earth's surface is termed the boundary layer; above it is the free troposphere. The boundary layer is normally bounded at its upper extreme by a temperature inversion (a horizontal band in which temperature increases with height) through which little exchange of air can occur with the free troposphere above. The depth of the boundary layer is typically around 100 m at night and 1 000 m during the day, although these figures can vary greatly. Pollutant emissions are generally into the boundary layer and are mostly constrained within it. Free tropospheric air contains the longer-lived atmospheric components, together with contributions from pollutants which have escaped the boundary layer, and from some downward mixing stratospheric air.

The average composition of the unpolluted atmosphere is given in Table 1. Some of the concentrations are very uncertain since:

(i)   analytical procedures for some components have only recently reached the stage where good data can be obtained.

(ii) some components such as CH_4 and N_2O are known to be increasing in concentration at an appreciable rate.

(iii) it is questionable whether any parts of the atmosphere can be considered entirely free of pollutants.

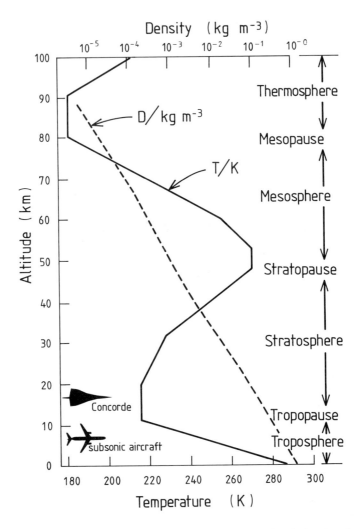

Figure 1 *The vertical temperature structure of the atmosphere showing the location of troposphere and stratosphere*

Table 1 also includes estimates of the lifetime of the various components. Those with lifetimes of a few days or less are cycled mainly within the boundary layer. Components with longer lifetimes mix into the free troposphere more substantially and those with lifetimes of a year or more will penetrate the stratosphere to a significant degree. An indication of polluted air concentrations of some of these components is given in Chapter 7.

Many of these components are chemically reactive and are transformed to other chemical species within the atmosphere. In some instances the products of such relations, termed secondary pollutants, are more harmful than the primary pollutants from which they are formed. Thus an appreciation of atmospheric

Table 1 *Average composition of the dry unpolluted atmosphere* (based upon Seinfeld[1] and Brimblecombe[2])

Gas	Average concentration (ppm)	Approx. residence time
N_2	780 840	10^6 years
O_2	209 460	5 000 years
Ar	9 340	)
Ne	18	) not
Kr	1.1	) cycled
Xe	0.09	)
CO_2	332	15 years
CO	0.1	65 days
CH_4	1.65	7 years
H_2	0.58	10 years
N_2O	0.33	20 years
O_3	0.01–0.1	100 days
NO/NO_2	10^{-6}–10^{-2}	1 day
NH_3	10^{-4}–10^{-3}	5 days
SO_2	10^{-5}–10^{-4}	10 days
HNO_3	10^{-5}–10^{-3}	1 day

chemical processes is fundamental to any attempt to limit the adverse effects of air pollutant emissions.

Table 1 includes lifetimes (also termed residence times) of the atmospheric gases. In this context, lifetime, τ, is defined as

$$\tau = \frac{A}{F}$$

where A = global atmospheric burden (Tg) (1 Tg = 10^{12} g)
F = global flux into and out of atmosphere (Tg year^{-1})
This treatment assumes a steady-state between the input and removal fluxes.

Sources of trace gases are not normally evenly spaced over the surface of the globe. Thus spatial variability in airborne concentrations occurs. For gases with an atmospheric lifetime comparable with the timescale of mixing of the entire troposphere (a year, or more), there is little spatial variation in concentration, as mixing processes outweigh the local variability in source strengths. For gases with short lifetimes, atmospheric mixing cannot prevent a substantially variable concentration. High spatial variability will also be associated with high temporal variability at one point as differing air mass sources and different mixing conditions will affect different concentrations of trace gas to the fixed receptor. An example of a rather well mixed gas of long lifetime is carbon dioxide which shows little variation in concentration over the globe and only small fluctuations at a given site (except very close to major combustion sources). At the other

[1] J.H. Seinfeld, 'Atmospheric Chemistry and Physics of Air Pollution,' Wiley, New York, 1986.
[2] P. Brimblecombe, 'Air Composition and Chemistry,' Cambridge University Press, Cambridge, 1986.

extreme, ammonia, which is chemically reactive and subject to efficient dry and wet deposition processes, is highly variable on both spatial and temporal scales. The spatial variability may be described by the coefficient of variation (equal to the standard deviation of the mean concentration divided by the mean) which relates to residence time as shown in Figure 2.

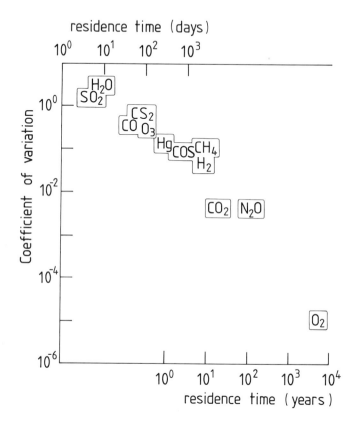

Figure 2 *The relationship between coefficient of variation in spatially averaged concentration and residence time for atmospheric gases*

Early books on tropospheric chemistry tended to consider the cycle of one substance in isolation from those of others. It is now well recognized that many of the important atmospheric chemical cycles are closely inter-linked and that a more integrated approach to study is appropriate. In this context, the hydroxyl radical, a species not recognized as significant until recent years, has been recognized as having an immensely important role. It is responsible for the breakdown of many atmospheric pollutants, whilst its formation is dependent upon others. It is appropriate to commence a description of tropospheric chemistry with this short-lived free radical species.

2 ATMOSPHERIC CHEMICAL TRANSFORMATIONS

2.1 The importance of the Hydroxyl Radical (OH)

The primary source of OH in the background troposphere is photolysis of ozone by light of short wavelengths ($\lambda < 315$ nm) to form singlet atomic oxygen $O(^1D)$, which may either relax to the triplet state, $O(^3P)$, or may react with water vapour to form OH:

$$O_3 + h\nu \qquad \rightarrow \quad O(^1D) + O_2 \quad \lambda < 315 \text{ nm} \qquad (1)$$

$$O(^1D) + M \qquad \rightarrow \quad O(^3P) + M \qquad (2)$$

$$O(^1D) + H_2O \qquad \rightarrow \quad 2OH \qquad (3)$$

Minor sources are available also through reactions of $O(^1D)$ with CH_4 and H_2:

$$CH_4 + O(^1D) \qquad \rightarrow \quad CH_3 + OH \qquad (4)$$

$$H_2 + O(^1D) \qquad \rightarrow \quad H + OH \qquad (5)$$

Photolysis of both HONO and H_2O_2 produces OH directly:

$$HONO + h\nu \qquad \rightarrow \quad OH + NO \quad \lambda < 400 \text{ nm} \qquad (6)$$

$$H_2O_2 + h\nu \qquad \rightarrow \quad 2OH \quad \lambda < 360 \text{ nm} \qquad (7)$$

Formation from nitrous acid may be of significance in polluted air. The route from hydrogen peroxide is not likely to represent a net source of OH since the main source of H_2O_2 is from the HO_2 radical:

$$HO_2 + HO_2 \qquad \rightarrow \quad H_2O_2 + O_2 \qquad (8)$$

In polluted atmospheres, however, HO_2 is able to give rise to OH formation by a more direct route:

$$HO_2 + NO \qquad \rightarrow \quad NO_2 + OH \qquad (9)$$

A review of tropospheric concentrations of the hydroxyl radical[3] found considerable variations in concentrations estimated by direct spectroscopic measurement, indirect measurement, and modelling. There are also genuine variations with latitude, season, and the presence of atmospheric pollutants. The overall consensus was of tropospheric concentrations within the ranges $(0.5–5) \times 10^6$ cm^{-3} daytime mean and $(0.3–3) \times 10^6$ cm^{-3} 24 h mean. A seasonal variation of about threefold is suggested by model studies.[3]

As the following sections will demonstrate, the hydroxyl radical plays a central role, via the peroxy radicals with which it is intimately related, in the production of ozone and hydrogen peroxide. It also itself contributes directly to formation of sulphuric and nitric acids, in the atmosphere, as well as indirectly contributing via ozone and hydrogen peroxide. Thus atmospheric processes leading to ozone formation will also tend to favour production of other secondary pollutants, including the strong acids HNO_3 and H_2SO_4.

[3] C. N. Hewitt and R. M. Harrison, *Atmos. Environ.*, 1985, **19**, 545.

3 ATMOSPHERIC OXIDANTS

3.1 Formation of Ozone

Mid-latitude northern hemisphere sites show background ozone concentrations typically within the range 20–50 ppb. These concentrations were for many years attributed solely to downward transport of stratospheric ozone and the seasonal fluctuation, with a pronounced spring maximum and broad winter minimum,[4,5] relating to adjustments in the altitude of the tropopause.

In the late 1970's, Fishman and Crutzen showed that ozone could be formed from oxidation of methane and carbon monoxide in the troposphere in processes involving the hydroxyl radical:

$$CH_4 + OH \qquad \rightarrow \quad CH_3 + H_2O \qquad\qquad (10)$$
$$CH_3 + O_2 + M \qquad \rightarrow \quad CH_3O_2 + M \qquad\qquad (11)$$
$$CO + OH \qquad \rightarrow \quad CO_2 + H \qquad\qquad (12)$$
$$H + O_2 + M \qquad \rightarrow \quad HO_2 + M \qquad\qquad (13)$$

(M is an unreactive third molecule such as N_2)
In the presence of NO:

$$CH_3O_2 + NO \qquad \rightarrow \quad CH_3O + NO_2 \qquad\qquad (14)$$
$$CH_3O + O_2 \qquad \rightarrow \quad CHBI2O + HO_2 \qquad\qquad (15)$$
$$HO_2 + NO \qquad \rightarrow \quad OH + NO_2 \qquad\qquad (9)$$

Thus via reactions of HO_2 and RO_2 (R = alkyl), NO is converted to NO_2. Then:

$$NO_2 + h\nu \qquad \rightarrow \quad NO + O(^3P) \quad \lambda < 435\,nm \qquad\qquad (16)$$
$$O(^3P) + O_2 + M \qquad \rightarrow \quad O_3 + M \qquad\qquad (17)$$

The magnitude of this source of ozone is presently controversial. However, recent evidence suggests that background northern hemisphere tropospheric ozone concentrations have approximately doubled since the turn of the century.[6] If this is indeed the case, the cause is almost certainly enhanced formation from the oxidation cycles of methane (whose concentration is known to be increasing), other less reactive hydrocarbons, and carbon monoxide. Such a source is also consistent with a spring maximum in ozone caused by reaction of hydrocarbons accumulated through the less reactive winter months. Figure 3 shows near-surface ozone concentrations measured at Montsouris in France from 1876–1905 and Arkona in the German Democratic Republic from 1956–1984. Although there are some remaining doubts over the validity of the measurements, they provide quite strong evidence of an increase in ground-level tropospheric ozone by a factor of approximately two since the turn of the century.

[4] J. Fishman and P. J. Crutzen, *J. Geophys. Res.*, 1977, **82**, 5897.
[5] J. Fishman and P. J. Crutzen, *Nature*, 1978, **274**, 855.
[6] United Kingdom Photochemical Oxidants Review Group, 'Ozone in the United Kingdom,' Department of Environment, London, 1987.

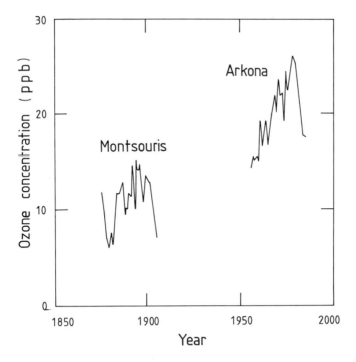

Figure 3 *Near-surface ozone concentrations measured at Montsouris, France 1876–1905 and Arkona, GDR 1956–1984 (from ref. 6)*

In urban air there is abundant NO_2 whose photolysis leads to ozone formation. However, fresh emissions of NO lead to ozone removal; the full cycle of reaction is:

$$NO_2 + h\nu \xrightarrow{J_1} NO + O(^3P) \quad \lambda < 435\,nm \tag{16}$$

$$O(^3P) + O_2 + M \rightarrow O_3 + M \tag{17}$$

$$NO + O_3 \xrightarrow{k_3} NO_2 + O_2 \tag{18}$$

J_1 and k_3 are the rate constants for the NO_2 photolysis and O_3 removal reactions respectively. All three reactions are rapid and an equilibrium is reached when the rate of ozone formation (equal to the rate of NO_2 photolysis if all $O(^3P)$ leads to O_3 formation) equals the rate of O_3 removal. Then:

$$J_1[NO_2] = k_3[O_3][NO_2] \tag{19}$$

$$\text{and} \quad [O_3] = \frac{J_1[NO_2]}{k_3[NO]} \tag{20}$$

This is termed the photostationary state, and thus the ozone concentration is determined by the value of J_1, highest at peak sunlight intensity, and the ratio of NO_2/NO. Hydrocarbons, and to a lesser extent CO, play a crucial role in producing HO_2 and NO_3 radicals which convert NO to NO_2 *without* consumption of O_3. For example, propene is attacked by hydroxyl:

$$CH_3CH=CH_2 + OH \quad \rightarrow \quad CH_3CH-CH_2OH \tag{21}$$

$$CH_3CH-CH_2OH + O_2 \quad \rightarrow \quad \underset{\underset{OO}{|}}{CH_3CH}-CH_2OH \tag{22}$$

$$\underset{\underset{OO}{|}}{CH_3CH}-CH_2OH + NO \quad \rightarrow \quad \underset{\underset{O}{|}}{CH_3CH}-CH_2OH + NO_2 \tag{23}$$

$$\underset{\underset{O}{|}}{CH_3CH}-CH_2OH \quad \rightarrow \quad CH_2OH + CH_3CHO \tag{24}$$

$$CH_2OH + O_2 \quad \rightarrow \quad HCHO + HO_2 \tag{25}$$

In this case atmospheric photochemistry acts as a source of O_3 and aldehydes. It also leads to formation of HO_2, a source of hydrogen peroxide. None of the above processes is an effective free radical sink and hence, during hours of daylight, reactive free radicals such as OH and HO_2 are constantly recycled.

3.2 Formation of PAN

Peroxyacetyl nitrate (PAN) is of interest as a characteristic product of atmospheric photochemistry, as a probable reservoir of reactive nitrogen in remote atmospheres and because of its adverse effects upon plants. The formation route is via acetyl radicals (CH_3CO) formed from a number of routes, most notably acetaldehyde oxidation.

$$CH_3CHO + OH \quad \rightarrow \quad CH_3CO + H_2O \tag{26}$$

$$CH_3CO + O_2 \quad \rightarrow \quad CH_3C(O)OO \tag{27}$$

$$CH_3C(O)OO + NO_2 \quad \rightarrow \quad \underset{\underset{O}{\|}}{CH_3C}-OONO_2 \tag{28}$$
$$\text{(PAN)}$$

In heavily polluted urban areas, subject to photochemical air pollution (*e.g.*, Los Angeles), the concentrations of NO, NO_2, O_3, and other secondary pollutants tend to follow characteristic patterns, illustrated in Figure 4. Primary pollutants NO and hydrocarbons tend to peak with heavy traffic around 7 a.m. This is followed by a peak in NO_2 some time later when the atmosphere has developed sufficient oxidizing capability to oxidize the NO emissions of motor vehicles. The NO_2/NO ratio is now high, favouring ozone production, and ozone peaks with peak sunlight intensity around noon, or a little later. In the United Kingdom, such diurnal profiles have been observed in urban areas during summer anticyclonic conditions, but peak ozone concentrations are generally rather later, around 3 p.m. The situation is more complex, however, as in the UK much of the ozone is generated during long-range transport of air pollution, often from continental European sources.

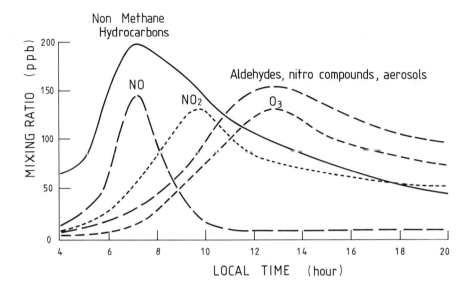

Figure 4 *Typical mixing ratio profiles, as a function of time of day, in the photochemical smog cycle*

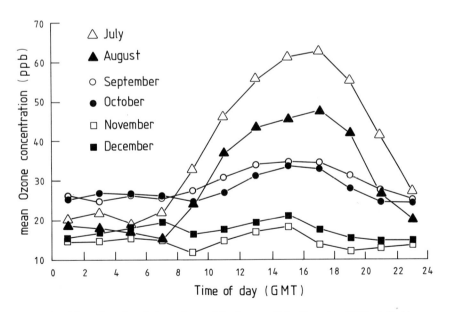

Figure 5 *Mean diurnal variations of ground-level ozone, July–December, 1983 at Stodday, near Lancaster[7]*

[7] I. Colbeck and R. M. Harrison, *Atmos. Environ.*, 1985, **19**, 1577.

Perhaps surprisingly, UK rural sites exhibit diurnal variations in ozone which are remarkably similar to those at urban sites. Figure 5 shows average variations at Stodday, near Lancaster, for various months in 1983. This diurnal change is not due to the same causes as that in urban areas, where fresh NO emissions destroy ozone in the night-time. Measurement of vertical profiles of ozone and temperature (Figure 6) show that nocturnal depletion of ozone is a surface phenomenon due to dry deposition (see Chapter 7) with little change in airborne concentrations at 50 m to 100 m altitude. The temperature profiles show the depletion occurs from a non-turbulent stable layer at the surface indicated by temperature rising with height (termed a temperature inversion). During the winter months, little diurnal change in ozone is seen (see Figure 5) since the atmospheric temperature structure shows a less marked change between night and day.

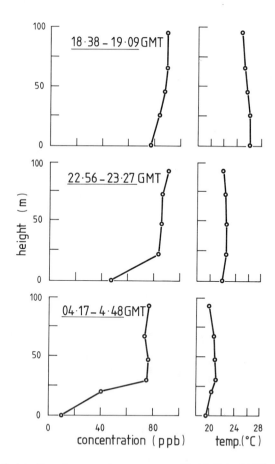

Figure 6   *Vertical profiles of ozone and temperature on* 11–12 *July,* 1983[8]

[8] I. Colbeck and R. M. Harrison, *Atmos. Environ.,* 1985, **19**, 1807.

4 ATMOSPHERIC ACIDS

4.1 Weak Acids

Well-known weak acids in the atmosphere which may contribute to corrosion processes and influence the pH of precipitation are carbon dioxide and sulphur dioxide. Atmospheric CO_2 at a concentration of 340 ppm leads to an equilibrium pH of 5.6 at 15 °C in otherwise unpolluted rainwater. This is normally taken as the boundary pH below which rain is considered acid. Sulphur dioxide is a stronger acid than CO_2 and at a concentration of only 5 ppb in air will, at equilibrium, cause a rainwater pH of 4.6 at 15 °C.[2] In many instances, this pH is not attained due to severe kinetic constraints upon achievement of equilibrium, as is the case with many atmospheric trace gases. Dissolved SO_2 may contribute appreciably to total sulphate and acidity in urban rainwater.

Organic acids such as formic acid may be formed in the atmosphere from formaldehyde:

$$HO_2 + HCHO \rightleftarrows HOOCH_2O \rightleftarrows OOCH_2OH \qquad (29)$$
$$OOCH_2OH + NO \rightarrow OCH_2OH + NO_2 \qquad (30)$$
$$OCH_2OH + O_2 \rightarrow HCOOH + HO_2 \qquad (31)$$

Carboxylic acids are believed to contribute significantly to rainwater acidity in remote areas, although their contribution is not at all well quantified. In polluted regions they are unlikely to be of much importance.

4.2 Strong Acids

Strong acids of major importance in the atmosphere are as follows:
H_2SO_4
HNO_3
HCl
CH_3SO_3H (methyl sulphonic acid)

4.3 Sulphuric Acid

Atmospheric oxidation of SO_2 proceeds via a range of mechanisms. Consequently, dependent upon the concentrations of the responsible oxidants, the oxidation rate is extremely variable with space and time but typically lies around $1\% \, h^{-1}$.

Several gas phase mechanisms have been investigated including photo-oxidation, reaction with hydroxyl radical, Criegee biradical, ground state atomic oxygen, $O(^3P)$, and peroxy radicals. Table 2, based upon the treatment of Finlayson-Pitts and Pitts,[9] summarizes the rate data and clearly indicates the overwhelming importance of the hydroxyl radical reaction in the gas phase.

---

[9] B. J. Finlayson-Pitts and J. N. Pitts Jr., 'Atmospheric Chemistry,' Wiley, New York, 1986.

Table 2 *Homogeneous oxidation mechanisms for SO_2 (based on ref 9)*

Oxidizing species	Concentration* (cm^{-3})	Rate constant $(cm^3\,molec^{-1}\,s^{-1})$	Loss of SO_2 $(\%\,h^{-1})$
OH	5×10^6	9×10^{-13}	1.6
O_3	2.5×10^{12}	$<8 \times 10^{-24}$	$<7 \times 10^{-6}$
Criegee biradical	1×10^6	7×10^{-14}	3×10^{-2}
$O(^3P)$	8×10^4	6×10^{-14}	2×10^{-3}
HO_2	1×10^9	$<1 \times 10^{-18}$	$<4 \times 10^{-4}$
RO_2	3×10^9	$<1 \times 10^{-18}$	$<1 \times 10^{-3}$

* Assuming a moderately polluted atmosphere

The mechanism of the SO_2—OH reaction is not entirely clear, but most probably is as follows:

$$SO_2 + OH \rightarrow HOSO_2 \tag{32}$$
$$HOSO_2 + O_2 \rightarrow HO_2 + SO_3 \tag{33}$$
$$SO_3 + H_2O \rightarrow H_2SO_4 \tag{34}$$

In the presence of water droplets, which can take the form of fogs, clouds, rain, or hygroscopic aerosols, sulphur dioxide will dissolve opening the possibility of aqueous phase oxidation. Upon dissolution, the following equilibria operate.

$$SO_{2(g)} + H_2O \rightleftarrows SO_2.H_2O \tag{35}$$
$$SO_2.H_2O \rightleftarrows HSO_3^- + H^+ \tag{36}$$
$$HSO_3^- \rightleftarrows SO_3^{2-} + H^+ \tag{37}$$

These equilibria are sensitive to pH, and HSO_3^- is the predominant species over the range pH 2–7. The other consequence of these equilibria is that the more acidic the droplet, the greater the degree to which the equilibria move towards gaseous SO_2 and limit the concentrations of dissolved S(IV) species. Some rate constants are pH-dependent in addition.

The major proposed oxidation mechanisms for SO_2 in liquid droplets include the following:

(i) uncatalysed oxidation by O_2
(ii) transition metal-catalysed oxidation by O_2
(iii) oxidation by dissolved oxides of nitrogen
(iv) oxidation by ozone
(v) oxidation by H_2O_2 and organic peroxides.

Relative rates for typical specified concentrations of reactive species are indicated in Table 3.

At high pH, all mechanisms in Table 3 are capable of oxidizing SO_2 at appreciable rates, generally far in excess of those observed in the atmosphere over a significant averaging period. Slower rates are observed, because of mass transfer limitations to the rate of introduction of SO_2 and oxidant to the water droplet, and the previously noted effect of pH reduction (due to H_2SO_4

formation) on the solubility of SO_2. At pH 3, only the reaction with hydrogen peroxide is very fast, due to an increasing rate constant at reduced pH compensating for the lower solubility of SO_2. Recent experimental studies indicate that this reaction can be very important in the atmosphere, but is limited by the availability of H_2O_2 which rapidly becomes depleted. The rate of formation of H_2SO_4 is therefore a function more of atmospheric mixing processes than of chemical kinetics in this case.

Several studies have emphasized the importance of SO_2 oxidation upon the surface of carbonaceous aerosols.[10,11] Our own work,[12] including experiments at high relative humidity and in aqueous suspension, indicates that this reaction pathway is likely to make a negligible contribution to atmospheric sulphate formation under most conditions.

Table 3 *Aqueous phase oxidation mechanisms for SO_2 (adapted from ref 9)*

Oxidant	Concentration*	Oxidation rate $(\% \, h^{-1})$ pH 3	pH 6
O_3	50 ppb (g)	3×10^{-2}	5×10^3
H_2O_2	1 ppb (g)	8×10^2	5×10^2
Fe-catalysed	3×10^{-7} M (l)	2×10^{-2}	5×10^1
Mn-catalysed	3×10^{-8} M (l)	3×10^{-2}	7
HNO_2	1 ppb (g)	5×10^{-4}	3

* g and l denote gas and liquid phase concentrations respectively. Conditions are 5 ppb gas phase SO_2 at 25 °C with no mass transfer limitations.

4.4 Nitric Acid

The main daytime route of nitric acid formation is from the reaction:

$$NO_2 + OH \rightarrow HNO_3 \qquad (38)$$

The rate constant is $1.1 \times 10^{-11} \, cm^{-3} \, molec^{-1} \, s^{-1}$ at 25 °C[9] implying a rate of NO_2 oxidation of $19.8\% \, h^{-1}$ at an OH concentration of $5 \times 10^6 \, cm^{-3}$ (*c.f.* Table 2). This is thus a much faster process than gas phase oxidation of SO_2. This process is not operative during hours of darkness due to near zero OH radical concentrations.

At nighttime, reactions of the NO_3 radical become important which are not operative during daylight hours due to photolytic breakdown of NO_3. The radical itself is formed as follows:

$$NO_2 + O_3 \rightarrow NO_3 + O_2 \qquad (39)$$

It is converted to HNO_3 by two routes. The first is hydrogen abstraction from hydrocarbons or aldehydes:

$$NO_3 + RH \rightarrow HNO_3 + R \qquad (40)$$

[10] P. Middleton, C. S. Kiang, and V. A. Mohnen, *Atmos. Environ.*, 1980, **14**, 463.
[11] S. G. Chiang, R. Toosi, and T. Novakov, *Atmos. Environ.*, 1981, **15**, 1287.
[12] R. M. Harrison and C. A. Pio, *Atmos. Environ.*, 1983, **17**, 1261.

Typical reaction rates imply formation of HNO_3 at about $0.3\,ppb\,h^{-1}$ in a polluted urban atmosphere by this route.[9] This is modest compared to daytime formation from NO_2 and OH.

The other nighttime mechanism of HNO_3 formation is via the reaction sequence:

$$NO_3 + NO_2 \overset{M}{\rightleftarrows} N_2O_5 \qquad (41)$$
$$N_2O_5 + H_2O \rightleftarrows 2HNO_3 \qquad (42)$$

The reaction involving water is rate determining and may be fairly slow at low relative humidity, contributing HNO_3 at $\sim 0.3\,ppb\,h^{-1}$.[9] As humidity increases, and especially in the presence of liquid water, more rapid reactions may be observed. This is presently supported only by indirect evidence and requires further experimental investigation.

Aqueous phase oxidation of NO_2 is of little importance due primarily to low aqueous solubility of NO_2.

4.5 Hydrochloric Acid

Hydrochloric acid differs from sulphuric and nitric acids in that it is emitted into the atmosphere as a primary pollutant and is not dependent upon atmospheric chemistry for its formation.

Some perspective of its significance as an acidic pollutant may be gained from Table 4 which shows the estimated emissions of SO_2, NO_x, and HCl in the UK and the potential hydrogen ion equivalent. At only 4.3%, the HCl contribution appears small until it is realized that this is immediately available acidity. Using diurnally averaged concentrations of reactive species, oxidation of SO_2 and NO_2 may be proceeding at only $\sim 1\%$ and $10\%\,h^{-1}$ respectively, or less, and thus over appreciable distances from a major source of all three pollutants (*e.g.*, a power plant), HCl may be the predominant strong acid.

Table 4 *Total UK acid emission*[13]

Species	Emission $(kt\,a^{-1})$	Potential H^+ equivalent $(kt\,a^{-1})$	% total potential acidity
SO_2	3 690	115.3	70.9
NO_x	1 856	40.4	24.8
HCl	250	7.0	4.3

Another source, not considered in the above table, may also contribute to HCl in the atmosphere. HCl is a more volatile acid than either H_2SO_4 or HNO_3 and thus may be displaced from aerosol chlorides, such as sea salt:

$$2NaCl + H_2SO_4 \rightarrow Na_2SO_4 + 2HCl \qquad (43)$$
$$NaCl + HNO_3 \rightarrow NaNO_3 + HCl \qquad (44)$$

[13] P. J. Lightowlers and J. N. Cape, *Atmos. Environ.*, 1988, **22**, 7.

This can be an appreciable source of HCl in areas influenced by maritime air masses.

4.6 Methanesulphonic Acid (MSA)

This strong acid is unlikely to be of importance in polluted regions but can contribute appreciably to acidity in remote areas. It is a major product of the oxidation of dimethylsulphide, $(CH_3)_2S$. A likely mechanism[9] is:

$$CH_3SCH_3 + OH \xrightarrow{M} CH_3SCH_3 \qquad (45)$$
$$| \qquad\qquad$$
$$OH$$

$$CH_3SCH_3 \rightarrow CH_3SOH + CH_3 \qquad (46)$$
$$|$$
$$OH$$

$$CH_3SOH + O_2 \xrightarrow{M} CH_3SO_3H \qquad (47)$$

At nighttime, the main breakdown mechanism of dimethylsulphide is reaction with NO_3 radicals, but the mechanism and products are presently unclear. Dimethylsulphide is a product of biomethylation of sulphur in seawater and coastal marshes and its production shows a strong seasonal pattern.

5 ATMOSPHERIC BASES

Carbonate rock such as calcite or chalk, $CaCO_3$, or dolomite, $CaCO_3.MgCO_3$, exist in small concentrations in atmospheric particles and provide a small capacity for neutralizing atmospheric acidity. In western Europe, however, the major atmospheric base is ammonia. Although not emitted to any major extent by industry or motor vehicles, ammonia is arguably a man made pollutant as it arises primarily from the decomposition of animal wastes; atmospheric concentrations relate closely to the density of farm animals in the locality. Release from chemical fertilizers can also be significant.[1]

In areas with a moderate or high ammonia source strength, ground-level atmospheric acidity is generally low. Sulphuric acid is present as highly neutralized $(NH_4)_2SO_4$ and HNO_3 and HCl predominantly as NH_4NO_3 and NH_4Cl. The latter two salts are appreciably volatile and may release their parent acids under conditions of low atmospheric ammonia, high temperature, or reduced humidity.

$$NH_4NO_3 \rightleftarrows HNO_3 + NH_3 \qquad (48)$$
$$NH_4Cl \rightleftarrows HCl + NH_3 \qquad (49)$$

The relative neutrality of ground-level air at such locations may, however, not be reflective of far greater acidity at greater heights above the ground, as is shown from simultaneous sampling of ground-level air and rainwater (see later).

Additionally, once deposited in soils, ammonium salts are slowly oxidized to release strong acid in a process which may be represented as:

$$(NH_4)_2SO_4 + 4O_2 \rightarrow H_2SO_4 + 2HNO_3 + 2H_2O \tag{50}$$

Thus the neutralization process has only a temporary influence and ultimately causes additional acidification.

Deposition of ammonia and ammonium may also contribute directly to damage to vegetation, which has proved to be a particular problem in the Netherlands. Since the reaction of ammonia with the OH radical is slow, the main sinks lie in wet and dry deposition processes.

6 ATMOSPHERIC AEROSOLS

Much of the atmospheric aerosol over populated areas is termed secondary as it is formed in the atmosphere from chemical reactions affecting primary pollutants. Water-soluble materials account typically for about 60% of the aerosol by mass[14] and comprise nine, or ten, major ionic components:

Anions: SO_4^{2-}; NO_3^-; Cl^-; (CO_3^{2-})
Cations: Na^+; K^+; Mg^{2+}; Ca^{2+}; NH_4^+; H^+

If concentrations are expressed in gram equivalents per cubic metre of air, some interesting relationships appear:

$$\text{Concentration (g equiv m}^{-3}) = \text{Concentration (g m}^{-3}) \times \frac{Z}{M}$$

where Z = Ionic charge and M = Molecular weight.
This is an expression as charge equivalents and hence if all major components are accounted for:

$$\Sigma \text{ anions} = \Sigma \text{ cations} \tag{51}$$

Typically for UK aerosol:

$$Na^+ \simeq 4.35 \, Mg^{2+} \tag{52}$$

and

$$Na^+ + Mg^{2+} \simeq Cl^- \tag{53}$$

when expressed in gram equivalents, relationships very similar to those pertaining in seawater, suggesting the latter as the major source of these components. In some instances land-derived Mg^{2+} may be significant and upsets this relationship.

Another relationship commonly observed is:

$$SO_4^{2-} + NO_3^- \simeq NH_4^+ + H^+ \tag{54}$$

In UK air, the ratio H^+/NH_4^+ is normally very low, although in other parts of the world it can be much higher. This indicates the existence of NH_4NO_3 and $(NH_4)_2SO_4$ arising from ammonia neutralization of nitric and sulphuric acids respectively. Close examination of a large data set[14] revealed that the NH_4^+, which is not accounted for by the above relationship, and Cl^-, not accounted for by association with Na^+ and Mg^{2+} in seawater, are approximately equal indicating the presence of NH_4Cl.

K^+ and Ca^{2+} are mainly soil-derived, although some seawater contribution is likely, whilst CO_3^{2-}, not always observed, arises from carbonate minerals such as $CaCO_3$ in rocks and soils.

The water-insoluble fraction of atmospheric aerosols represents the minor part in the UK, but may be far more significant in other countries, especially those with a dry climate. It comprises such materials as soil (*e.g.*, clay materials), rock fragments (*e.g.*, α-quartz), and elemental carbon from combustion processes.

Table 5 *Abundance of major crystalline compounds identified in the atmosphere of Toronto, Canada (from ref 15)*

Compound	Concentration ($\mu g\ m^{-3}$)	
	Summer	*Winter*
Total suspended particulate	26.2 ± 14.0	19.4 ± 8.1
α-SiO_2 (quartz)	1.37 ± 0.67	0.51 ± 0.39
$CaCO_3$ (calcite)	2.43 ± 1.42	1.19 ± 0.86
$CaMg(CO_3)_2$ (dolomite)	1.41 ± 0.89	0.71 ± 0.64
$CaSO_4.2H_2O$ (gypsum)	0.13 ± 0.21	0.32 ± 0.45
$NaCl$ (halite)	<d.l.	1.40 ± 1.00
$Al_{1-2}\ Si_{2-3}\ O_8$ (K,Na,Ca) (feldspar)	1.07 ± 0.07	0.21 ± 0.26
$(NH_4)_2SO_4$	3.36 ± 2.90	2.70 ± 1.49
$3(NH_4)_2SO_4.NH_4HSO_4$	0.38 ± 0.60	0.07 ± 0.18
$CaSO_4.(NH_4)_2SO_4.H_2O$	0.58 ± 0.51	0.43 ± 0.50
$PbSO_4.(NH_4)_2SO_4$	0.14 ± 0.13	0.07 ± 0.12
$(NH_4)_2SO_4.2NH_4NO_3$	0.25 ± 0.22	0.56 ± 0.55
$(NH_4)_2SO_4.3NH_4NO_3$	<d.l.	0.86 ± 1.67
NH_4Cl	0.01 ± 0.02	0.12 ± 0.12

Another approach to identifying the chemical compounds within the atmospheric aerosol is to use *X*-ray powder diffraction analysis. This technique is capable only of identifying crystalline components representing a significant proportion of the total mass; thus minor crystalline components and amorphous materials are not quantified. Table 5 shows the result of application of this method to seven winter samples and six spring/summer samples collected in Toronto, Canada.[15] A rather similar composition has been observed at other mid-latitude northern hemisphere sites, including some in the UK, although the mineral composition is influenced by local geology. In the winter samples, a total of $52 \pm 4\%$ of the total suspended particulate matter is accounted for by the compounds listed; in the spring/summer samples it is $46 \pm 10\%$.

[14] R. M. Harrison and C. A. Pio, *Environ. Sci. Technol.*, 1983, **17**, 169.
[15] W. T. Sturges, R. M Harrison, and L. A. Barrie, *Atmos. Environ.*, 1989, **23**, 1083.

As Table 5 shows, X-ray powder diffraction is capable of identifying mixed salts (*e.g.*, $(NH_4)_2SO_4.2NH_4NO_3$), which could not be distinguished from a mixture of $(NH_4)_2SO_4$ and NH_4NO_3 in the ionic balance work mentioned above. Lead, in the form of ammonium lead sulphate, $PbSO_4.(NH_4)_2SO_4$ is observed in most localities where leaded gasoline is utilized. This is formed by a liquid phase reaction process occurring after coagulation of vehicle-emitted lead bromochloride, $PbBrCl$, with ambient ammonium sulphate, present at higher relative humidities as solution droplets:

$$2PbBrCl + 2(NH_4)_2SO_4 \rightarrow PbSO_4.(NH_4)_2SO_4 + PbBrCl.(NH_4)_2BrCl \quad (55)$$

The other lead compound produced in this reaction, $PbBrCl.(NH_4)_2BrCl$ has also been identified in roadside air.[16]

The mixed salts of ammonium sulphate with ammonium nitrate and calcium sulphate are assumed to arise from coagulation mechanisms similar to that proposed above. Halite, $NaCl$, arises from marine sources at sites within 100 km or so of the sea. In Toronto, its presence in the winter, but not the summer, aerosol is suggestive of a source in road deicing salt. Quartz, calcite, dolomite, gypsum, and feldspar are very commonly observed in atmospheric aerosols, arising from local rocks and soils and in some instances also from building materials and roads.

Recent years have seen a realization that aerosol particles can play a substantial active role in atmospheric chemistry, rather than being simply an inert product of atmospheric processes. They can act as a surface for heterogeneous catalysis, or probably more important, can provide a liquid phase reaction medium. Soluble particles deliquesce at humidities well below saturation and many compounds, such as the abundant $(NH_4)_2SO_4$, probably exist predominantly in solution droplet form in humid climates such as that of the UK. As indicated above these droplets can form a medium for reaction of solutes. They can also provide a medium for reaction with gaseous components, for example, the dissolution of nitric acid vapour in sodium chloride aerosol, with release of hydrochloric acid vapour:

$$HNO_3 + NaCl \rightarrow NaNO_3 + HCl \quad (44)$$

One fascinating aspect of aerosol chemistry is the fact that at typical tropospheric temperatures both ammonium nitrate and ammonium chloride aerosols are close to dynamic equilibrium with their gaseous precursors:

$$NH_4NO_3(aerosol) \rightleftharpoons NH_3(g) + HNO_3(g) \quad (48)$$
$$NH_4Cl(aerosol) \rightleftharpoons NH_3(g) + HCl(g) \quad (49)$$

The position of these equilibria may be predicted from chemical thermodynamics, both for crystalline particles, and at higher humidities for solution droplets. A comparison of atmospheric measurements of the concentration product $[HNO_3]$ $[NH_3]$ with theoretical predictions appears in Figure 7. The diagonal lines indicate the theoretical values of concentration product as a function of inverse Kelvin temperature at the indicated relative humidity. The data points

[16] P. D. E. Biggins and R. M. Harrison, *Environ. Sci. Technol.*, 1979, **13**, 558.

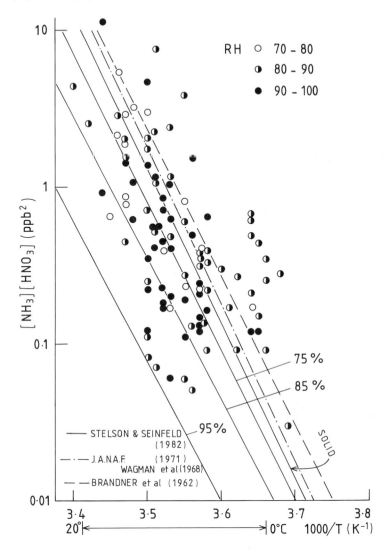

Figure 7 *A comparison of theoretically predicted values of the concentration product* [NH$_3$] [HNO$_3$] *(lines) with measurements from eastern England (points)*[17]

correspond to 24 h-average measurements. There is clearly a rather good agreement between theory and practical measurements. Deviations probably arise from kinetic constraints upon the achievement of equilibrium. The implication for the UK, where ammonia levels are fairly high, is that the ammonia concentration at a given atmospheric temperature and relative humidity controls the nitric acid concentration via the ammonium nitrate dissociation equilibrium. It appears also that very similar considerations apply to the concentration of HCl vapour in air which is subject to control by the ammonium chloride dissociation equilibrium.

[17] A. G. Allen, R. M. Harrison, and J. W. Erisman, *Atmos. Environ.*, 1989, **23**, 1591.

Atmospheric aerosols are responsible for the major part of the visibility reduction associated with polluted air. In the UK and other heavily industrialized regions, visibility reduction tends to correlate closely with the concentration of aerosol sulphate, a component tending to be present in sizes optimum for scattering of visible light. In very dry climates, visibility reduction is primarily associated with dust storms and is thus correlated with airborne soil.

7 THE CHEMISTRY OF RAINWATER

Substances become incorporated in rainwater by a number of mechanisms. The major ones are as follows:

(a) particles act as cloud condensation nuclei, *i.e.*, they act as centres for water condensation in clouds and fogs when the relative humidity reaches saturation

(b) particles are scavenged by cloudwater droplets or falling raindrops as a result of their relative motion

(c) gases may dissolve in water droplets either within or below the cloud.

As mentioned in Chapter 7, in-cloud scavenging is referred to as rainout, whilst below-cloud processes are termed washout. Incorporation in rain is a very efficient means of cleansing the atmosphere as evidenced by the substantial improvements in visibility occasioned by passage of a front.

The dissolved components of rainwater are the same nine or, at high rainwater pH, ten major ions listed earlier for atmospheric aerosol. There are also insoluble materials, again similar to the insoluble components of the atmospheric aerosol. When the composition of rainwater collected in the Lancaster, UK area was compared with that of aerosol collected over the same periods close to ground-level, the average composition shown in Figure 8 as percentages of total anion or cation load is very similar for many components, but is markedly different for some ions in aerosol and rainwater. The most obvious difference is in the H^+/NH_4^+ ratio: rainwater has a far higher ratio and is thus much more acidic. This arises for two main reasons:

(a) the neutralizing agent, ammonia, has a ground-level source and is thus more abundant at ground-level where the aerosol is sampled than at cloud level where the major pollutant load is incorporated into the rain

(b) in-cloud oxidation processes for SO_2 (see above) may lead to appreciable acidification of the cloudwater.

The regional distribution of rainwater pH over the UK is shown by Figure 9. The trend of increasing acidity from west to east arises from the prevailing south-westerly circulation bringing relatively clean air into the west of the country. Primary pollutants emitted over the country and the secondary pollutants formed from them lead to acidification which increases progressively as the air moves eastwards. The quantitative deposition of acidity in rain (as $g\,H^+\,m^{-2}$ per year) shows a rather different pattern (Figure 10) which arises from the fact

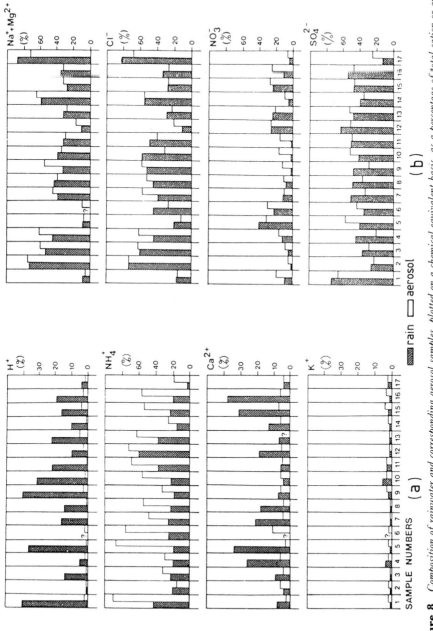

Figure 8 *Composition of rainwater and corresponding aerosol samples, plotted on a chemical equivalent basis, as a percentage of total cation or anion concentrations. Samples collected at a rural site near Lancaster, UK[18]*

[18] R. M. Harrison and C. A. Pio, *Atmos. Environ.*, 1983, **17**, 2539.

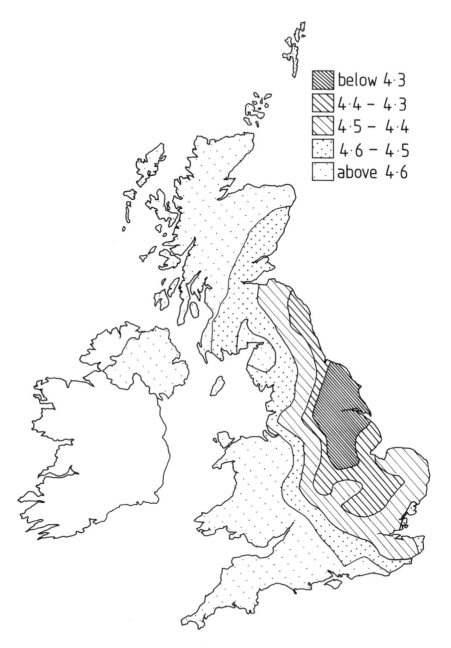

Figure 9   *Precipitation-weighted mean hydrogen ion concentration expressed as pH for 1986*[19]

that the annual deposition field is derived from the product of the volume-weighted mean concentration and the annual rainfall. Thus areas of high rainfall (*e.g.*, Scottish Highlands) exhibit high hydrogen ion inputs despite a relatively

[19] Warren Spring Laboratory, 'United Kingdom Acid Rain Monitoring,' Stevenage, 1988.

high pH. One feature of acidic inputs is the substantial influence of episodicity: a very small number of rainfall events in any year combining high acidity with a large rainfall amount can contribute a large proportion of the total annual hydrogen ion input.

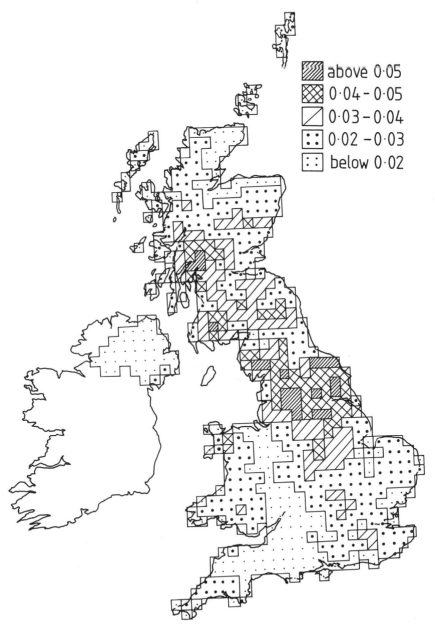

above 0·05
0·04 – 0·05
0·03 – 0·04
0·02 – 0·03
below 0·02

Figure 10 *Wet deposited acidity* (g H$^+$ m^{-2}) *in the UK in 1986*[19]

CHAPTER 9

Chemistry and Pollution of the Stratosphere

I. COLBECK AND J. C. FARMAN

1 INTRODUCTION

In the early 1970's the possibility of polluting the stratosphere and, in particular, depleting the ozone layer allowing biologically harmful ultraviolet radiation to reach the ground with potential adverse effects on human health and on the aquatic and terrestrial ecosystems, was first raised. Among the chemical agents considered harmful were nitrogen oxides, produced by the detonation of nuclear weapons or from the exhaust of supersonic aircraft, and chlorofluorocarbons (CFC), especially $CFCl_3$ and CF_2Cl_2, which are used mainly as refrigerants and aerosol propellants. Ten years later, concern that man's activities could affect significantly stratospheric ozone had diminished. Observations and models agreed; observed and predicted changes were less than 1 per cent per decade. However, concern over ozone resurfaced abruptly in 1985 with the publication by Farman et al. who reported that the total ozone column over Antarctica in September and October had decreased by up to 40% in the past decade. Such a change had not been predicted by any model of future stratospheric composition. This depletion precipitated a great deal of research on the chemistry and dynamics of the stratosphere. Recent data leave no doubt that man made chlorofluorocarbons are primarily responsible for this depletion. This ozone decrease is not confined to the Antarctic. Ozone concentrations generally, over the Northern Hemisphere, have fallen several per cent between 1969 and 1986. This again has been attributed to an increased abundance of chlorofluorocarbons.

Recent measurements have shown that conditions in the Arctic stratosphere are ideal for a significant depletion of ozone. Preliminary indications of ozone

181

depletion come from Alert Bay, high in the Canadian Arctic, which reported a 5% decrease in ozone. The consequences of an ozone decline over the Arctic are potentially greater than that over the Antarctic. An Arctic ozone hole could endanger some of the most populated parts of the globe, including northern Europe, North America, and the Soviet Union.

Intensive theoretical and experimental investigations of stratospheric ozone have been underway for many years. Our present understanding of the stratosphere has grown out of the research motivated by the perceived threat of man's activities. Laboratory studies have improved the accuracy of reaction rate data and have identified significant species and reactions which had been omitted from numerical models. Many trace constituents of the atmosphere are now measured from balloon-borne platforms and satellites, providing much needed data on the temporal and spatial variability of chemical species. If we want to protect our environment, it is essential that we know more about the various atmospheric trace species and their impact on chemical processes in the atmosphere. Whereas acid rain and photochemical oxidant pollution are somewhat localized environmental problems, modification of stratospheric ozone is a global phenomenon and hence affects everyone.

2 STRATOSPHERIC OZONE

The Earth's atmosphere may be divided into a series of 'spherical' layers each characterized by a distinctive vertical gradient of temperature. Figure 1 shows a typical mid-latitude temperature profile. The stratosphere is characterized by a strong temperature inversion and dynamically stable air. This distinguishes it from the troposphere where vertical mixing is rapid. The lower limit of the stratosphere is at an altitude of 8 to 16 km, depending on latitude, and it extends to about 45 km or 50 km. Whereas the troposphere is heated from below by solar radiation absorbed at the surface, the stratosphere is heated from above by absorption of ultraviolet radiation by ozone. Hence, the distribution of ozone has a profound influence on the atmospheric temperature structure and consequent motions of the stratosphere. It should also be noted, that due to its low density and its stability against vertical mixing, pollutants in the stratosphere have a lifetime for removal by transport of several years, and hence could build up to globally damaging levels.

It is just over 100 years ago that Hartley (1881) pointed out that ozone is a normal constituent of the higher atmosphere and is in a larger proportion there than near the earth's surface. Fabry and Dobson were the pioneers in monitoring the abundance of atmospheric ozone, using spectroscopic techniques. We now know that ozone occurs in trace amounts throughout the atmosphere with a peak concentration in the lower stratosphere between about 20 and 25 km altitude (Figure 1). Observations of the total amount of ozone in a vertical column have established the average column abundance and the general pattern of latitudinal, longitudinal, seasonal, and meteorological variations. Total ozone is usually stated in Dobson Units or matm cm, related to the thickness of an equivalent layer of pure ozone at standard pressure and temperature. If the ozone in a column of the atmosphere were concentrated into a thin shell surrounding the earth at atmospheric pressure, it would be about 3 mm thick, *i.e.*, the average total amount of ozone is 300 DU. Seasonal and latitude variations at sub-polar latitudes are about $\pm 20\%$ of this value. The annual average total ozone is a minimum of approximately 260 DU at equatorial latitudes and increases poleward in both hemispheres to a maximum at sub-polar latitudes of about 400 DU. The high latitude maximum results from transport of ozone from the equatorial middle and upper stratosphere, the region of primary production, to the lower stratosphere in polar regions. About 90% of all the ozone in the atmosphere resides in the stratosphere.

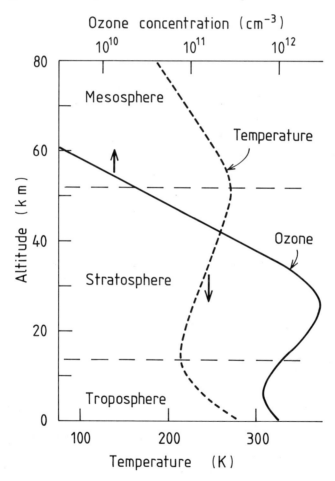

Figure 1 *Ozone distribution and temperature profile in the atmosphere*

How is this ozone produced? Sixty years ago Chapman (1930) proposed a static pure oxygen photochemical steady state model that could explain the formation of the ozone layer. Chapman's predictions agreed well with observations available at the time. The reactions were:

$$O_2 + h\nu \rightarrow O + O \quad \lambda < 243 \text{ nm} \tag{1}$$
$$O + O_2 + M \rightarrow O_3 + M \tag{2}$$
$$O_3 + h\nu \rightarrow O + O_2 \tag{3}$$
$$O + O_3 \rightarrow O_2 + O_2 \tag{4}$$

Ozone is produced by the absorbance of solar ultraviolet radiation by oxygen molecules (O_2) which then photodissociate to form oxygen atoms (O). These O atoms in turn react with O_2 via a three-body reaction to form O_3. The O_3 produced is subsequently photodissociated by both u.v. and visible light. Finally the chemical loss for odd oxygen is the reaction between O and O_3 to reform two O_2 molecules. Since reaction (2) becomes slower with increasing altitude, while reaction (3) becomes faster, atomic oxygen predominates at high altitudes and O_3 is favoured at lower ones. Reactions (2) and (3), in which odd oxygen is interchanged are much faster than reactions (1) and (4) in which odd oxygen is created or destroyed. If n_1, n_2, n_3, and n_m are the concentrations of O, O_2, O_3, and all molecules, respectively and k_2 and k_4 are the reaction rates of reaction (2) and (4) and J_1 and J_3 the photolysis rate for reaction (1) and (3) then the equilibrium ozone concentration is given by:

$$n_3 = n_2 \left[\frac{J_1 k_2 n_m}{J_3 k_4} \right]^{\frac{1}{2}}$$

The reaction rate for loss of O_3 (reaction 4) was unknown at the time of Chapman's proposal. Laboratory measurements have since revealed that reaction (4) is too slow to destroy ozone at the rate it is produced globally. There exist chemical processes which provide a more efficient route than reaction (4) for the loss of odd oxygen. The catalytic cycle:

$$X + O_3 \rightarrow XO + O_2 \tag{5}$$

$$\frac{XO + O \rightarrow X + O_2}{\text{net} \quad O + O_3 \rightarrow 2O_2} \tag{6}$$

achieves the same result as reaction (4), where the reactive species X is not consumed in the process and can effectively destroy many ozone molecules. Several species have been suggested for the catalyst X, among these are X = H, OH, NO, Cl and Br.

The significance of these catalytic cycles is that small quantities of the catalytic species can affect ozone in a substantial manner. These species can, with varying degrees of efficiency, control the abundance and distribution of ozone in the stratosphere. The rate parameters for reactions (5) and (6) and different catalysts are given in Table 1. It can be seen that these catalytic cycles are faster than the direct reaction (4), for a temperature of 220 K, typical of the stratosphere.

Table 1 *Rate constants [and uncertainty factor*] at a temperature of 220 K for some catalytic cycles*

X	Reaction $X + O_3$	Reaction $XO + O$
	$k(cm^3 \text{ molecule}^{-1} s^{-1})$	$k(cm^3 \text{ molecule}^{-1} s^{-1})$
H	1.7×10^{-11} [1.59]	3.8×10^{-11} [1.35]
OH	2.2×10^{-14} [1.86]	7.4×10^{-11} [1.35]
NO	3.4×10^{-15} [1.52]	1.1×10^{-11} [1.27]
Cl	8.9×10^{-12} [1.30]	4.1×10^{-11} [1 30]
Br	4.5×10^{-13} [1.52]	3.0×10^{-11} [4.04]
	$O + O_3$ (reaction 4)	6.9×10^{-16} [1.55]

* Plausible range = mean ÷ uncertainty factor to mean × uncertainty factor.

Ozone may also be destroyed in cycles not involving atomic oxygen which hence are important in the lower stratosphere:

$$OH + O_3 \rightarrow HO_2 + O_2 \tag{7}$$

$$\text{net} \quad \frac{HO_2 + O_3 \rightarrow OH + 2O_2}{2O_3 \rightarrow 3O_2} \tag{8}$$

$$NO + O_3 \rightarrow NO_2 + O_2 \tag{9}$$

$$NO_2 + O_3 \rightarrow NO_3 + O_2 \tag{10}$$

$$\text{net} \quad \frac{NO_3 + h\nu \rightarrow NO + O_2}{2O_3 + h\nu \rightarrow 3O_2} \tag{11}$$

Species X and XO may be involved in cycles without odd oxygen removal. For X = N we have:

$$NO + O_3 \rightarrow NO_2 + O_2 \tag{9}$$

$$\text{net} \quad \frac{NO_2 + h\nu \rightarrow NO + O}{O_3 + h\nu \rightarrow O_2 + O} \tag{12}$$

This cycle is in competition with the catalytic cycle and the NO_x tied up in this cycle is ineffective as a catalyst.

The removal of radicals from the stratosphere occurs by chemical conversion to more stable oxidation products. For example:

$$OH + NO_2 + M \rightarrow HNO_3 + M \tag{13}$$

The HNO_3 may be transported down into the troposphere and removed in rain or may be photolysed to regenerate $OH + NO_2$. This latter process is relatively slow and HNO_3 is said to act as a reservoir of NO_x. About half the stratospheric load of NO_x is stored in this HNO_3 reservoir. For Cl_x about 70% of the stratospheric load is present as HCl via:

$$Cl + CH_4 \rightarrow CH_3 + HCl \tag{14}$$

The chlorine tied up as stable HCl may be released by:

$$OH + HCl \rightarrow H_2O + Cl \tag{15}$$

This ClO_x can then participate in catalytic cycles leading to the removal of ozone.

Extensive research has emphasized the importance of temporary reservoir species such as $ClONO_2$, $HOCl$, N_2O_5, and HO_2NO_2 which act to lessen the efficiency of ClO_x and NO_x species in destroying ozone in the lower stratosphere. These reservoir species may be formed by:

$$NO_2 + O_3 \rightarrow NO_3 + O_2 \tag{10}$$
$$NO_3 + NO_2 + M \rightarrow N_2O_5 + M \tag{16}$$
$$ClO + HO_2 \rightarrow HOCl + O_2 \tag{17}$$
$$HO_2 + NO_2 + M \rightarrow HO_2NO_2 + M \tag{18}$$
$$ClO + NO_2 + M \rightarrow ClONO_2 + M \tag{19}$$

These reactions emphasize the coupling between the HO_x, NO_x, and ClO_x families. The effects of these catalytic families are not additive since members of one family can react with members of another, *e.g.*,

$$NO + ClO \rightarrow NO_2 + Cl \tag{20}$$
$$HO_2 + NO \rightarrow NO_2 + OH \tag{21}$$

These reactions interconvert NO_x, HO_x, and ClO_x without participation of odd oxygen and hence reduce the efficiency of the catalytic cycles.

The recycling of HO_x, NO_x, and ClO_x from the reservoirs N_2O_5, $HOCl$, and $ClONO_2$ is by photolysis. Removal of odd hydrogen in the lower stratosphere occurs mainly by:

$$OH + HNO_3 \rightarrow H_2O + NO_3 \tag{22}$$
$$OH + HO_2NO_2 \rightarrow H_2O + NO_2 + O_2 \tag{23}$$

These reactions also release NO_x from the reservoirs.

Bromine chemistry is in many respects similar to that for chlorine, but differs in several important aspects. The reaction of Cl with CH_4 (reaction 14) to produce HCl limits the abundance of active chlorine in the stratosphere. The reaction of Br with CH_4 is endothermic and can be neglected. As a result BrO is the dominant form of BrO_x in the stratosphere. Despite relatively low BrO_x concentrations in the lower stratosphere O_3 may be destroyed by:

$$BrO + BrO \rightarrow 2Br + O_2 \tag{24}$$
$$BrO + ClO \rightarrow Br + Cl + O_2 \tag{25}$$

and

$$3(Br + O_3 \rightarrow BrO + O_2) \tag{26}$$

net

$$\frac{Cl + O_3 \rightarrow ClO + O_2}{4O_3 \rightarrow 6O_2} \tag{5}$$

Since HBr and $BrONO_2$ are rapidly photolysed there are no known efficient reservoirs for Br and BrO. On a molecule for molecule basis, bromine is a much more efficient sink for O_3 than chlorine.

Hydrocarbon oxidation is closely related to all other reactive trace gas species and hence ozone photochemistry. Methane is the dominant stratospheric hydrocarbon and is the main source of water in the stratosphere, via:

$$OH + CH_4 \rightarrow CH_3 + H_2O \tag{27}$$

The reaction:

$$Cl + CH_4 \rightarrow HCl + CH_3 \tag{14}$$

is the primary loss process for active chlorine. The CH_3 radical may be further oxidized to CO_2 through intermediate products CH_3O_2, CH_3O, HCHO, and CO. Below 35 km there is sufficient NO for CH_4 oxidation to be a net source of O_x by the reactions:

$$CH_2O_2 + NO \rightarrow CH_3O + NO_2 \tag{28}$$
$$NO_2 + h\nu \rightarrow O + NO \tag{12}$$
$$O + O_2 + M \rightarrow O_3 + M \tag{2}$$

This scheme is also a source of HO_x through the reactions:

$$CH_3O + O_2 \rightarrow HO_2 + HCHO \tag{29}$$
$$HCHO + h\nu \rightarrow H + HCO \tag{30}$$
$$HCHO + h\nu \rightarrow H_2 + CO \tag{31}$$

Figures 2, 3, 4, and 5 summarize the current state of understanding of the atmospheric reactions involving HO_x, NO_x, ClO_x, and CH_4 oxidation. Improvements in our knowledge of interfamily coupling have resulted in many changes in calculated ozone depletions over the past decade.

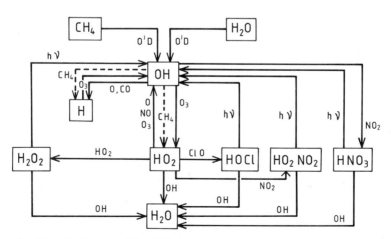

Figure 2 *Schematic diagram of the odd-hydrogen cycle*

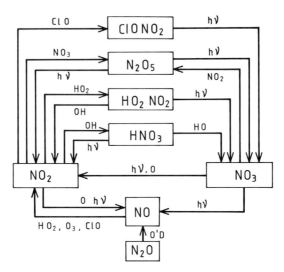

Figure 3   *Schematic diagram of the odd-nitrogen cycle*

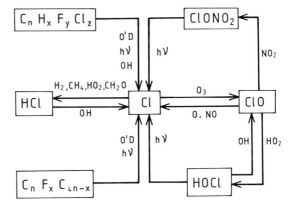

Figure 4   *Schematic diagram of the odd-chlorine cycle*

3   NATURAL SOURCES OF TRACE GASES

Anthropogenic sources undoubtedly contribute to natural background concentrations of many trace gases which have a profound effect on the ozone layer. From a knowledge of these natural sources it is then possible to say how man's activities have perturbed the atmosphere.

Nitrous oxide is the dominant precursor of stratospheric NO_x. A major natural source of N_2O is nitrification in aerobic soils and waters. Soils in tropical forests emit N_2O at rates far in excess of soils in most other environments. This source of N_2O is likely to be changing since the world's tropical forests are being rapidly modified. During the polar night, cosmic rays contribute to NO production in the altitude range 10–30 km. This source is not important on the global scale.

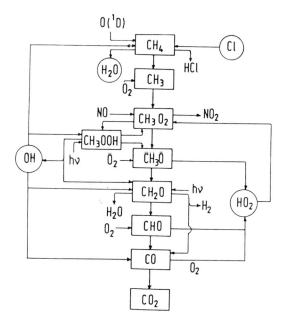

Figure 5 *Schematic diagram of methane oxidation chemistry*

Natural sources of Cl_x, providing enough to have a marked influence on atmospheric ozone, are few. Only three compounds, CH_3Cl, CH_2Cl_2, and $CHCl_3$ are thought to have substantial natural sources. Industrial sources of CH_3Cl are far outweighed by releases from the oceans and from the burning of vegetation. Volcanic eruptions can also result in CH_3Cl and HCl being released into the atmosphere.

The main source of OH radicals is the reaction of $O(^1D)$ with water vapour and methane:

$$O(^1D) + H_2O \rightarrow OH + OH \tag{32}$$
$$O(^1D) + CH_4 \rightarrow OH + CH_3 \tag{33}$$

The atmospheric concentration of CH_4 has been increasing at an average annual rate of about 1%. The major natural sources are wet lands, rice paddies, termites, wild fires, and enteric fermentation in ruminants. These sources may account for up to 35% of the total annual production.

4 ANTHROPOGENIC SOURCES

Man's capability to produce nitrogen and chlorine compounds at levels equalling or exceeding natural sources has led to numerous investigations aimed at assessing the impact of human activities on the ozone layer.

Johnston (1971) postulated that NO_x, emitted in the exhausts of supersonic aircraft (SST), would result in large ozone depletions. Early models predicted depletions of approximately 10% in total ozone. The altitude of greatest depletion occurred between 20 and 30 km and incremental decreases tended to be larger with each additional unit of NO_x injection. As the kinetics were improved by laboratory measurements the predicted depletion gradually fell and by mid-1978 the total ozone was forecast to increase. The effect of SST's is largely academic, with no prospect of large fleets of SST's becoming operational in the near future.

The space shuttle launch and re-entries provide a minor source of strato-spheric NO_x. However, the solid-fuel booster engines release about 100 tons of HCl vapour to the stratosphere. The effect of 60 shuttle launches a year would reduce ozone over the Northern Hemisphere by about 0.2%. If the shuttle stopped flying or if the booster rockets were changed so that chlorine compounds were not emitted the ozone layer would recover in 4 to 6 years. Again, the earlier predictions of flight numbers has never materialized.

Nuclear explosions produce copious quantities of NO_x which may be injected into the stratosphere. Early work predicted O_3 reductions of up to 70% in the Northern Hemisphere. With the move towards large numbers of smaller yield weapons the amount of NO_x injected into the stratosphere has been greatly reduced. Calculated ozone perturbations are very sensitive to the vertical distribution of NO_x injections and therefore the assumed yields of individual weapons. Hence, calculated ozone changes may vary widely among scenarios, estimated to produce roughly equivalent quantities of smoke. For a 6500 Mt scenario, the baseline case, ozone is predicted to be reduced by 17% over the Northern Hemisphere, one year after the war. Recovery to one-half of the peak reduction would require an additional 2 years (SCOPE, 1986).

There are two significant sources of N_2O which result from human activity. These are combustion and the use of nitrogenous fertilizers that would enhance biogenic emissions of N_2O. A doubling in N_2O concentrations is predicted to give a 10–16% global ozone depletion. The depletions vary with altitude and are greatest at higher latitudes. For a more realistic increase of 20% in N_2O, total ozone decreases range from 1.1 to 2.6%. Small increases in O_3 are calculated in the upper troposphere and lower stratosphere due to increased efficiency of the CH_4–NO_x–smog reactions from the added NO_x (see Chapter 8).

Molina and Rowland (1974) suggested that free chlorine introduced into the upper atmosphere as a result of the photodissociation of man made chlorofluoro-methanes could greatly enhance the catalytic destruction of ozone, *e.g.*,

$$CF_2Cl_2 + h\nu \rightarrow CF_2Cl + Cl \qquad (34)$$
$$CFCl_3 + h\nu \rightarrow CFCl_2 + Cl \qquad (35)$$

$$Cl + O_3 \rightarrow ClO + O_2 \qquad (5)$$
$$\underline{ClO + O \rightarrow Cl + O_2} \qquad (6)$$
$$\text{net} \quad O + O_3 \rightarrow 2O_2$$

CFC's are used in a wide variety of industrial applications including aerosol propellants, refrigerants, solvents, and foam blowing. The CFC's which have attracted most attention in ozone depletion are CFC-11 ($CFCl_3$) and CFC-12 (CF_2Cl_2). The lifetime of CFC-11 is about 77 years, while that for CFC-12 is 139 years (SORG, 1988). Hence any reduction in CFC release will have little effect in the immediate future. Model predictions of ozone column changes due to increases in CFC's have varied widely during the last 10 years. If CFC emissions continue at the 1980 rates long enough to obtain a steady-state, 8 ppb of chlorine would result. Calculated changes in column ozone, when chlorine is increased from 1 ppb to 8 ppb, range from approximately 3 to 9%, depending on the model used. In all models, the major decrease occurs in the upper stratosphere, near 40 km. As chlorine is increased it exceeds a certain threshold where each additional chlorine atom destroys progressively more and more ozone. By increasing chlorine to 15 ppb, the models then predict ozone column changes of 9 to 22%. This nonlinearity for large chlorine perturbations can have significant implications for the interpretation of effects of changing CFC emissions. The most abundant organochlorine compounds in the atmosphere are CCl_2F_2, CCl_3F, CH_3Cl, CCl_4, and CH_3CCl_3. The total atmospheric content of organochlorine compounds has grown from about 0.8 ppb in 1950 to 1.0 ppb in 1970 and 3.5 ppb in 1987 (WMO–NASA, 1988). The current rate of increase exceeds 1 ppb per decade.

Wofsy *et al.* (1975) suggested that trace amounts of bromine might efficiently catalyse ozone destruction. The predominant anthropogenic source is CH_3Br and CF_3Br. Some sources, such as CF_2BrCl and $C_2H_4Br_2$, are entirely man made. CH_3Br is used as a soil fumigant and other compounds of bromine are used as fire retardants and fuel additives. An increase in CH_3Br from 20 to 100 ppt would result in a decrease in total ozone of 30%. The largest relative change in ozone occurs around 15 km with a secondary peak at 40 km.

Carbon dioxide concentrations are increasing largely as a result of burning fossil fuel. Increased CO_2 leads to lower stratospheric temperatures, and hence a slowing down of temperature dependent ($O + O_3$, $NO + O_3$) ozone-destruction reactions. This results in a net increase in stratospheric ozone concentrations. Doubling CO_2 concentrations leads to temperature decreases of 7 to 10 K near 40 km, and a corresponding increase in ozone between 9 and 19%. Increases in the total ozone column are in the range 1.2 to 3.5%.

Increases in total ozone between 0.9 and 3.0% are also predicted for a doubling of CH_4 concentrations. For such an increase in CH_4, stratospheric OH is predicted to increase by up to 50% and HO_2 by up to 100%. The resulting increase in HNO_3 and HNO_4 production reduces NO_2 concentrations and a subsequent reduction in the NO_x catalytic cycle. Methane is currently increasing due to the growth in rice paddy farming and cattle farming.

5 MULTIPLE PERTURBATIONS

Whereas single source perturbations define the relative importance and potential effects of individual pollutants they do not reflect reality. Interactions between the various pollutants are not additive. Simultaneous NO_x and Cl_x injections are less effective in reducing ozone than separate injections. Table 2 shows the results of one-dimensional model runs for single and multiple perturbations. It can be seen for each model with combined perturbations, the calculated changes in total ozone are much less than the sum of the individual perturbations involved. Calculated changes in O_3 with altitude show large decreases in the upper stratosphere primarily from chlorine and increases in the lower stratosphere and troposphere as a result of methane changes.

Table 2 *Changes in column ozone predicted by a variety of 1-D models, at steady-state, and scenarios for a perturbed atmosphere, relative to an atmosphere with 1.3 ppb chlorine and no CFC*

	% change in total ozone predicted by model:					
	A	B	C	D	E	F
8 ppb Cl_x	−5.1 (−5.7)	−2.9	−4.6		(−4.1)	(−9.1)
15 ppb Cl_x	−12.2 (−12.4)	−17.8	−15.0		(−8.8)	(−22.0)
2 × CH_4	+2.0 (+2.9)	+0.3	+0.9	+1.7	(+1.6)	(+1.4)
1.2 × N_2O	−2.1 (−1.7)	−2.6	−1.8	−2.3	(−1.1)	(−1.2)
2 × CO_2	(+3.5)		(+2.6)	(+2.8)	(+3.1)	(+1.2)
8 ppb Cl_x +2 × CH_4 +1.2 N_2O	−3.4 (−2.8)	−3.0	−3.3	−3.1	(−2.3)	(−6.0)
8 ppb Cl_x +2 × CH_4 +1.2 × N_2O +2 × CO_2	(+0.2)			(−1.4)	(0.0)	(−5.2)
15 ppb Cl_x +2 × CH_4 +1.2 × N_2O	−7.8 (−7.2)	−8.2	−8.8	−7.2	(−5.6)	−13.7
15 ppb Cl_x +2 × CH_4 +1.2 × N_2O +2 × CO_2	(−4.6)				(−3.5)	(−13.6)

Numbers in parentheses refer to model runs with temperature feedback. Adapted from Present State of Knowledge of the Upper Atmosphere, NASA reference publication 1162, 1986.

Two-dimensional models are able to model seasonal variations, explicitly as the sun angle is varied, as well as latitude and altitude variations. Such models indicate a significant seasonal effect but one which is less than the latitudinal effect. Large decreases of ozone are predicted around 40 km for chlorine increases of 7 ppb in April, which is in agreement with one-dimensional models. There is, however, substantial latitudinal variation below 25 km. An increase in ozone in equatorial latitudes and a decrease in polar regions is predicted.

During the early 1980's models were predicting rather small changes in average column ozone abundances. For a 3% growth rate per year of CFCs a 10% ozone depletion was calculated after 70 years. Significant vertical and latitudinal changes of ozone were also predicted. The observations of ozone decreases of up to 40% over Antarctica during the spring since the late 1970's were hence unexpected.

6 ANTARCTIC OZONE

Following the paper by Farman *et al.* (1985) other groups soon confirmed the report. Measurements from instruments aboard the Nimbus 7 satellite showed that the region of ozone depletion was larger than the Antarctic continent (Stolarski *et al.*, 1986), while balloon borne instruments indicated that greatest depletion occurred between 12 and 24 km in altitude (Hofmann *et al.*, 1987). Figure 6 shows the monthly mean values of the ozone column for October. Prior to the mid-1970's the concentration of ozone was approximately 300 DU. By 1985 total ozone had fallen to below 200 DU and yet greater depletions were observed in 1987. The variation of ozone with altitude above Halley Bay is shown in Figure 7. In mid-August the profile was near normal but in less than two months 97% of the ozone between 14 and 18 km in altitude has been destroyed.

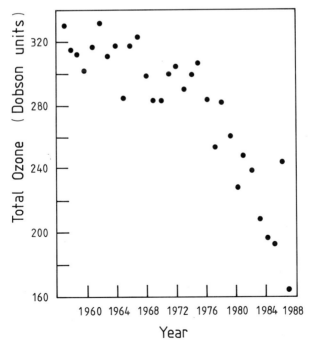

Figure 6   *Mean October value of column ozone over Halley Bay* (76 °S) *since* 1957 *(adapted from SORG, 1988)*

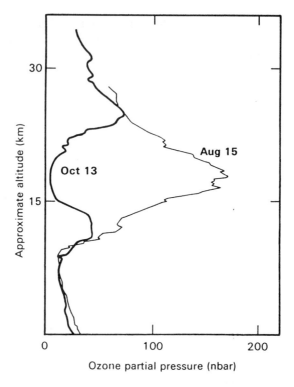

Figure 7   *Vertical ozone profiles above Halley Bay for* 15th *August,* 1987 *and* 13th *October,* 1987 *(adapted from SORG,* 1988)

Various theories have been postulated to explain this ozone depletion (Solomon, 1988) including dynamical air motions near the pole, a connection with the 11 year cycle in solar activity, and a chemical mechanism. It is now recognized that the depletion is caused by man made chlorine pollution of the atmosphere in combination with unique wintertime meteorological conditions.

The circulation of the winter stratosphere over Antarctica and the surrounding oceans is dominated by the polar vortex. This is a region of very cold air surrounded by westerly winds which may reach speeds of $\sim 100\ \mathrm{m\ s^{-1}}$ or more by the spring. Air within the vortex is sealed off from that at lower latitudes and the vortex acts as a containment vessel in which chemistry may occur in near isolation. Temperatures within the vortex fall to well below 190 K; cold enough in the lower half of the stratosphere for polar stratospheric clouds (PSC) to form. These clouds remove active nitrogen from the stratosphere that would otherwise remove active chlorine and also provide surfaces for heterogeneous chemical reactions.

Recent results from aircraft flights into the vortex have shown it to be very dry, very poor in compounds that contain nitrogen, and very high levels of both chlorine monoxide and chlorine dioxide. The low NO_2 concentrations may be explained by the conversion of NO_2 to N_2O_5 during the polar night via reactions

10 and 16. N_2O_5 is then converted to HNO_3 by a heterogeneous reaction which takes place on the surface of aerosols or in cloud particles. These PSCs are composed of submicron-size particles consisting of about 50% HNO_3. Below 193 K HNO_3 and H_2O can form PSCs. If temperatures drop a further 4 degrees, PSCs, consisting primarily of water ice, grow to much larger sizes ($\sim 10\,\mu m$). Such particles with a sedimentation velocity around 1 km day^{-1}, are soon removed from the stratosphere carrying with them nitrogen oxides.

The PSCs also act as sites of heterogeneous reactions that may free active chlorine atoms in reservoir compounds (HCl, $ClONO_2$) far more efficiently than homogeneous gas-phase reactions. The reactions are:

$$HCl + ClONO_2 \xrightarrow{\text{surface}} Cl_2 + HNO_3 \tag{36}$$

$$H_2O + ClONO_2 \xrightarrow{\text{surface}} HOCl + HNO_3 \tag{37}$$

The chlorine compounds are released as a gas and HNO_3 remains in the ice particles. During the winter and early spring forms of chlorine are then accumulated that quickly dissociate with the reappearance of sunlight. Molecular Cl_2 is photodissociated to release Cl atoms which may then be converted to ClO. Any NO_2 present can convert the ClO back to $ClONO_2$. Hence we have:

$$Cl_2 + h\nu \rightarrow Cl + Cl \tag{38}$$
$$Cl + O_3 \rightarrow ClO + O_2 \tag{5}$$
$$ClO + NO_2 + M \rightarrow ClONO_2 + M \tag{19}$$

This chlorine nitrate is then available to react with any HCl to release more Cl, which may then be converted to ClO. Hence within the polar vortex we would expect to find high concentrations of ClO.

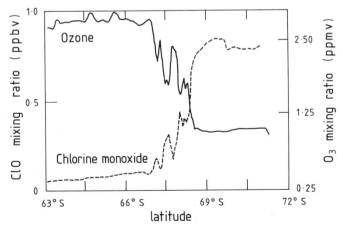

Figure 8 *Chlorine monoxide and ozone concentrations poleward of 63 °S at an altitude of 18 km in Mid-September, 1987 (adapted from SORG, 1988)*

Extensive measurements of O_3 and ClO have been made during the US Airborne Antarctic Ozone Experiment. During late August, at an altitude of 18 km, the ClO concentration showed a large increase at about 65 °S and relatively constant ozone levels. By mid September, when the sun was shining for a few hours a day, there was a sharp reduction in ozone concentration with 2.5 ppm north of 65 °S and about 1.0 ppm to the south (Figure 8). Concentrations of ClO within the Antarctic vortex are up to 500 times greater than those found at similar altitudes outside the vortex.

In addition to ozone destruction by chlorine (reactions 5 and 6) a catalytic process involving the ClO dimer, Cl_2O_2, is also important

$$ClO + ClO + M \rightarrow Cl_2O_2 + M \tag{39}$$

$$Cl_2O_2 + h\nu \rightarrow Cl + ClOO \tag{40}$$

$$ClOO + M \rightarrow Cl + O_2 + M \tag{41}$$

$$\mathrm{net} \quad \frac{2(Cl + O_3 \rightarrow ClO + O_2)}{2O_3 + h\nu \rightarrow 3O_2} \tag{5}$$

McElroy *et al.* (1986) suggested that bromine could make a significant contribution to ozone depletion via reaction (25) which liberates two halogen atoms. Measurement of BrO showed concentrations around 5 ppt at 18 km. These are too low to account for more than 10% of the ozone depletion. It should, however, be noted that atmospheric bromine concentrations are increasing rapidly. All these chains are able to continue throughout September since the usual chain-breaking steps of Cl reaction with CH_4, or ClO reaction with NO_2 are particularly weak, since methane concentrations are low and NO_x is tied up as HNO_3.

The severity and duration of the ozone hole has changed over the last ten years. As sunlight returns to the polar regions, atmospheric temperatures rise and the vortex collapses. The onset of depletion is largely fixed to the time of the first sunrises of the spring. As chlorine concentrations increase the rate of ozone depletion also increases. Hence the lowest ozone values may be reached progressively earlier in the year resulting in increased radiative cooling, lower stratospheric temperatures, a higher probability of PSC formation later in the season and a delayed breakup of the vortex. When the vortex breaks down, the ozone depleted air becomes distributed over the hemisphere, leading to ozone reductions at lower latitudes. In 1987, a large region of ozone depleted air was observed to drift over the southern ocean towards South Africa before finally being distributed over the hemisphere. Ozone depletions may also result from air sinking through the vortex during the period of ozone destruction and then being transported to lower latitudes. Hartman *et al.* (1988) have suggested that as much as one-third of the air in the vortex may reach lower latitudes and altitudes in late August and September.

Could this ozone depletion happen elsewhere? Whilst the Arctic has a warmer stratosphere and weaker vortex than the Antarctic, PSCs, containing ice crystals, have been observed in large numbers there. Additionally, researchers have measured substantial amounts of OClO, which is virtually absent in the normal atmosphere, and elevated concentrations of ClO. By early February, 1989, ClO mixing ratios were similar to those seen in the Antarctic, as high as 0.8 ppb inside the vortex. BrO radicals were also similar to levels in the Antarctic vortex, with mixing ratios of 2 to 8 ppt. Denitrification was not observed over the Arctic. Levels of nitric acid were much higher than over Antarctica. With the return of sunlight, HNO_3 could be converted to NO_2 which would reconvert the ClO into an inactive form (reaction 19). As yet there is no conclusive evidence of an ozone decline, but the ClO concentration is capable of destroying about 0.5 to 1 per cent of the region's stratospheric ozone every day in the presence of sufficient sunshine. When that occurs will depend on how long the vortex stays together. Canadian government scientists have provided preliminary evidence of a 5% decline in stratospheric ozone over Alert Bay. The cause of this depletion is currently being investigated since ozone over the Arctic is naturally highly variable. Any major decrease in ozone over the Arctic could have serious environmental consequences because of the North Pole's great proximity to populated areas.

Table 3 *Substances controlled by the Montreal Protocol*

Formula	Number	Ozone depletion potential	Atmospheric lifetime in years
$CFCl_3$	CFC- 11	1.0	77
CF_2Cl_2	CFC- 12	1.0	139
$C_2F_3Cl_3$	CFC-113	0.8	92
$C_2F_4Cl_2$	CFC-114	1.0	180
C_2F_5Cl	CFC-115	0.6	380
CF_2BrCl	Halon 1211	2.7	12.5
CF_3Br	Halon 1301	11.4	101
$C_2F_4Br_2$	Halon 2402	5.6	unknown

Data from Hammitt *et al.*, 1987; SORG, 1988.

By 1978, CFC-11 and CFC-12 had been banned as aerosol propellants in the United States, Canada, and Scandinavia. Leading industrial nations signed the Montreal Protocol in September, 1987 which aims to limit emissions of certain CFCs and halons. The controlled compounds are shown in Table 3. The Protocol has the following essential characteristics:

(i)   consumption of the five CFCs will be frozen at 1986 levels, beginning in 1990
(ii) reduction in consumption by 20% below the 1986 level by 1994
(iii) reduction in consumption by 50% below the 1986 level by 1999
(iv) developing countries that use less than 0.3 kg of CFCs per person annually are exempt from the restrictions for an extra decade.

These measures will be reviewed regularly, with the first review in May, 1989 in Helsinki. Scarcely more than 30 countries ratified the Montreal Protocol, although many more countries have now agreed to abide by it. It is possible that the remaining countries, mainly Third World ones, might undo all the environmental gains made by the signatories to the protocol. If there is only 92% compliance in the world in phasing out CFCs there will be no reduction in the amount of chlorine. The Montreal Protocol as it stands will result in an increase in stratospheric CFCs by about three times the present level by 2020. In order to stabilize atmospheric concentrations at 1989 levels an 85% reduction in the amount of CFCs being released is required. The European Community has committed themselves to a total ban on CFCs by the year 2000. Not mentioned in any of the reductions are carbon tetrachloride and methylchloroform, both of which contribute more to present day ozone removal than CFC-114, CFC-115, and the halons, which are included in the protocol.

7 SUMMARY

The role of human activity may seem insignificant compared with the enormous number of natural chemical cycles. However, many substances are already being emitted in greater quantities by anthropogenic sources than natural sources. Even minor anthropogenic sources of certain species pose a significant threat to the environment. The destruction of the ozone layer is a prime example of a serious environmental problem of global extent. The Antarctic ozone hole has forced the international community to co-operate and has resulted in the world's first global treaty for the protection of the environment. It has also spurred investigators to study atmospheric chemistry and dynamics in new detail.

8 BIBLIOGRAPHY

S. C. Chapman, 'A theory of upper atmospheric ozone,' *Mem. R. Meteorol. Soc.*, 1930, **3**, 103–125.

J. C. Farman, B. G. Gardiner, and J. D. Shanklin, 'Large losses of total ozone reveal seasonal ClO_x/NO_x interactions,' *Nature*, 1985, **315**, 207–210.

J. K. Hammitt, F. Camm, P. S. Connell, W. E. Mooz, K. A. Wolf, D J. Wuebbles, and A. Bamezai, 'Future emission scenarios for chemicals that may deplete stratospheric ozone,' *Nature*, 1987, **330**, 711–716.

W. N. Hartley, 'On the absorption spectrum of ozone,' *J. Chem. Soc.*, 1881, **39**, 57–60.

D. Hartmann, L. Heidt, M. Loewenstein, J. R. Podolske, W. Starr, and J. Vedder, 'Transport into the south polar vortex in early Spring,' Polar Ozone Workshop, NASA Conference Publication 10014, Washington DC, 1988.

D. J. Hofmann, J. W. Harder, S. R. Rolf, and J. M. Rosen, 'Balloonborne observations of the development and vertical structure of the Antarctic ozone hole in 1986,' *Nature*, 1987, **326**, 59–62.

H. S. Johnston, 'Reduction of stratospheric ozone by nitrogen oxide catalysts from supersonic transport exhaust,' *Science*, 1971, **173**, 517–762.

M. B. McElroy, R. J. Salawitch, S. C. Wofsy, and J. A. Logan, 'Reductions of Antarctic ozone due to synergistic interactions of chlorine and bromine,' *Nature*, 1986, **321**, 759–762.

M. J. Molina and F. S. Rowland, 'Stratospheric sink for chlorofluoromethanes: chlorine atom catalysed destruction of ozone,' *Nature*, 1974, **249**, 810–812.

SCOPE, 'Environmental Consequences of Nuclear War — Volume 1, Physical and Atmospheric Effects,' ed. A. B. Pittock, T. P. Ackerman, P. J. Crutzen, M. C. Mac-Cracken, C. S. Shapiro, and R. P. Turco, John Wiley and Sons, Chichester, 1986.

S. Solomon, 'The mystery of the Antarctic ozone hole,' *Rev. Geophys.*, 1988, **26**, 131–148.

SORG, 'United Kingdom Stratospheric Ozone Review Group — Second Report,' HMSO, London, 1988.

R. S. Stolarski, A. J. Krueger, M. R. Schoeberl, R. D. McPeters, P. A. Newman, and J. C. Alpert, 'Nimbus 7 satellite measurements of the springtime Antarctic ozone decrease,' 1986, **322**, 808–811.

WMO–NASA, 'Atmospheric ozone 1985: Assessment of our understanding of the processes controlling its present distribution and change,' World Meteorological Organization, Geneva, Switzerland, 1986, Report No. 16.

S. C. Wofsy, M. B. McElroy, and Y. L. Ying, 'The chemistry of atmospheric bromine,' *Geophys. Res. Lett.*, 1975, **2**, 215–218.

CHAPTER 10

Atmospheric Dispersal of Pollutants and the Modelling of Air Pollution

M. L. WILLIAMS

1 INTRODUCTION

Considerable resources are often devoted to the measurement of air pollutant concentrations in the ambient atmosphere but measurements on their own provide little information on the origin of the pollutants in question, on the dispersal process in the atmosphere, and on the impact of new sources or the benefits of controls. There is frequently the need, therefore, for detailed knowledge of the characteristics and quantities of pollutants emitted to the atmosphere and on the atmospheric processes which govern their subsequent dispersal and fate. This knowledge must then be built into an appropriate dispersion model, whether the problem to be addressed is the emissions from a single chimney or the emissions from a large multi-source urban/industrial area, or, on a larger scale, from a region or country.

The trend within the EEC towards formal air quality guildelines or standards is creating the need for formal air quality management systems and for a more strategic approach to air pollution control. Neither can be accomplished without the use of atmospheric dispersion modelling techniques. A wide variety of techniques are available, ranging from the most simple 'box' model through to numerical solutions of the basic equations of fluid flow, *etc*. For the most chemically reactive pollutants, it is also necessary to incorporate the relevant atmospheric chemistry. The spatial and temporal resolution and accuracy of the model output ideally must match the questions being posed. Where emissions vary greatly, both in space and time, or it is necessary to predict the time series of concentrations at specified locations, then the modelling task is extremely difficult, even without the complications of a very chemically reactive pollutant species or substantial topographical effects on the dispersal pattern. On the other hand, if it is not necessary to know when a specified concentration will occur but rather it is the probability of occurrence during a given period (*e.g.*, a year) which is of interest, then the modelling task is generally much less demanding. However in general, long-period (*e.g.*, annual) average concentrations can be modelled more accurately than can shorter averaging times where the turbulent fluctuations in the atmosphere can result in agreement within factors of 2 to 3 with observed values.

2 DISPERSION AND TRANSPORT IN THE ATMOSPHERE

A pollutant plume emitted from a single source is transported in the direction of the mean wind. As it travels it is acted upon by the prevailing level of atmospheric turbulence which causes the plume to grow in size as it entrains the (usually) cleaner surrounding air. There are two main mechanisms for generating atmospheric turbulence. These are mechanical and convective turbulence, and will be discussed in Sections 2.1 and 2.2 below.

2.1 Mechanical Turbulence

This is generated as the air flows over obstacles on the ground such as crops, hedges, trees, buildings, and hills. The intensity of such turbulence increases with increasing wind speed and with increasing surface roughness and decreases with height above the ground. If there is only a small heat flux, either to or from the surface, in the atmosphere so that most of the turbulence is mechanically generated the atmosphere is said to be neutral or in a state of neutral stability. In this case the wind speed will vary logarithmically with height z:

$$u(z) = (u_*/k) \ln (z/z_o) \qquad (1)$$

where k is von Karman's constant (~ 0.4), z_o is the so called surface roughness length (~ 1 m for cities and ~ 0.3 m for 'typical' countryside in the UK), and u_* is the friction velocity and is a measure of the flux of momentum to the surface.

2.2 Turbulence and Atmospheric Stability

As solar radiation heats the earth's surface, the lower layers of the atmosphere increase in temperature and convection begins driven by buoyancy forces. The motion of air parcels from the surface is unstable as a parcel in rising finds itself warmer than its surroundings and will continue to rise. Convective circulations are set up in the boundary layer and this form of turbulence is usually associated with large eddies the effects of which are often visible in 'looping' plumes from stacks. At night when there is no incoming solar radiation and the surface of the earth cools, temperature increases with height, and turbulence tends to be suppressed. During calm clear nights, when surface cooling is rapid and little or no mechanical turbulence is being generated, turbulence may be almost entirely absent.

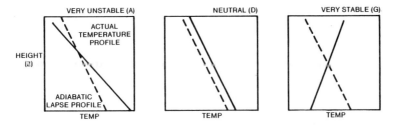

Figure 1 *Typical atmospheric temperature profiles and corresponding stabilities*

Table 1 *Pasquill's stability categories**

Surface wind speed m s^{-1} ($\equiv U_{10}$)	Insolation Strong	Insolation Moderate	Insolation Slight	Night Thickly overcast or $\geqslant 4/8$ low cloud	Night $\leqslant 3/8$ Cloud
<2	A	A–B	B	—	G
2–3	A–B	B	C	E	F
3–5	B	B–C	C	D	E
5–6	C	C–D	D	D	D
>6	C	D	D	D	D

Strong insolation corresponds to sunny midday in midsummer in England. Slight insolation in similar conditions in midwinter. Night refers to the period from 1 hour before sunset to 1 hour after dawn. A is the most unstable category and G the most stable. D is referred to as the neutral category and should be used, regardless of wind speed, for overcast conditions during day or night.
* Based on Figure 6.10 in reference 1.

When considering most dispersion problems it is convenient to classify the possible states of the atmosphere into what are usually referred to as stability categories. The typing scheme developed by Smith from the original Pasquill formulation[1,2] is widely used because of its relative simplicity yet dependence on sound physical principles. Stability is classified according to the amount of incoming solar radiation, wind speed, and cloud cover. A semi-quantitative guide is given in Table 1 and Figure 1 shows typical temperature profiles corresponding to the unstable, neutral, and stable cases. The adiabatic lapse profile in Figure 1 is the vertical temperature gradient for the atmosphere in a

[1] F. Pasquill and F. B. Smith, 'Atmospheric Diffusion,' Chichester, Ellis Horwood, 1983.
[2] 'A Model for Short and Medium Range Dispersion of Radionuclides Released to the Atmosphere,' First Report of a UK Working Group on Atmospheric Dispersion, ed. R. H. Clark. London, HMSO, 1979, NRPB Report R91.

state of adiabatic equilibrium when a parcel of air can rise and expand, or descend and contract, without gain or loss of heat; the temperature of the air parcel is always the same as that of the level surrounding air and the conditions correspond to neutral stability. The numerical value of the adiabatic lapse rate is $\sim 1\,°C/100$ m. For wind speeds in excess of about 6–8 m s^{-1} mechanical turbulence dominates irrespective of the degree of insolation and neutral stability prevails. Table 2 gives typical annual frequencies of occurrence of the different stability categories in Great Britain. For other regions quite different frequencies might apply. In central Continental regions at lower latitudes, for example, the greater incidence of solar radiation would probably result in smaller incidence of neutral conditions and increased frequencies of unstable and stable categories.

Table 2   *Typical annual frequency of occurrence of stability categories in Great Britain*

Stability category	Frequency of occurrence %
A	0.6
B	6.0
C	17.0
D	60.0
E	7.0
F	8.0
G	1.4

2.3 Mixing Heights

The stable atmosphere depicted in Figure 1 is an example of a ground based temperature inversion, *i.e.*, the temperature increases with height unlike the normal decrease. An elevated inversion is often observed where a region of stable air caps an unstable layer below. Pollutants emitted below the inversion can be mixed up to, but not through, the inversion, the height of which is referred to as the mixing height. This term can be used more generally to describe the height of a boundary between two stability regimes. In the case of an emission above an elevated inversion, the pollutant will be prevented from reaching the ground so that for both surface and elevated sources, inversions can have a significant effect on ground level concentrations. The variation of mixing heights throughout the day due to solar heating and atmospheric cooling can have profound effects on ground level concentrations of pollutants. At night the atmosphere is typically stable with a shallow (~ 1–300 m) layer formed by surface cooling. As the sun rises the surface heating generates convective eddies and the turbulent boundary layer increases in depth, reaching a maximum in the afternoon, at a depth of $\sim 1\,000$ m. As the solar input decreases and stops, the surface cools and a shallow stable layer begins to form again in the evening. In this idealized day, concentrations from surface sources will thus be at a maximum in the periods when the stable layers (with low wind speeds and mixing heights) are present and minimized during the afternoon, emission rates remaining the same. Sources emitting above the stable overnight layer will not contribute to ground level concentrations until the height of the growing convective layer reaches the plume and brings the pollutants to ground level, a process known as fumigation. The pattern of ground level concentrations from elevated sources can therefore be quite different from that of surface releases. In reality the diurnal pattern of emissions can of course play a significant role.

In assessing air quality impacts, particularly of elevated sources such as power stations, estimates of mixing heights and their frequency of occurrence and variability throughout the day are therefore essential. WSL has used acoustic sounding (SODAR) to determine mixing heights[3] and an example of the diurnal and seasonal variation of mixing heights measured at Stevenage from 1981–1983 is given in Figure 2. The broad features described above are apparent in this diagram.

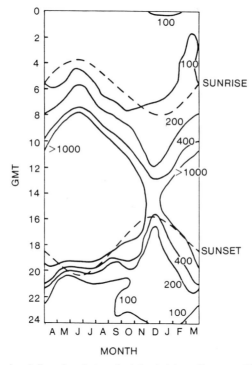

Figure 2   *Annual and diurnal variation of mixing height at Stevenage, 1981–3* (metres)

2.4   Building and Topographical Effects

Hills or buildings can have significant adverse effects on plume dispersion if their dimensions are large in comparison with the dimensions of the plume or if they significantly deflect or disturb the flow of the wind. Figure 3 shows a simplified and idealized representation of the flow over a building. There is a zone on the immediate downwind (or leeward) side of the building which is to some extent

[3] A. M. Spanton and M. L. Williams, 'A Comparison of the Structure of the Atmospheric Boundary Layers in Central London and a Rural/ Suburban Site Using Acoustic Sounding,' *Atmos. Environ.*, 1988, **22**, 211–223.

isolated from the main flow and within which there is a reversal of the air flow. Further downstream the air flow is highly turbulent. Waste gases escaping through a relatively short chimney attached or adjacent to the building, will be entrained in this characteristic flow pattern and will not disperse according to the conventional equations (2) and (3). Recent wind tunnel studies demonstrate that up-wind buildings can have a significant effect on emissions from a chimney located within a few, say 5, building heights; for example, to maintain the same maximum ground level concentration the chimney height required in the presence of one building type studied would be between $1\frac{1}{2}$ and 2 times the height of the chimney required if the buildings was not present. Further downwind, beyond roughly 10 building heights, the near-field effects of the buildings can be incorporated into dispersion models in a parameterized way as discussed in Section 3.1.

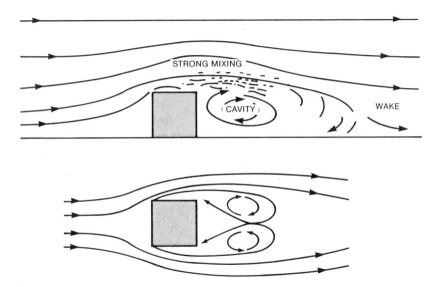

Figure 3 *Simplified schematic flow patterns around a cubical building*

Topographical features such as hills and sides of valleys can have similar effects on dispersion to those described above for buildings. Valleys are also somewhat more prone to problems arising from emissions fairly close to the ground. The incidence of low level or ground based temperature inversions can be greater, either because solar heating of the ground is somewhat delayed in the morning or because during the night cold air drains down the valley sides (katabatic winds) thus creating a 'pool' of cold air on the valley floor. Any low level emissions will therefore disperse very slowly and may even accumulate to some extent. The relatively undiluted emissions can also drift along and across the valley thus affecting areas other than the immediate surroundings of the source. Emissions from high chimneys located on the valley floor may not be detected at all on the valley floor while the ground based inversion persists.

However, considerable horizontal spreading of the plume aloft can occur and during the morning fumigation period large parts of the valley may experience relatively high concentrations at much the same time.

The effects of hills on the flow of pollutant plumes and the resulting concentrations are complex and will not be discussed here.

2.5   Removal Processes — Dry and Wet Deposition

When considering pollution impacts from nearby sources, *e.g.*, within say 10 km or so, the various losses of pollutants are generally not important (unless one happens to be interested specifically in such issues as the short range washout of HCl or the deposition of a particularly toxic species in the near field). However in considering impacts over long ranges and especially on the international scale then consideration of the removal processes is essential.

Dry deposition takes place continuously in a turbulent boundary layer as a result of turbulent flux towards the surface. The efficiency of the process is determined by the deposition velocity which in general is a function of the prevailing level of turbulence (high levels of which result in increasing deposition other things being equal) and of the nature of the gas and the surface (for example a reactive gas such as HNO_3 will be deposited more readily than a less reactive species such as NO). The flux to the surface is given by the product of the surface concentration and the deposition velocity, v_g, so that the process is linear. The equivalent first-order rate constant for this process is given by (v_g/H) where H is the depth of the mixing layer through which deposition is taking place. Typical half-lives for this process are ~ 1–2 days for species such as SO_2 and NO_2, but ~ 5 days or more for sulphate aerosols, which is one reason why 'acid rain' is a continental rather than purely a local phenomenon.

Wet deposition is the term given to the removal of gases or particles from the atmosphere in clouds and/or in rain. For species such as SO_2 and aerosols this process is relatively efficient. Typical lengths of dry and wet periods in the UK are such that if transport times are of the order of the dry period duration (~ 70 hours in the UK), wet removal processes should be included in the model.

3   MODELLING OF AIR POLLUTION DISPERSION

In Section 2 we discussed the underlying physical principles of air pollution dispersion, transport and transformation in the atmosphere. In this Section we summarize the methods used to apply these principles in a quantitative way to model the processes mathematically. The techniques used depend on the distance scales involved in the transport from the source to the receptor or 'target' area. If this distance is small, say of the order of tens or hundreds of metres, then very often buildings and local topography are important and the mathematical description of the ensuing complex flows and turbulence may not be tractable. In such cases (and others such as the dispersion of dense gases or longer distance problems in complex topography) a physical model in a wind tunnel may be the only practicable solution. These complications will be neglected in all that follows and we will deal with situations of ideal flat terrain.

It is fairly clear that as a plume is transported in the direction of the mean wind, it grows through the effect of atmospheric turbulence producing, very roughly, a cone shaped plume with the apex towards the stack. Now clearly the plume will continue to expand until, in the vertical, it fills the atmospheric boundary layer (~ 1 km deep in neutral conditions). Beyond this point vertical dispersion has no further effect; concentrations are thence reduced only by horizontal dispersion, and by the deposition processes and, if appropriate, by chemical reactions. It can be shown that in neutral conditions this point is reached at downwind distances from a source of very roughly 50–100 km, so for source–receptor distances less than this value, vertical dispersion should be included in a model for an accurate representation of the dispersion. Beyond this region, plumes generally fill the mixing layer and uniformly mixed 'box-models' can be used with some confidence.

We will firstly discuss modelling on scales where vertical dispersion is important, before discussing problems involving longer range transport.

3.1 Modelling in the Near Field

In this section we will discuss modelling of pollutant dispersion from 0 to ~ 100 km, using the Gaussian plume approach. This is not to condemn more sophisticated methods, but for most practical applications, the quality of the available emission and meteorological data does not justify the increased resources required to set up and run more complex models. In many cases a sound knowledge of meteorology and aerodynamics can be used to parameterize the Gaussian model to simulate adequately, for example, the effects of buildings on dispersion.

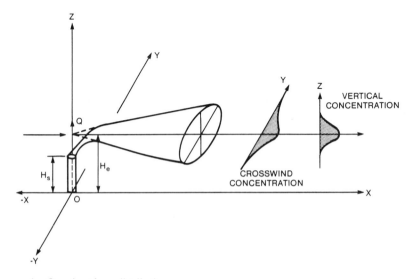

Figure 4 *Gaussian plume distribution*

In the Gaussian plume approach the expanding plume has a Gaussian, or Normal, distribution of concentration in the vertical (z) and lateral (y) directions as shown in Figure 4. The concentration C (in units of $\mu g\,m^{-3}$ for example) at any point (x, y, z) is then given by:

$$C(x,y,z) = \frac{Q}{2\pi\sigma_y\sigma_z U}\,\exp\left[-\frac{y^2}{2\sigma_y^2}\right]\left\{\exp\left[-\frac{(z-H_e)^2}{2\sigma_z^2}\right]+\exp\left[-\frac{(z+H_e)^2}{2\sigma_z^2}\right]\right\} \quad (2)$$

where Q is the pollutant mass emission rate in $\mu g\,s^{-1}$, U is the wind speed, x, y, and z are the along wind, crosswind, and vertical distances, H_e is the effective stack height given by the height of the stack plus the plume rise defined below. The parameters σ_y and σ_z measure the extent of plume growth and in the Gaussian formalism are the standard deviations of the horizontal and vertical concentrations respectively in the plume. When $y = z = 0$, this equation reduces to the familiar ground level concentration below the plume centreline:

$$C(x) = \frac{Q}{\pi\sigma_y\sigma_z U}\,\exp\left[-\frac{H_e^2}{2\sigma_z^2}\right] \quad (3)$$

Equation (3) is rather less cumbersome to deal with and several points of interest emerge. Firstly, concentrations are directly proportional to the emission rate, Q, so it is essential that this is known accurately in any practical application. Secondly, unless $H_e = 0$ (*i.e.*, unless the source is at ground level) the maximum concentration will occur at a point downwind and this downwind distance will increase with increasing H_e and furthermore the value of C_{max} will decrease with increasing H_e. In fact, C_{max} is roughly proportional to H_e^{-2}. This is the mathematical statement of the so-called 'tall stacks' policy which underpinned air pollution control in the UK until relatively recently.

The specification of H_e involves calculating the plume rise, which is the height above the point of emission reached by the plume due to its buoyancy (if it is warmer than the surrounding air, as most combustion emissions are) or momentum (plumes may be driven up stacks at relatively high velocities). For most plumes, buoyancy is the dominating force and there have been a large number of studies of methods of determining plume rise. A widely used method however is that due to Briggs,[4] where the plume rise H is given by:

$$\Delta H = 3.3\,(Q_H)^{1/3}\,(10H_s)^{2/3}\,U^{-1} \text{ for } Q_H \geqslant 20\,\text{MW}$$

or $\qquad\qquad\qquad\qquad\qquad\qquad\qquad\qquad\qquad\qquad\qquad\qquad\qquad (4)$

$$\Delta H = 20.5(Q_H)^{0.6}\,(H_s)^{0.4}\,U^{-1}\ Q_H \leqslant 20\,\text{MW}$$

where Q_H is the sensible heat emission from the stack and U is the wind speed at the stack height (H_s). The dependence on Q_H is not strong and, if measured values of this quantity are not available, an approximation often used is to assume Q_H is equal to one-sixth of the total heat generated in combustion of the

[4] G. A. Briggs, 'Plume Rise,' US Atomic Energy Commission, 1969.

fuel. An expression for ΔH due to Moore[5] has been developed for power stations in the UK and is:

$$\Delta H = a\, Q_H^{\frac{1}{4}}/U \tag{5}$$

where $a = 515$ in unstable and neutral conditions and $a = 230$ in stable atmospheres, with Q_H in MW and U in m s^{-1}. More recent (and more complex) formulations of Moore's and Briggs' formulae have been summarized by the UK Dispersion Modelling Working Group.[6]

The standard deviations of the plume in the vertical and lateral directions, σ_z and σ_y, are extremely important quantities. They are determined by the prevailing atmospheric turbulence in the boundary layer.

Turbulent motions or eddies in the atmosphere vary in size and intensity; the greater their size and/or intensity the more rapid is the plume growth and hence the dilution of the pollutants. Small scale turbulent motions tend to dominate the plume growth close to the point of emission where the plume is still relatively small and the larger scale eddies dominate at greater distances. Furthermore, the small and larger eddies are associated respectively with short and larger time scales. Consequently, σ_y and σ_z increase in value with distance from the source (see Figure 4); also they increase with the time or sampling period over which they have been measured. This latter point means that it is essential to state the sampling period to which σ_y and σ_z apply, especially if comparisons are being made between calculated and measured concentration; ideally the two periods should be identical. Most values of σ_y and σ_z to be found in the literature are for sampling periods in the range 3–60 minutes. It is also evident that σ_y and σ_z are dependent on atmospheric stability, being smallest when the atmosphere is most stable (category G) *i.e.*, when atmospheric turbulence is least, increasing to their greatest values in highly turbulent very unstable conditions (category A). The underlying surface roughness elements also play a part, σ_y and σ_z increasing with increasing surface roughness so that for a given distance downwind of a chimney σ_y and σ_z will be larger in, for example, an urban area than in an area of open relatively flat agricultural land.

In general, lateral (horizontal) motion is less constrained than vertical motion with the result that there are larger scale eddies in the horizontal than in the vertical. Fluctuations in wind direction also become important for longer sampling periods. Consequently, σ_y increases more rapidly with increasing sampling or averaging period than does σ_z. This dependence of σ_y on wind direction fluctuations also means that for longer sampling periods, say greater than 1 hour, σ_y values can increase with increasing atmospheric stability because during low wind speed stable conditions plume meandering can be significant.

[5] D. J. Moore, 'A Comparison of the Trajectories of Rising Buoyant Plumes with Theoretical/Empirical Models,' *Atmos. Environ.*, 1974, **8**, 441–457.
[6] 'Models to Allow for the Effects of Coastal Sites, Plume Rise, and Buildings on Dispersion of Radionuclides and Guidance on the Value of Deposition Velocity and Washout Coefficients.' Fifth Report of a UK Working Group on Atmospheric Dispersion, ed. J. A. Jones, London, HMSO, 1983, NRPB Report R157.

Ignoring for the moment the plume meandering component of σ_y the parameters are often conveniently expressed in the form:

$$\sigma_y = \sigma_{yo} + ax^b$$

$$\sigma_z = \sigma_{zo} + cx^d \tag{6}$$

where a, b, c, and d are constants dependent on atmospheric stability, x is the downwind distance from the source and σ_{yo}, σ_{zo} are the initial plume spreads generated by, for example, building entrainment. To incorporate plume meander into σ_y, an extra term is added so that:

$$\sigma_y^2 = \sigma_{yt}^2 + 0.0296\, T x^2 / U \tag{7}$$

where σ_{yt} is given by equation (6) and T is the averaging time in hours.

A simple expression for σ_z, based on Smith's[1] work is:

$$\sigma_z = \sigma_{zo} + 0.9(0.83 - \log_{10}P)x^{0.73} \tag{8}$$

which gives a good representation out to ~ 30 km. Here P is Smith's stability parameter equal to 3.6 for neutral conditions and ranging from 0–1 (stability A) through to 6–7 in stability G.

Values of coefficients specifying σ_{yt}, the so-called microscale σ_y, i.e., not including any plume meander effects, are given in Table 3. This table also includes typical values of mixing heights in the stability categories A–G. The effect of the mixing height on vertical plume dispersion can be taken into account in the following modification of equation (2) for ground level concentrations:

$$C(x) = \frac{Q}{\pi \sigma_y \sigma_z U} \exp\left[-\frac{y^2}{2\sigma_y^2}\right]\left\{\exp\left[-\frac{H_e^2}{2\sigma_z^2}\right] + \exp\left[-\frac{(2L-H_e)^2}{2\sigma_z^2}\right]\right\} \tag{9}$$

where L is the mixing height. This equation is not valid for $H_e > L$ when the concentration is zero (i.e., the pollutant is emitted above the mixing height).

Table 3 *Typical mixing heights and coefficients in* $\sigma_y = cx^d$ *(x in km) for different stabilities*

	Stability					
	A	B	C	D	E	F/G
Mixing height (m)	1300	900	850	800	400	100
c	213	156	104	68	50.5	34
d	0.894	0.894	0.894	0.894	0.894	0.894

Where long period averages (e.g., annual) are of concern the detailed dependence on σ_y is of much less importance and the pollutant concentrations can be assumed to be uniformly distributed cross-wind within each wind sector.

For a 30° sector the first two terms of equation (9) become:

$$\frac{1.524\, Q}{U\, \sigma_z\, x} \tag{10}$$

The contribution of this wind sector to the overall annual average is then given by the equation (9) modified as in equation (10) multiplied by the combined frequency of occurrence of that wind sector and stability category.

In section 2.1 we introduced the concept of the logarithmic wind speed profile with height in conditions of neutral stability. In different atmospheric stability conditions the variation will be different, but, in general the wind speed will increase with height because of the surface drag. As would be expected intuitively this variation is smallest in unstable conditions (since in such boundary layers there is a considerable degree of vertical mixing) and greatest in stable conditions (for the opposite reason). In general one can write:

$$U(z) = U_{10}\,(z/10)^{\alpha} \tag{11}$$

where U_{10} is the 10 metre wind speed and α is ~ 0.15 in unstable conditions, ~ 0.2 in neutral, and ~ 0.25 in stable conditions.

There are several interesting derivations from the standard Gaussian equation and one particularly useful one is the formula giving the concentrations from a line source (of infinite length) obtained by integrating, for simplicity, equation (2) over y to yield:

$$C(x) \; = \; \sqrt{\frac{2}{\pi}}\, \frac{Q}{\sigma_z U}\, \exp\left[-\frac{H_e^2}{2\sigma_z^2} \right] \tag{12}$$

which can be used to estimate the concentrations downwind of roads for example. In this application, Q is the mass emission rate per unit length of road ($\mu g\, m^{-1}\, s^{-1}$).

3.2   Emission Inventories

We have already seen how important it is to specify the emission rate of a single source in order to model concentrations with confidence. In single stack applications this is often relatively straightforward. However, in multiple source applications such as in the use of an urban air quality model, there can in principle be literally thousands of individual sources. It would be clearly impracticable to attempt to quantify the emission rate of every house, office, shop, and car in, say, London so methods have to be devised of making the problem tractable yet retaining as accurate a description of reality as possible.

The usual way of achieving this is to apportion the area to be modelled into a grid and to combine all the numerous small emitters within each grid square (such as individual houses, cars, *etc.*) into so-called 'area sources.' Major sources are usually treated explicitly as individual point sources. The size of the grid square used will usually be determined by the size of the area, or domain, to be modelled and the computing resources available. Typical grid scales are 1 km (or smaller) for urban areas, 20 km for nation-wide modelling in a country the size of the UK, to 50–100 km for European or other international scale long range transport.

The specification of emissions is therefore fundamental to modelling and, apart from single source problems, is a difficult task. The usual approach is to collect information on fuel consumption in particular sectors (such as power generation, domestic heating, *etc.*) and multiply this by appropriate emission factors which ideally will have been measured over a range of representative fuels, appliances, and combustion conditions.

Very often such data on fuel consumption (or some other measure of industrial commercial activity) are available only on a large scale, *e.g.*, at national level, when the area to be modelled is much smaller. The modeller then has to use some means of spatially disaggregating the total domain emissions over the individual grid square of the model. This usually involves the introduction of another level of uncertainty as surrogate statistics have to be employed—for example domestic heating emissions may be assumed to have the same spatial pattern over the model grid as population for which data are often fairly readily available. Other surrogates which can be used are office floor space for emissions from the commercial sector and population for motor vehicle emissions. If the domain is of an appropriate size then questionnaires and other, often labour intensive, techniques can be used to assess the magnitude and spatial pattern of emissions in a particular town or city.[7] Some examples of urban and national emission inventories are given in Figures 5 and 6. The emissions shown in Figure

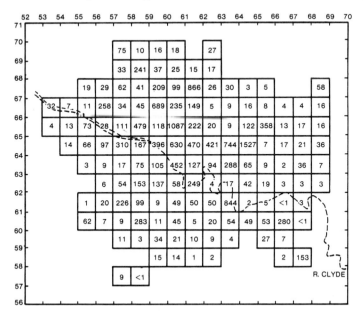

Figure 5   *Total* SO_2 *emissions for the Glasgow area* (tonnes/yr) *in* 1 × 1 km *grid squares*

---

[7] D. J. Ball and S. W. Radcliffe, 'An Inventory of SO_2 Emissions to London's Air,' GLC Research Report No. 23, London, 1979.

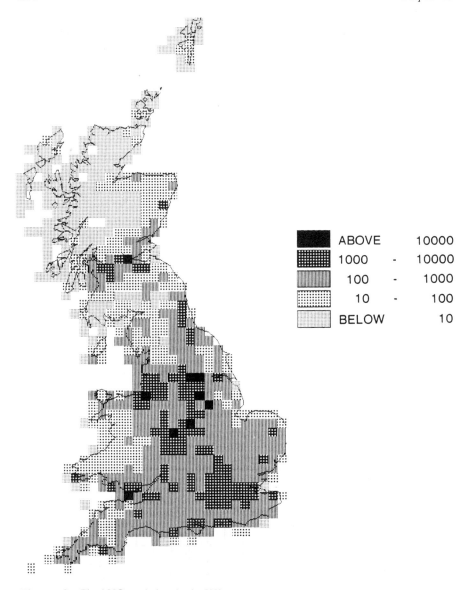

	ABOVE	10000
	1000 -	10000
	100 -	1000
	10 -	100
	BELOW	10

Figure 6   *Total* NO$_x$ *emissions in the UK*

5 formed part of a study of the East Strathclyde area by WSL[8] which used some 1200 1 × 1 km grid squares as area sources and about 1900 individual point sources. Figure 5 illustrates the total SO$_2$ emissions in the City of Glasgow, *i.e.*, the central part of the area. Table 4 is a disaggregation of the total emissions for the 1200 km^2 area by fuel and use type. Figure 6 is a map of total NO$_x$ emissions

[8] M. L. Williams *et al.*, 'The Measurement and Prediction of Smoke and Sulphur Dioxide in the East Strathclyde Region: Final Report,' Warren Spring Laboratory Report LR 412 (AP), Stevenage, 1981.

for the UK at a resolution of 20×20 km grid squares and has been derived by WSL[9] on the basis of national and regional fuel use and other statistics.

We have already seen, in Section 3.1, how the averaging time inherent in the dispersion model structure and parameters should match that of the concentrations being modelled. Similarly it is vital that the emissions should be of the appropriate timescale. This is often straightforward in the case of annual averages when annual emissions must be used, but can be more difficult if seasonal or diurnal variations are being modelled. However, various approaches are possible.

Table 4 *Estimated emissions for the East Strathclyde* 1200 km^2 *inventory*

Fuel	Emission rate (tonnes/yr)					
	Indust/Commerc		*Domestic*		*Motor traffic*	
	Smoke	SO$_2$	Smoke	SO$_2$	Smoke	SO$_2$
Bituminous Coal	69	1 040	6 375	2 035	—	—
Solid Smokeless	61	3 406	367	754	—	—
Domestic Oil	—	—	1	284	—	—
Ind/Comm Oil	1 556	46 624	—	—	—	—
Motor Spirit	—	—	—	—	480	320
DERV	—	—	—	—	2 726	909
Totals	1 686	51 070	6 742	3 073	3 206	1 229
Relative Emission (%)	14	92	58	6	28	2

Fuel requirements for space heating purposes (residential, commercial, and a proportion of industry) depend on the ambient temperature and therefore vary with the season of the year. The Degree–Day principle[10] can be used to calculate these temperature dependent emission rates, $E(T)$, from the annual average emission rate Eo:

$$E(T) = Eo[0.33 + 0.11(14.5 - T)] \text{ for } T \leqslant 14.5 \,^{\circ}\text{C}$$

$$E(T) = 0.33 \, Eo \qquad\qquad \text{for } T > 14.5 \,^{\circ}\text{C}$$

(13)

A further factor can be introduced if required to take account of the typical diurnal variation in emissions.

When modelling traffic pollution diurnal variations in traffic flow are often available so that emissions can be scaled accordingly. One very important feature in dealing with traffic pollution is the variation of emissions with speed. This is particularly important for the pollutants carbon monoxide (CO) and hydrocarbons both of which, being products of incomplete combustion are formed in the biggest quantities at low speeds. Recent work at WSL using on the

[9] H. S. Eggleston and G. McInnes, 'Methods for the Compilation of UK Air Pollutant Emission Inventories,' Warren Spring Laboratory Report LR 634 (AP), 1987.
[10] 'Degree Days,' London, Gas Council, Technical Handbook No. 101.

road measurements in actual driving conditions[11] has begun to quantify this effect. For example emissions of CO can vary from 25–30 g km^{-1} at 20 kph to 5–10 g km^{-1} at 100 kph.

3.3 Long Range Transport

With the increasing interest in the problems of acid rain and photochemical ozone in recent years, the modelling of these phenomena has become important. There are several important features which must be considered in modelling long (in this context, greater than 100 km) as opposed to shorter range transport. Firstly the time scales of transport are such that the removal processes of wet and dry deposition must be incorporated. Secondly large scale meteorological features must be taken into account which involve specifying the movement of an air mass on a synoptic scale. Thirdly chemical reactions will generally be important and must be included. To incorporate all these effects in detail, demands immense computing resources as well as raising the questions over how accurately the input data on emissions and meteorology can be known and how well one can describe the detailed physics and chemistry of the processes. In practice, therefore, simplified models have been used. These have generally been of two types, Lagrangian, where a series of air mass trajectories is followed, or Eulerian, where the governing equations are solved for every grid point in the domain at each time step. In many ways Lagrangian models are the simpler in concept and application, as Eulerian models generally require greater complexity of input data and computing requirements, as well as suffering from the numerical 'pseudo-diffusion' if not appropriately constructed. Lagrangian models in widest use in the UK take two forms, statistical models and trajectory/box models. The former are typically applied to long period averages of concentration and deposition fields of acidic pollutants such as SO_2 and sulphate, NO_x and nitrate, and use climatological data or annual frequencies of wind directions, speeds, stability categories, *etc.* Rainfall may be taken into account in several ways, the simplest but not necessarily the most satisfactory being to assume continuous constant rainfall (at an annual average rate). Alternatively the sporadic or stochastic nature of rainfall can be incorporated in a probabilistic way.[12] Trajectory/box Lagrangian models are usually applied to a succession of air mass trajectories arriving at a receptor at relatively short intervals (*e.g.*, every 6 hours in the case of the most recent UN/ECE EMEP model[13]). The specification of the trajectories is a fundamental step in using these models and they are usually obtained from the detailed models used in national meteorological services. In practice back-track trajectories up to 96 hours are used; longer timescales would

---

[11] C. J. Potter and C. A. Savage, 'A Summary of Gaseous Pollutant Emissions from Tuned In-service Gasoline Engined Cars Over a Range of Road Operating Conditions,' Warren Spring Laboratory Report LR 447 (AP), 1983.

[12] H. Rodhe and J. Grandell, 'On the Removal Time of Aerosol Particles from the Atmosphere by Precipitation Scavenging,' *Tellus XXIV*, 1972, **5**, 442–454.

[13] Ø. Hov, A. Eliassen, and D. Simpson, 'Calculation of the Distribution of NO_x Compounds in Europe,' In Tropospheric Ozone, Regional and Global Scale Interactions, ed. I. S. A. Isaksen, Reidel, Dordecht, 1988.

introduce unacceptable errors. As it is, errors in trajectories increase with time back along the path, and particularly in slack pressure areas these can be very large even at relatively short times. These models work by moving the box or air parcel along the trajectory and at each time step the appropriate emissions are introduced from the underlying grid, pollutant is lost by dry deposition at the rate given by $v_g c$ where the deposition velocity v_g is appropriate to the underlying surface type. Wet deposition removes pollutant according to the rainfall field at the particular location of the air parcel, and chemical transformations of reactive species are updated since the previous time step.[13] A good summary of the use of Lagrangian and Eulerian long range transport models of acid rain has been given by Pasquill and Smith.[1]

Much use has been made of Lagrangian box models in modelling photochemical ozone formation in the UK[14] and elsewhere.[15] Because the computational demands of the physical and meteorological aspects of the problems are relatively small, particularly if a one or two layer box model is used, quite complex chemical schemes can be used to describe the chemical processes involved. The model developed by Derwent,[14] for example, uses 339 chemical reactions describing the fate of 40 species. Explicit chemistry is used rather than so-called 'lumped' schemes where for example one hydrocarbon is used as a surrogate for its class (*e.g.*, propene could be used to describe all alkenes, *etc*).

In attempting to simulate observed ozone concentrations with Lagrangian models, the correct specification of the air parcel's trajectory is of paramount importance. This may not be easy particularly, as we have noted, when anticyclonic conditions with slack pressure gradients exist. Then, although conditions may be optimal for ozone formation, the uncertainties in the calculated trajectories are often at their greatest. Large differences in the calculated ozone concentration can result, depending on the quantity of precursors (nitrogen oxides and hydrocarbons) picked up along each trajectory.

3.4 Accuracy of Models

In using models such as those described in sections 3.2 and 3.3 for air pollution control purposes, for assessing air quality impacts or for elucidating chemical and transport mechanisms, it is important to assess the accuracy with which the model can reproduce observed concentrations. Detailed model validation exercises can be expensive, involving considerable resources in measurements of pollutants at many locations and timescales, and the appropriate meteorological variables over the domain of the model. In general the more complex the model, the more complex is the validation required. However, the confidence one has in the output of a model depends very much on the questions one is attempting to answer. The prediction of a peak concentration at a specific location will place

[14] R. G. Derwent and A. M. Hough, 'The Impact of Possible Future Emission Control Regulations on Photochemical Ozone Formation in Europe,' London, HMSO, Harwell Report AERE R 12919, 1988.
[15] A. Eliassen, Ø. Hov, I. S. A. Isaksen, J. Saltbones, F. Stordal, 'A Lagrangian Long-Range Transport Model with Atmospheric Boundary Layer Chemistry,' *J. Appl. Met.*, 1982 **21**, 1645–1661.

different demands on a model from answering a question such as whether or not controlling a particular category of source in a region would have a beneficial or an adverse effect.

In general, long term (*e.g.*, annual) average concentrations can be predicted with greater confidence than short term (hourly or less) averages. Some assessments of the likely accuracy of dispersion models in predicting the concentrations of non-reactive pollutants have been given by Jones[16] for single source situations. In summarizing work of several authors, he suggests that annual averages from a low level release can be predicted within a factor of about 2. At larger distances the factor increases with concentrations at about 100 km being predicted to within a factor of four with high probability. Factor of two accuracy for peak hourly concentrations is also suggested by Jones with the indication that if specification of the time and location of the peak are also required then the accuracy is likely to be worse.

In urban areas where there are usually numerous sources on all wind directions around receptors, annual average concentrations of relatively inert pollutants, such as SO_2, can generally be predicted to better than a factor of two accuracy on most occasions. A summary of applications of a climatological Gaussian plume model to several urban areas of the UK[17] as shown in Figure 7. This diagram shows the frequency distribution of the percentage error of the calculation of annual average SO_2 concentrations. The percentage error here is defined as $100 \times$ (modelled value $-$ observed) /observed. For London, for example, the modelled results were within $\pm 30\%$ of observed for about 75% of the 20 receptors considered.

In the context of the evaluation of air pollution control strategies it should be noted that the Gaussian plume model and the behaviour of the primary pollutants to which it is usually applied are linear in emission rates so that such models will predict concentration reductions proportional to emission reductions which might arise from any postulated control technology. This is very straightforward in single source problems; however in urban areas a large number of sources will be present and only one sector (*e.g.*, domestic sources under smoke control or motor vehicles under emission regulations) may be subject to controls. The accuracy of prediction of the effects of these controls will then depend on the accuracy with which the proportional contribution of the particular sources is predicted by the model, and this may be difficult to assess.

Turning to the larger scale effects such as acid deposition and photochemical oxidant formation, the question of the linearity, or proportionality, is more important, that is, whether or not for a given reduction in emissions of a particular species (such as SO_2 for example) there is likely to be a proportional reduction in deposition of sulphur. Total sulphur deposition over annual time-

[16] J. A. Jones, 'What is Required of Dispersion Models and Do They Meet the Requirements?' Paper to 17th NATO/CCMS International Technical Meeting on Air Pollution Modelling and its Applications, Cambridge, 1988.
[17] M. L. Williams, 'Models as Tools for Abatement Strategies,' in Acidification and its Policy Implications, ed. T. Schneider, Elsevier, Amsterdam, 1986.

scales has been shown to be approximately proportional[18] even though in some circumstances the wet deposition component may be non-linear.

The problem of the evaluation of photochemical ozone formation is more complex in that ozone is formed from the atmospheric reactions of nitrogen oxides and many individual hydrocarbon species. The governing reaction schemes are overall non-linear and can be further complicated by the fact that NO_x (and some hydrocarbons) can act as both sources and sinks of ozone, on different scales.

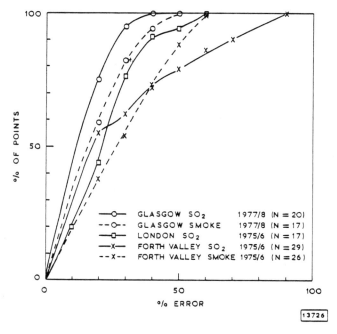

Figure 7 *Percentage of Receptor Points within percentages of observed concentrations for annual average SO_2 concentrations*

With increasing interest in global problems such as the effects of so-called greenhouse gases on climatic change, and on the effects of man made pollutants on stratospheric ozone, modelling techniques are now being used to advise pollution control policies over a very wide range of atmospheric problems.

---

[18] 'Acid Deposition in the UK 1981–1985,' Second Report of the UK Review Group on Acid Rain, Warren Spring Laboratory, 1987.

CHAPTER 11

Catalyst Systems for Emission Control from Motor Vehicles

G. J. K. ACRES

1 INTRODUCTION

When the amendments to the American Clean Air Act were made in 1970, requiring carbon monoxide, hydrocarbon, and nitrogen oxide emissions from motor vehicles to be reduced by 90% it is estimated that there were some 200 million cars in the world of which a majority were in use in that country. Since then the number of cars in the world has increased by 100% to 400 million vehicles. It is therefore not surprising that the problem of photochemical smog, which originated on the West Coast of America, and attributable to a large extent to emissions from motor vehicles, Table 1, has become more widespread and of increasing concern to the principal industrialized countries of the world. In addition to photochemical smog, global environmental problems such as acid rain, the greenhouse effect, and depletion of the ozone layer are also attributable in part to emissions from motor vehicles. As the environmental effects of vehicle emissions have now become a world issue, interest in the catalyst technology which has been developed for use on American vehicles is now being extended to the rest of the world and is the subject of this chapter.

The need for a rapid improvement in air quality in America was reflected by the inclusion of technology forcing standards in the amendments to the Clean Air Act since it was judged that established engine technology was not amenable to the reduction in emissions sought by the American Government. After an intense four-year research and development programme, involving the major car manufacturers, the oil and chemical industries, many advanced concepts were evaluated with the prime objective of meeting the emission standards without impairing fuel economy and driveability and at minium additional cost. The technologies may be divided into two categories. Those designed to eliminate the emissions problem within the engine, such as the stratified charge and lean burn engine concepts, and those such as thermal and catalytic reactors whose role is the elimination of emissions post the combustion chamber. In the event, catalyst technology was chosen and is now universally adopted in America for controlling carbon monoxide, hydrocarbon, and nitrogen oxide emissions from gasoline fuelled vehicles and is used in those countries such as Japan, Australia, Switzerland, and Austria which have adopted standards similar to those in force in America.

Table 1   *Air pollution in Los Angeles in 1973 (Tons/Day)*

Source	Hydrocarbons		Nitrogen oxide		Sulphur dioxide		Carbon monoxide	
	A	B	A	B	A	B	A	B
Vehicles	1 930	520	490	—	30	18	10 330	950
Oil refining	220	990	45	—	45	810	170	1 465
Solvents	495	20	—	—	—	—	—	—
Chemicals	55	20	—	—	65	25	—	—
Heat and power	14	1	270	95	310	330	1	—

A: Actual emissions to atmosphere.
B: Pollutants prevented from reaching the atmosphere.

Although catalyst systems have now been in use in America for some fifteen years, involving more than 150 million cars, only now is catalyst technology widely accepted as the most cost effective approach to achieving standards necessary for the maintenance of, or preferably, the improvement in air quality standards. Current American and EEC standards are given in Chapter 7.

2   EXHAUST EMISSIONS FROM MOTOR VEHICLES

The two principal power sources for motor vehicles, namely the petrol and diesel fuelled internal combustion engines, are projected to remain in use for the foreseeable future and certainly for the rest of this century. To many this may be surprising in view of their low energy conversion efficiency, limited fuel capability, and exhaust emission levels. However, alternatives, such as the Sterling engine, gas turbine, rotary engine, and fuel cell or battery powered vehicle, do not offer sufficient benefits at the present time to be considered for volume production. Trends such as future needs to reduce imports of hydrocarbon fuels while at the same time substantially reducing exhaust emissions, including carbon dioxide, may further stimulate the development of alternative power systems as outlined later.

    In an ideal power system, emissions other than water would be absent and carbon dioxide would be minimized. In practice, incomplete combustion leads to the emission of carbon monoxide and a wide range of hydrocarbons including aromatics and oxygenated species such as aldehydes. In addition, nitrogen oxides which arise as a result of nitrogen and oxygen reacting at the high temperatures produced in the combustion chamber, are also emitted in the exhaust gases. The use of lead and other metal based compounds as octane improvers is no longer permitted in countries such as America and Japan and is being phased out in most countries. Thus emissions of lead and other compounds are largely eliminated.

The emission levels of carbon monoxide, hydrocarbons, and nitrogen oxides are a function of fuel composition, engine type, and the power/load conditions on the engine. As far as a gasoline engine is concerned however, the overriding factor is the air/fuel ratio under which the engine is operating as shown in Figure 1. Under no conditions are the three principal emissions at a minimum level and, further, the maximum power output of the engine corresponds with maximum nitrogen oxide emissions. While significant improvements to fuel economy, power output, and emission have been made in recent years by engine modification and control, none of them has resulted in an engine capable of meeting American 1983 standards while maintaining satisfactory driveability, power output, and fuel economy without the use of a catalyst.

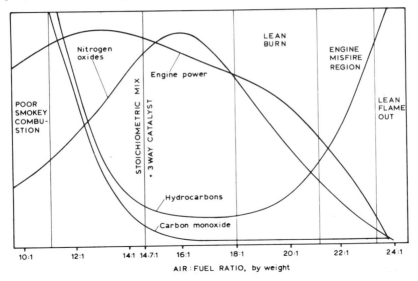

Figure 1 *The effect of air/fuel ratio on engine operations*

In the case of the diesel engine, as the fuel composition and ignition process are different to that of a petrol engine, exhaust emissions differ significantly as shown in Table 2. Exhaust emission standards for diesel engines are therefore different. Standards applicable to American vehicles are shown in Table 3.

Table 2 *A comparison of diesel and petrol exhaust emissions*

	Carbon monoxide %	Hydrocarbons ppm	Nitric oxide ppm	Sulphur dioxide ppm	Particulates g/m³
Diesel engine	0.1	300	4 000	200	0.5
Petrol engine	10.0	1 000	4 000	60	0.01

Table 3   *Particulate emissions from diesel engines*
American Standards

Automobiles	grams/mile
Current	0.2
1990	0.08

Light duty trucks	grams/mile*
Current	0.26
1990	0.08

Heavy duty trucks	grams/brake horse power. hour.
1991	0.25
1994	0.10

* EPA proposed relaxation.

3   THE USE OF CATALYSTS FOR EMISSION CONTROL

The concept of using a catalyst to convert carbon monoxide, hydrocarbons, and nitrogen oxides to less environmentally active compounds such as nitrogen, water, and carbon dioxide was well established practice prior to the need arising on motor vehicles.

The principal reactions are shown in Table 4. However, the demands put upon catalyst technology by the need to match the performance of the catalyst to the engine as a result of rapid changes in exhaust gas temperature, volume, and composition were features not previously encountered in chemical and petroleum industry applications of catalysis. Other unique requirements were the control of emissions such as ammonia, hydrogen sulphide, and nitrous oxide which could result from secondary catalytic reactions and for the catalyst system to maintain its performance after high temperature excursions up tp 1000 °C, in the presence of trace catalyst poisons such as lead and phosphorus.

Table 4   *Reactions occuring on automobile exhaust catalysts*

Oxidation reactions
$$2CO + O_2 \rightarrow 2CO_2$$
$$HC + O_2 \rightarrow CO_2 + H_2O$$

Reduction reactions
$$2CO + 2NO_2 \rightarrow 2CO_2 + N_2$$
$$HC + NO \rightarrow CO_2 + H_2O + N_2$$

By the nature of the oxidation and reduction reactions which are involved in the removal of carbon monoxide, hydrocarbons, and nitrogen oxides and the operating characteristics of the preferred catalysts, Figure 2, several combinations of engine/catalyst system have been used since catalysts were introduced on American cars in 1975.

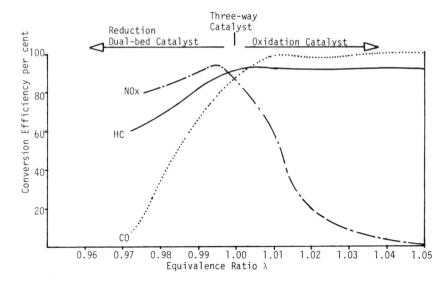

Figure 2 *The effect of exhaust gas stoichiometry on the conversion efficiency of a rhodium/platinum catalyst*

3.1   The Carbon Monoxide/Hydrocarbon Oxidation Catalyst Concept

As shown in Figure 2, which illustrates the best available technology to date, only a limited region of the air/fuel ratio span exists in which carbon monoxide, hydrocarbons, and nitrogen oxides can be reacted on the catalyst with greater than 90% conversion. Hence, in situations where carbon monoxide and hydro-carbons but not high nitrogen oxide emission control is required, *e.g.*, the European 'Euronorm' standards, oxidation catalysts are used. A Schematic of this system is shown in Figure 3.

Key features of this system are the use of a secondary air supply to the exhaust gas stream to ensure oxidizing conditions under all engine operating loads and the use of exhaust gas recirculation (EGR) to limit nitrogen oxide emissions from the engine.

This system was used initially in America to meet interim emission standards and is likely to be adopted to meet similar standards on medium and smaller engined cars (less than 2 litres) in Europe.

3.2   Dual Bed and Threeway Catalyst Concepts

In order to overcome the limitations imposed by the use of EGR and to meet severer nitrogen oxide standards, such as those imposed in America in 1981, catalysts capable of reducing nitrogen oxide emissions are necessary. Initially, as a result of the difficulty of controlling air/fuel ratios to the tolerances required by a single catalyst unit, Figure 2, a dual catalyst bed was used. In order to ensure reducing conditions in the first catalyst bed, where nitrogen oxides were reacted,

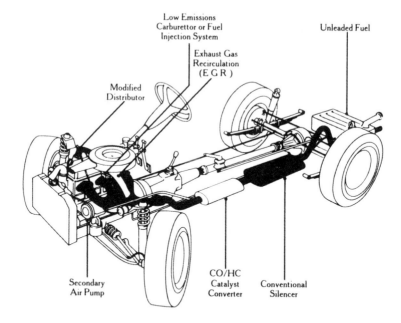

Figure 3   *The oxidation catalyst*

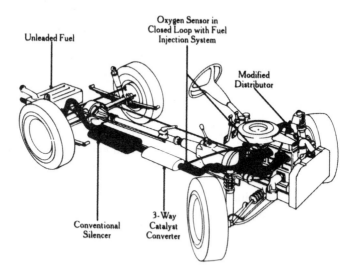

Figure 4   *The 'three way' catalyst*

the engine was tuned slightly rich of the stoichiometric ratio. Secondary air was then injected into the exhaust stream ahead of the second catalyst bed, the oxidation bed, to complete the removal of carbon monoxide and hydrocarbons. With developments in engine control and catalyst technology, in the latter case involving widening the air/fuel operating window for 90% removal of hydrocarbons, carbon monoxide, and nitrogen oxides, the dual bed system has been replaced with a single threeway catalyst unit. A schematic of this system is shown in Figure 4.

Key features of this system in addition to the catalyst unit are an electronically controlled air/fuel management system incorporating, in its most advanced form the use of an oxygen censor to monitor and control exhaust gas composition. Systems such as this are now universal on American and Japanese cars and in those countries that have adopted similar emission standards. Cars having engines of greater than 2 litres which will be required by EEC legislation to meet standards similar to those in America will use TWC systems from 1989.

3.3 Lean Burn Catalyst Systems

The concept of operating engines under very lean, greater than 20/1 air/fuel ratios is attractive in terms of reducing nitrogen oxide emissions and improving fuel economy. However, with current engine technology, in order to achieve nitrogen oxygen emissions consistent with US legislation requirement, the engine must operate in the very lean region where, as shown in Figure 1, hydrocarbon emissions increase to levels which may exceed current American standards. In these situations an oxidation catalyst is incorporated into the exhaust system to control hydrocarbon emissions.

A feature of the European test cycle as presently defined as its low average speed as it is intended to be representative of city driving. The emissions that result are therefore typical of low speed, low acceleration conditions. A more representative cycle incorporating higher speeds and accelerations has been proposed so as to assess emissions under other conditions including urban and motorway driving.

In order to develop and maintain a higher speed more power is required from the engine which, in the case of the lean burn system, entails decreasing the air/fuel ratio. This in turn increases nitrogen oxide emissions, Figure 1, to levels where current engine technology is likely to exceed the American 1983 standards and equivalent European standards if a high speed, acceleration requirement is incorporated into the test cycle.

It is therefore desirable that catalysts used on lean burn engines should in addition to having a hydrocarbon oxidation capability also have a nitrogen oxide reduction capability when fuel enrichment occurs for increased power. The effect on the reduction of hydrocarbons and nitrogen oxide emissions which can be achieved on a lean burn engine using a catalyst with oxidation and reduction capabilities is shown in Table 5 for a Volkswagen Jetta Series 1, powered with a 1.4 litre Ricardo HRCC (high ratio compact chamber) lean burn engine.

A further illustration of the effectiveness of lean burn catalysts having oxidation and reduction capabilities has been demonstrated by incorporating such catalysts into the exhaust system of small, *i.e.*, less than 1.4 litre cars. No other modifications were made to the engines or associated systems. These cars

Table 5   *Lean burn engine emissions (ECE 15 cold start cycle. g/test.)*

	Hydrocarbons	*Carbon monoxide*	*Nitrogen oxides*
Without catalyst	11.7	15.9	5.9
With catalyst	1.7	12.4	4.2

were selected as being representative of cars likely to be manufactured in Europe in the early 1990's. Tables 6 and 7 show the results of ECE 15 cold cycle emission tests with and without catalysts fitted. In most cases these cars are capable of meeting the EEC Stage One standards without a catalyst. However, by incorporating a lean burn dual function catalyst and with no further modifications, emission levels consistent with the EEC Stage Two standard for small cars can be achieved.

Table 6   *Emission levels from small vehicles without catalysts. (Cold ECE 15 cycle)*

	$HC + NO_x$	NO_x	CO
Peugeot 205	18.3	7.8	26.3
Fiat UNO 45	15.2	6.2	26.7
VW Golf C	16.1	5.7	50.5
Rover 213	12.3	3.6	46.7

Table 7   *Emission levels from small vehicles with catalysts. (Cold ECE 15 cycle)*

	$HC + NO_x$	NO_x	CO
Peugeot 205	8.5	5.8	8.8
Fiat UNO 45	4.1	2.7	9.8
VW Golf C	6.4	2.0	42.7
Rover 213	5.2	1.4	27.5
Stage 2 EEC standard	8	—	30.0

4   AIR QUALITY STANDARDS

In some countries, and notably America, significant and progressive reductions in tail pipe emissions from motor vehicles have been required by legislation for the past twenty years. As the effectiveness of such legislation and particularly the means by which it has been achieved are questioned by some groups, it is relevant to evaluate air quality data which has been collected over this period and compare it with countries having less stringent exhaust emission legislation.

According to recent EPA surveys, carbon monoxide levels in America have averaged a fall of 5% per year between 1975 and 1984 giving an overall reduction of 33%. In terms of air quality standards there has been an 87%

reduction in the number of violations of the standard. As the motor vehicle accounts for most of the carbon monoxide emitted to the atmosphere, these results provide the clearest evidence of the success of the American programme. Further evidence is provided in Figure 5 for tail pipe emissions of carbon monoxide, hydrocarbons, and nitrogen oxides from American motor vehicles. The improvement in air quality concerning nitrogen oxides is not so impressive. Overall, nitrogen oxide concentrations have only recently started to decline after decreasing in the period 1975 to 1979. In the period 1975 to 1983, composite average nitrogen oxide levels fell by 6%. This data indicates that air quality is better than it would have been in the absence of motor vehicle control but the overall gains are less due to increases in the number of vehicles, vehicle miles travelled, the delayed introduction of severe standards as well as increased emissions of nitrogen oxides from stationary power sources.

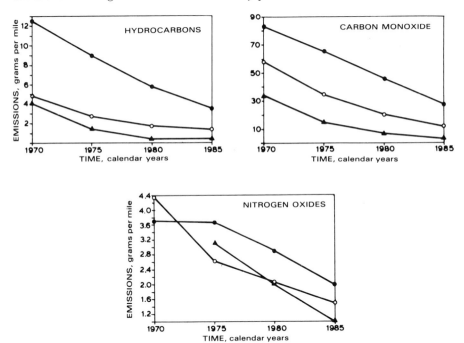

Figure 5 *Tail pipe emission reductions since the introduction of the Amendments to the Clean Air Act in America in 1970.* ▲ *Model Year Standard,* ○ *Model Year Average Emissions, and* ● *Fleet Average Emissions*

Unlike carbon monoxide and nitrogen oxide emissions, hydrocarbon emissions from motor vehicles encompass a wide range of compounds having different environmental impacts. Whilst current legislation does not require emissions of individual hydrocarbons to be monitored, the effectiveness of catalyst control systems in dealing with the most biochemically reactive species is significant. Aldehydes are the most prevalent oxygenated species in gasoline engine exhausts. In addition to their high photochemical reactivity, they contribute directly to eye irritation. As shown in Table 8, aldehyde emissions from gasoline fuelled vehicles are significantly reduced with a catalyst converter.

Table 8   *Aldehyde emission*

Vehicle type	Emissions grams per mile
Average of 10 non-catalyst gasoline cars	0.141
Average of 3 catalyst gasoline cars	0.023

Aromatic hydrocarbons are either produced in the refining process used to manufacture gasoline or added to it as a means of achieving the desired octane rating. Exhaust emissions of aromatic and polyaromatic hydrocarbons in vehicle exhaust gases are of particular interest as a result of their established mutagenic and carcinogenic activity. As illustrated in Tables 9 and 10, systems designed to control carbon monoxide and hydrocarbon emissions and, notably catalytic systems, are effective in controlling polyaromatic hydrocarbons and most notably benzo(a)pyrene and five ring aromatic compounds which are particularly bio-active.

Table 9   *Benzo(a)pyrene emissions*

Vehicle type	Emissions micrograms per mile
Pre-emissions control	12.04
1968 emissions controlled	2.77
1970 emissions controlled	1.62
Catalyst equipped	0.88

Table 10   *Polycyclic aromatic hydrocarbon emissions from a programmed combustion engine*

Polycyclic aromatic hydrocarbon	Emissions, micrograms per mile	
	Without catalyst	*With catalyst*
phenanthrene	1.85	0.16
anthracene	0.61	0.04
fluoranthrene	2.27	0.23
pyrene	2.91	1.50
perylene	1.21	0.40
benzo(a)pyrene	0.94	0.17
benzo(e)pyrene	2.76	0.41
dibenzopyrenes	0.28	0.23
coronene	0.41	0.27

5 ADVANCED EMISSION CONTROL TECHNOLOGY

Although some countries and notably some in the EEC have not yet adopted standards equivalent to those in force in America since 1983, an increasing need is arising for even lower emissions from motor vehicles and hence the need to consider other technical approaches to the problem.

The need for lower emissions from motor vehicles arises for two reasons. Firstly, the increase in the motor vehicle population resulting in a deterioration of lower atmosphere air quality as illustrated by effects such as photochemical smog and acid rain. Secondly, an increasing realization that changes in the upper atmosphere resulting in a greenhouse effect and depletion of the ozone layer may have to be tackled by reducing ground level emissions. The principal gases responsible for these effects, and to which the motor vehicle makes a significant contribution, are hydrocarbons, carbon monoxide, carbon dioxide, nitrogen oxides, and chlorofluorocarbons.

For example, proposals in America to reduce hydrocarbon and NO_x emission standards from motor vehicles from 0.4 to 0.25 grams per mile and 1 to 0.4 grams per mile respectively are now considered to be important steps in at least maintaining air quality. Included in these proposals are stricter standards for trucks and buses including those powered by diesel engines. Developments in air quality thinking such as this are resulting in the need to reassess current approaches to emission control from motor vehicles.

5.1 Diesel Exhaust Emission Control

Although diesel engines emit relatively low concentrations of carbon monoxide and hydrocarbons compared to gasoline engined vehicles, particulate emissions are now estimated to account for 40% of smoke emissions in Britain and 80% of the estimated 140 000 tonnes of carbon particulates emitted in Europe each year. However associated with the carbon particulates which are produced during the combustion process are a range of aromatic hydrocarbons having mutagenic activity. It is principally for this reason that the EPA proposals in America for limiting particulate emissions were first proposed.

The carbon and associated organics collected on the filter can be removed by oxidation so that the filter regenerates and is effective for the life of the vehicle. As the particulates are not oxidized at a significant rate below 600 °C, and these temperatures are only achieved in the exhaust system when the engine is near or at full power, the more promising systems incorporate a catalyst into the filter which reduces the oxidation temperature to approximately 300 °C. A schematic representation of a catalyst trap oxidizer is shown in Figure 6. Table 11 lists data on the degree of emission control which can be achieved with this system when fitted to a Volkswagen Golf 1.6 litre vehicle.

Catalyst developments aimed at reducing the temperature at which particulate oxidation commences and sulphate emissions are minimized are desirable for the system to be effective in meeting the proposed American standards on motor vehicles, trucks, and buses.

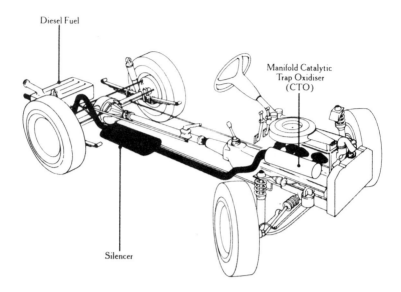

Figure 6   *The diesel catalyst trap oxidizer system*

Table 11   *Catalytic control of diesel exhaust emissions*

g/mile	HC	CO	NO_x	*Particulate*
Without catalyst	0.24	1.01	0.90	0.23
With catalyst	0.05	0.16	0.79	0.11

5.2 Catalytic Combustion

Nitrogen oxide emissions arise principally as a result of the reaction between oxygen and nitrogen at the temperatures arising from the combustion of fuel whether it is initiated by spark, as in the petrol engine, or compression as in the diesel engine. Lean burn operation of a gasoline engine, as described earlier, offers a partial solution to the problem but is limited by hydrocarbon emissions as the non-flammability limit for spark ignition is approached. While the diesel engine does not have these disadvantages it is limited by high particulate emissions and a high cetane fuel requirement.

An alternative approach is to use a catalyst to ignite the air/fuel mixture and in this way overcome the constraining factors of the gasoline and diesel engines. Having removed this constraint a further benefit is offered, namely, the operation of the engine at a compression ratio of 12 to 1 where engine efficiency and mechanical friction are optimized and fuel economy maximized.

The principle of the new catalytic engine is that during the engine operating cycle the fuel is injected into the combustion chamber just before the start of combustion is required. This fuel is then mixed with the air already in the cylinder and then passed through the catalyst, where heat release occurs. The use of this principle overcomes the fundamental problem of the stratified charge engine, which was outlined above, because since all the charge is passed through a catalyst, oxidation can occur even at low temperatures and very lean mixtures. Thus all the fuel should be oxidized, and the engine can be run unthrottled which should give good economy.

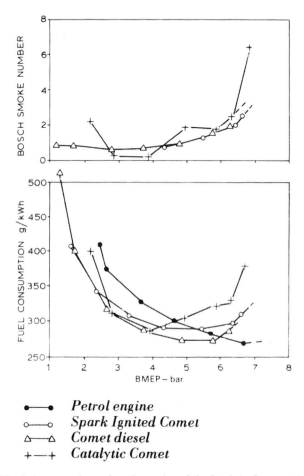

●——● *Petrol engine*
○——○ *Spark Ignited Comet*
△——△ *Comet diesel*
+——+ *Catalytic Comet*

Figure 7 *The fuel consumption and smoke number of the Catalytic Comet engine compared with conventional engines*

The formation of nitrogen oxides and carbon monoxide in the combustion chamber is also strongly dependent on the air/fuel ratio, and lean operation gives reduced emissions of these pollutants in the exhaust of the engine. The catalyst enables oxidation of hydrocarbons at much lower temperatures than normally possible, so these emissions are also reduced. Another important advantage of the catalytic engine concept is that it is capable of operating on many different liquid fuels. The engine does not require high octane fuel because: (a) since oxidation occurs on or immediately downstream of a catalyst, rather than in a flame front, there is less tendency for the unburnt gases to be heated above their spontaneous combustion temperature, and (b) since the fuel is injected late into the combustion chamber there is no possibility of pre-ignition; therefore the engine can be operated on fuels like diesel oils and paraffins. It has no requirement for fuels of high cetane number, this being a measure of the fuel's resistance to knock in a compression ignition engine, because the start of combustion is controlled by the catalyst instead of by the heat of compression; therefore it can be operated on fuels like petrols or alcohols.

The fuel consumption, smoke number, and gaseous emissions of the four engine systems using gasoline as fuel are shown in Figures 7 and 8. Although only limited development of the catalytic Comet engine has been done, of particular note are the low nitrogen oxide and hydrocarbon emissions which are achieved by this system. On alcohol fuels all of these parameters are significantly improved.

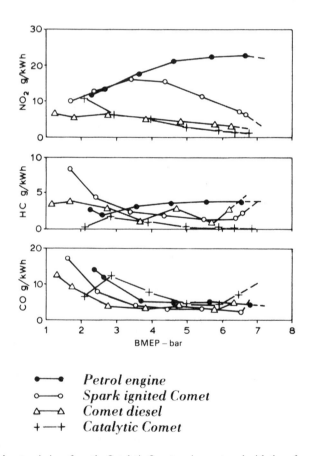

—•—•   *Petrol engine*
o——o   *Spark ignited Comet*
△——△   *Comet diesel*
+——+   *Catalytic Comet*

Figure 8   *Exhaust emissions from the Catalytic Comet engine compared with those from conventional engines*

5.3 Electrically Powered Vehicles

In principle the emissions problems from motor vehicles can be overcome by using electric power provided by rechargeable batteries. Apart from the range and limited acceleration provided by existing battery technology, recharging of a vehicle's battery pack is dependent upon electricity from nuclear or fossil fuel power plants which, in the latter case, are responsible for a significant part of the environmental carbon dioxide, nitrogen oxides, and sulphur oxides burden. Thus while some benefit may be achieved with ground level emissions it is outweighed both environmentally and economically by the other disadvantages.

The disadvantages of the battery powered vehicle may be overcome, if electric power is provided by a fuel cell since by this means fuel can be converted directly into electric power. The fuel cell, which in addition to emission levels of hydrocarbons, carbon monoxide, and nitrogen oxides substantially below those of internal combustion engines, even in their low emissions form, also has the benefit of significantly improved fuel economy. This results from the conversion of fuel directly into electric power by an electrocatalytic process which is not subject to efficiency limitations imposed by the Carnot Cycle. The resulting fuel economy benefits of the fuel cell over the internal combustion engine could be up to a factor of two with a corresponding reduction in carbon dioxide emissions from transport sources. Current interest in America and Japan in fuel cell powered vehicles centres on the low emissions and fuel economy practice.

Table 12 *Fuel cell systems for motor vehicles*

Type	Electrolyte	Electrode
PAFC	Phosphoric acid	Platinum
PEM	Solid polymer	Platinum
MCFC	Potassium carbonate	Nickel
SOFC	Zirconia	Nickel/Cobalt

Table 13 *Design parameters for a 200V, 60 kWe fuel cell system*

	PAFC	PEM	MCFC	SOFC
Parameter				
Cell voltage, V	0.67	0.63	0.70	0.63
Current density, A/cm^2	0.27	0.75	0.25	0.75
Stack temperature, °C	190	80	650	1 000
Gross efficiency, %	45.40	40.30	47.40	42.70
Active stack volume, L	109.00	38.00	86.00	22.00
Cell specific power, W/L	302.00	1 575.00	700.00	0.78
System specific power, W/L	56.00	62.00	36.00	77.00

The principle of a fuel cell was first demonstrated by Grove in 1849 but it was not until Bacon's pioneering work on alkaline fuel cells in the 1950's that it was shown that significant power could be generated by this means. In the 1960's notably Shell and British Petroleum looked to apply this technology to mobile applications including motor vehicles but were unable at that time to demonstrate a commercially viable system with the resources and materials technology at their disposal. Since then work in America on fuel cell powered system for manned space vehicles and, more recently, power generating plant, has significantly advanced the technology.

Fuel cell systems that are potential power units for motor vehicles are listed in Table 12. In Table 13 are listed key operating parameters for these fuel cell systems based upon a 200 volt, 60 kWe gross power unit.

6 CONCLUSIONS

Since the introduction of legislation in America in 1975 requiring a substantial reduction in emissions from motor vehicles, catalyst technology has played a major part in maintaining air quality. With the introduction of similar standards in other countries, this application now represents the largest single use of catalyst systems.

While the emission standards for cars set by the 1970 Clean Air Amendments Act were considered adequate at the time, air quality standards have not improved as projected largely as a result of the expanding car population in America and other industrialized countries. Advanced catalyst technology, including fuel cell systems, may be required in the future to achieve further reductions in exhaust emissions including carbon dioxide.

CHAPTER 12

Effects of Gaseous Pollutants on Crops and Trees

T. A. MANSFIELD and P. W. LUCAS

1 INTRODUCTION

Prior to the Clean Air Act of 1956 the concentrations of atmospheric pollutants, such as SO_2, in many towns and cities in the UK were high enough to cause visible damage to vegetation. The problem of very high pollutant concentrations was not confined to the UK but was a common occurrence throughout Europe and N. America in regions where industrial activity or large urban centres produced localized sources of pollutants from fossil fuel combustion. In N. America, especially in cities such as Los Angeles, topographic and climatic factors combined to produce photochemical smogs, and ozone damage to trees is well documented for areas such as the San Bernardino Forest to the east of the city.

Because these effects on vegetation were so clearly associated with high concentrations of air pollutants, the consensus view amongst researchers during the 1950's and early 1960's was that air pollutants had no effect on the yield of plants unless the concentrations were large enough to cause visible markings or damage to the foliage.[1,2] It was even considered that in the range 0.10 ppm– 0.20 ppm, pollutants such as SO_2 might actually exert a beneficial effect on plant growth.

In the last two decades, emission control legislation, increased urbanization, and new abatement technologies, such as the shift to the use of tall stacks in power stations, have resulted in changes in the quantity and quality of atmospheric pollutants. Associated with these changes has been a gradual shift of

[1] M. Katz, *Ind. Eng. Chem.*, 1949, **41**, 2450.
[2] M. D. Thomas, *Annu. Rev. Plant Physiol.*, 1951, **2**, 293.

emphasis in research not only in the species of plants studied but also in the concentration and type of pollutants used to carry out these studies. In the 1960's and 1970's the research into effects of air pollutants on plants was mainly concerned with crops and grasslands. In Europe SO_2 was the main pollutant of interest but in the USA the greatest concern was about the possible widespread effects of O_3. The two decades saw some important advances in our understanding of the modes of action of these two pollutants and, in particular, it was becoming recognized that affected plants could show physiological changes that were not immediately expressed as readily visible injury. These forms of 'invisible injury' could be revealed as reductions of growth and productivity or in the ability of plants to survive under periods of environmental stress.

During the 1980's several important further changes have taken place. First, the emphasis of research has shifted from herbaceous plants to trees, particularly the economically important conifers grown in Europe and North America. Second, there has been better information available on the distribution of pollutants in the atmosphere and this has led to more interest in their combined action, for example, of SO_2 and NO_x, and SO_2 and O_3. Third, there has been much more financial support for research and the facilities available for experimental studies are now greatly improved.

The switch in emphasis from herbaceous plants to trees came about because of worrying reports of damage to forests in central Europe and N. America, apparently worsening year by year, from about 1980 onwards.[3] Despite the much-expanded research efforts, the precise cause of this decline in tree health is still unknown. Some effects of individual pollutants are now better defined but these advances have been made mainly using herbaceous plants and not trees. It is still very difficult to perform realistic experiments on trees and there is often uncertainty about applying our knowledge of herbaceous plants to trees.

2 CURRENT KNOWLEDGE OF THE MECHANISM OF ACTION OF SOME MAJOR POLLUTANTS

2.1 Sulphur Dioxide

Sulphur is one of the important mineral nutrients of plants. It is normally taken into the roots from the soil in the form of sulphate then transported to the leaves. Enzymes for the reduction of sulphate are located in the leaves and reduced forms of sulphur are used in the synthesis of amino acids. During the reduction of sulphate ions there is transient formation of sulphite, but there is unlikely to be any significant accumulation of this ion. Sulphur can also be taken up directly by the leaves. Exchanges of carbon dioxide and water vapour between plants and the atmosphere are fundamental to many physiological processes such as photosynthesis, respiration, and transpiration carried out by plants. The exchange of gases with the atmosphere is regulated by opening and closing of stomata, and the uptake of pollutants, such as SO_2, is also thought to occur

[3] L. W. Blank, *Nature*, 1985, **314**, 311.

primarily via these pores at the surface of the leaf.[4] There is, however, some recent evidence which suggests that some gaseous pollutants may be absorbed by the leaf cuticle which could have important implications in reducing cuticular integrity, leading to enhanced water loss.[5]

When SO_2 enters the leaves from the atmosphere, via the stomata, it dissolves in a film of water at the surface of the mesophyll cells inside the sub-stomatal cavity forming sulphite and bisulphite ions. There is little evidence that SO_2 as a dissolved gas persists within the leaf.[6] The aqueous layer of the mesophyll cells is part of the pathway by which water from the soil is distributed to individual cells of the leaf. Thus sulphate ions will normally be present but not sulphite or bisulphite in significant concentrations. It is at present technically impossible to measure these ions in the water layer within the cell wall. We know, however, that the proton-pumping activities of the outer cell membrane (the plasma-lemma), which are part of normal cellular functioning, alter the pH of the extracellular environment appreciably at different times of day. Thus the equilibrium between the solution products of SO_2 may change, the ratio between sulphite and bisulphite increasing as the pH increases.

We do not know to what extent the plasmalemma of the cells in a leaf can tolerate the presence of sulphite and bisulphite but damage to membranes is known to be one feature of cellular injury by SO_2. The permeability of the plasmalemma is known to be affected by exposure to SO_2 and essential ions, such as potassium, can then leak out of cells.[7] Nevertheless potassium-leakage rates for leaves are not well correlated with differences in SO_2 sensitivity between cultivars or species and consequently it is not believed that the plasmalemma is the most sensitive cellular component to SO_2.[8]

The thylakoid membranes of chloroplasts seem to be more sensitive to the presence of SO_2. Swelling of these structures, which are the locations of the light-harvesting complexes and electron transport components, is seen very soon after the commencement of fumigation with SO_2.[9] This may explain why the process of photosynthesis is inhibited by SO_2. Despite this obvious sensitivity of the chloroplasts there is strong evidence that the primary sites of injury within leaves are located elsewhere. Noyes[10] fed bean (*Phaseolus vulgaris*) leaves with $^{14}CO_2$ and found that the translocation of ^{14}C-labelled products out of the leaves was inhibited by SO_2 concentrations too small to reduce the rate of photosynthesis. Subsequent studies with several different species have confirmed the high sensitivity of assimilate translocation to SO_2.[11]

[4] T. A. Mansfield and P. H. Freer-Smith 'Gaseous Air Pollutants and Plant Metabolism,' ed. M. J. Koziol and F. R. Whatley, Butterworth Scientific, London, 1984, p. 131.
[5] K. J. Lendzian, *Aspects of Applied Biology.*, 1988, **17**, 97.
[6] V. Black and M. H. Unsworth, *J. Exp. Bot.*, 1980, **31**, 667.
[7] E. Nieboer, D. H. S. Richardson, K. J. Puckett, and F. D. Tomassini, in 'Effects of Air Pollutants on Plants,' ed. T. A. Mansfield, Cambridge University Press, 1976, p. 61.
[8] D. T. Tingey and D. M. Olszyk, in 'Sulfur Dioxide and Vegetation,' ed. W. E. Winner, H. A. Mooney, and R. A. Goldstein, Stanford University Press, 1985, p. 178.
[9] A. R. Wellburn, O. Majernik, and F. A. M. Wellburn, *Environ. Pollut.*, 1972, **3**, 37.
[10] R. D. Noyes, *Physiol. Plant Pathol.*, 1980, **16**, 73.
[11] S. B. McLauchlin and G. E. Taylor, in 'Sulfur Dioxide and Vegetation,' ed. W. E. Winner, H. A. Mooney, and R. A. Goldstein, Stanford University Press, 1985, p. 227.

Long distance transport of assimilates in plants involves their movement from sites of photosynthesis (principally the mesophyll cells) into the sieve tubes of the phloem. The final step is known as phloem loading, and this can take place against a very large concentration gradient, *e.g.*, 1 : 40. The energy required to transport sucrose against such a gradient is almost certainly generated by the consumption of ATP and extrusion of protons, which then return across the membrane 'carrying' sucrose molecules. Presumably the membrane site of this 'sucrose/proton co-transport' may be the point of particular sensitivity to the solution products of SO_2. Minchin and Gould[12] used the very short-lived isotope ^{11}C to show that phloem-loading was almost immediately reduced after exposure of the C_3 plant wheat to SO_2.* This was not the case, however, with maize, a C_4 plant, in which there is a tight ring of cells protecting the phloem from the intercellular air spaces of the leaf. There is no such protective layer in wheat and consequently SO_2 in the intercellular spaces may have almost direct access to the phloem tissue. Thus this interesting comparative study of two plant species with contrasting internal leaf anatomy may well have confirmed that the phloem is the most sensitive tissue in the leaf to SO_2 pollution. Too little is known of the subcellular mechanism of the sucrose transport process to enable us to identify the precise sites of injury.

Studies of the effects of SO_2 on plants have often been performed in fumigation chambers subjected to natural illumination and ambient temperatures. In some of these experiments the growth responses to the pollutant seemed to be affected by climatic changes. A study under controlled conditions with the grass *Phleum pratense* (Timothy) showed that when growth was rapid, in high irradiance and long days, 120 ppb SO_2 (higher than usually found in the most controlled situations nowadays) had no detectable effect on the plants. On the other hand the same concentration applied to plants in low irradiance and short days reduced growth by about 50%.[13] Similar changes in sensitivity were found when growth was reduced by reducing the temperature (Figure 1).

The results of these and other similar experiments have led to the conclusion that there is no critical concentration which can usefully be regarded as the threshold for injury. Different amounts of SO_2 can clearly be tolerated under different climatic conditions, a factor which has not been considered in many of the models so far produced which attempt to evaluate the effects of pollutants on cultivated or natural vegetation. Perhaps our attention should be focused on those effects that occur under the least favourable conditions, particularly if they involve reductions in survival ability under extreme climatic conditions. There is now quite a considerable amount of evidence that plants exposed to SO_2 become more sensitive to frost injury. Figure 2 shows the substantial reduction in survival of SO_2-polluted ryegrass upon subsequent exposure to sub-zero temperatures. It is not uncommon to see frost injury to pasture grasses during winter in the more polluted parts of the UK. The impact on the productivity of an established

[12] P. E. H. Minchin and R. Gould, *Plant Sci.*, 1986, **43**, 179.
[13] T. Jones and T. A. Mansfield, *Environ. Pollut. Ser. A.*, 1982, **27**, 57.

* C_3 and C_4 plants have different first products of CO_2 fixation in light, *viz.* 3-carbon and 4-carbon compounds respectively.

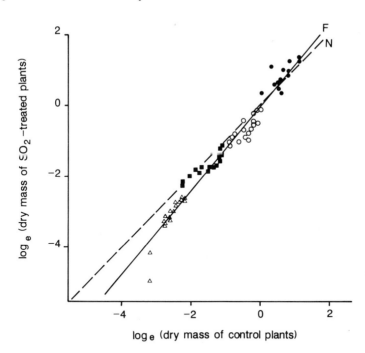

Figure 1 *Relationship between growth rate and effects of SO$_2$ on dry mass increase of the grass Phleum pratense (Timothy). N is the 'no effect' line and F is the line fitted to the data. Plants were fumigated for 44 days with 0.12 ppm SO$_2$ and the following environmental conditions were used to produce the different growth rates:* ●, *light of 400 μE m^{-2}s^{-1}, 19 °C (night) and 30 °C (day);* ○, *light of 400 μE m^{-2}s^{-1}, 12 °C (night) and 26 °C (day);* ■, *light of 100 μE m^{-2}s^{-1}, 19 °C (night) and 30 °C (day);* △, *light of 100 μE m^{-2}s^{-1}, 12 °C (night) and 26 °C (day)*

From Jones and Mansfield,[13] and used with the permission of Applied Science Publishers Ltd.

sward is, however, probably quite small because of recovery that can occur in spring and summer.[14] The enhancement by SO$_2$ of frost sensitivity in woody plants is more likely to be of concern. There is evidence that even dormant deciduous plants may be affected by SO$_2$ uptake into the shoots in winter, with subsequent death of the terminal buds.[15] This topic is covered in more detail in section 4.2.1.

In some agricultural areas the soil is deficient in sulphur and it is possible that SO$_2$ (and also H$_2$S) in the atmosphere might remedy this deficiency. The total sulphur deposition per unit area may exceed 6 g m^{-2} per year in parts of the UK, and more than 12 g m^{-2} per year in central and eastern Europe.[16] Such large inputs exceed the sulphur requirements of most plants. Cowling and Koziol[17]

[14] M. E. Whitmore and T. A. Mansfield, *Environ. Pollut. Ser. A.*, 1983, **31**, 217.

[15] T. Keller, *Environ. Pollut.*, 1978, **16**, 243.

[16] 'Acid Rain,' Report no. 14, Watt Committee on Energy, 1984.

[17] D. W. Cowling and M. J. Koziol, 'Effects of Gaseous Air Pollution in Agriculture and Horticulture,' ed. M. H. Unsworth and D. P. Ormrod, Butterworth Scientific, London, 1982, Chapter 17, p. 349.

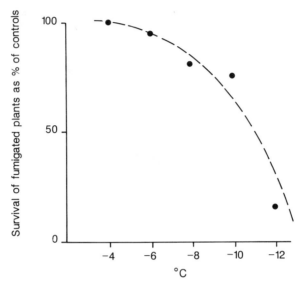

Figure 2 *Survival of ryegrass* (L. perenne *cv.* S23) *after exposure to sub-zero temperatures. Controls were subjected to less than* 15 µg SO_2 m^{-3}, *and the fumigated plants were exposed for 3 weeks to* 250 µg SO_2 m^{-3} (0.098 ppm)
From Davison and Bailey,[60] and reproduced with permission from Macmillan Journals Ltd.

reviewed the many experiments performed over a period of 50 years to ascertain whether atmospheric SO_2 can provide useful nutrition. There is much evidence that when supplies of sulphate in the soil are inadequate to support growth, then SO_2 in the atmosphere can make up the deficit. When the plants are growing rapidly, under favourable conditions, the SO_2 can be regarded as beneficial, but in most industrialized countries in Europe and N. America SO_2 pollution is greater in winter when plant growth is slowest. To benefit from SO_2 the leaves must be able to use the supply of sulphur as it enters the cells. If metabolism is proceeding too slowly for this to happen then toxic ions such as SO_3^{2-} and HSO_3^- may accumulate with resulting damage to cells and processes. It is thus possible to put mistaken emphasis on the beneficial effects of atmospheric SO_2. Sulphur deficiencies in the soil can be remedied by increases in the sulphate content of fertilizers at small cost to farmers and the economic advantage of sulphur supplied by SO_2 pollution is therefore very small.

2.2 Nitrogenous Pollutants

It has long been recognized that nitrogen is required by plants for normal growth as it is a major constituent of all proteins and nucleic acids. As a consequence, plants have developed a range of diverse mechanisms to enhance their acquisition of this important nutrient, for example, fixation of atmospheric nitrogen, associations with mycorrhizal fungi, and insectivory. The majority of plants obtain nitrogen through root absorption of the inorganic ions ammonium

(NH_4^+) and nitrate (NO_3^-) from the soil solution and have developed pathways of nitrogen assimilation which are now well characterized (Figure 3). (Soil processes such as nitrification and mineralization which influence the availability of nitrogen are complex and space does not permit a description of these processes here). Considerably less is known, however, about how plants adapted to low nitrogen inputs, particularly those growing in sensitive ecosystems such as forests or heathlands, react to additional nitrogen deposition from anthropogenic sources, either as gaseous nitrogen compounds (NO, NO_2, and NH_3) or from wet deposition $(NO_3^-$ and $NH_4^+)$

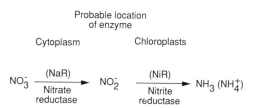

Figure 3 *The reduction pathway from nitrate to ammonia which occurs in plants prior to amino acid synthesis*

Because of the paucity of reliable information regarding the atmospheric concentration of NO_x and NH_3 in rural areas throughout Europe, only estimates of the amounts of nitrogen deposition in wet and dry forms to terrestrial ecosystems are available. In addition both the amount and form of deposited nitrogen will vary depending on the specific region under consideration.[18] Thus inputs of between 10 and 30 kg ha^{-1} a^{-1}, largely as NO_2, have been suggested for low altitude areas of Western Europe, and up to 20 kg ha^{-1} a^{-1} (predominantly as NO_3^- and NH_4^+ in cloud and rainfall) for high altitude regions of Britain and W. Germany. The long-term effects of high inputs of nitrogen to sensitive ecosystems are, however, only poorly understood, especially on soil processes,[19] the availability of other nutrients,[20] and the possible interaction with environmental stresses such as frost.[21,22] In the southern Pennines the atmospheric concentrations of SO_2 have declined considerably since the 1950's but there has been no recovery of the *Sphagnum* moss, the decline of which was attributed to pollution from the nearby urban areas. This is thought to be because the *Sphagnum* is now primarily affected by nitrogenous pollutants which have increased in concentrations over the period in which SO_2 has declined.[23]

2.2.1 Responses to Gaseous NO_x. Most research has concentrated on the effects of NO_2 rather than NO, although the latter is often a major component of NO_x, especially near sources such as busy roads. Anderson and Mansfield[24] found that

[18] F. T. Last, J. N. Cape, and D. Fowler, *Span*, 1986, **29**, 2.
[19] I. Kottke and F. Oberwinkler, *Trees*, 1986, **1**, 1.
[20] G. G. Gebauer and E. D. Schulze, *Aspects Appl. Biol.*, 1988, **17**, 123.
[21] B. Nihlgard, *Ambio*, 1985, **14**, 2.
[22] A. J. Friedland, G. J. Hawley, and R. A. Gregory, *Plant Soil*, 1988, **105**, 189.
[23] M. C. Press, S. J. Woodin, and J. A. Lee, *New Phytol.*, 1986, **103**, 45.
[24] L. S. Anderson and T. A. Mansfield, *Environ. Pollut.*, 1979, **20**, 113.

NO was five times more soluble in xylem sap than in distilled water. The xylem sap is continuous with the extracellular water in the leaf mesophyll, and hence it is likely that the uptake of NO into leaves may be greater than would be predicted from its water solubility. We know little about the fate of NO after it has entered the leaf, but some research has suggested that it can be more toxic than NO_2.[25] This is very difficult to explain on the basis of our present knowledge of the solution products of the two gases.

The nitrate and nitrite ions formed when NO_2 enters into solution might be expected to enter normal metabolism via the pathway that reduces them to ammonia (Figure 3). Nitrate and nitrite are the substrates for two reductase enzymes normally present in leaves, *viz.* nitrate reductase and nitrite reductase. The usual source of initial substrate is the nitrate (but not nitrite) transported into the leaves from the roots via the xylem.

Nitrites are known to be toxic to plant tissues and there is metabolic regulation within the reduction pathway to ammonia to prevent them accumulating. The necessary rate control is thought to be provided at the stage of nitrate reduction, the enzyme for which is located inside the cells but outside the chloroplasts,[26] while nitrite reductase is within the chloroplasts. There must be mechanisms for keeping the toxic NO_2^- ions away from sensitive sites during their movement within the cells but we know nothing of these. It is, however, likely that the arrival of NO_2 at the cell surface, leading to NO_2^- ions in the extracellular water, may pose problems because the movement of substantial numbers of these ions across the plasma membrane is probably not a normal requirement.

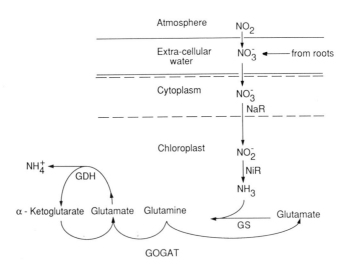

Figure 4 *The relationship between the synthesis of amino acids in chloroplasts and the nitrate formed from NO_2 entering the leaf. NaR and NiR are nitrate and nitrite reductase, respectively*

[25] H. Saxe, *New Phytol.*, 1986, **103**, 185.
[26] R. M. Wallsgrove, P. J. Lea, and B. J. Miflin, *Plant Physiol.*, 1979, **63**, 323.

Figure 4 shows diagrammatically the locations of nitrate and nitrite reduction and the incorporation of NH_3 into amino acids in chloroplasts. This is regarded as the normal pattern in the majority of plants but there are exceptions and these may be important in determining differences in sensitivity to NO_2 between species.

It has been suggested[27] that the susceptibility of different plant species to NO_x may depend on the precise location at which they normally carry out nitrate/nitrite reductions. Some woody plants, for example, might be particularly sensitive to NO_x because the reduction pathway is absent or poorly developed in the leaves. These species carry out the reductions of soil-derived nitrates in the roots and hence nitrogen in an already reduced form is transported to the shoots. Research has so far failed to indicate precisely how such plants are affected by NO_x entering the leaves from the atmosphere.

The exposure of plants to NO_x pollution has been shown to lead to changes in the activities of nitrate and nitrite reductases in leaves.[28] There are usually marked increases in nitrite reductase, a clear indication that the plant has some ability for metabolic adjustment to detoxify the nitrite ions as they enter the cells. In contrast the activity of nitrate reductase may decline in leaves exposed to NO_x. This is probably because the supply of nitrite ions from the atmosphere reduces the need for the production of nitrite by nitrate reductase. A consequence may be loss of the normal rate of control of the pathway, which could produce serious imbalances in subsequent metabolism. Figure 4 shows the routes for the incorporation of ammonium into amino acids, and it is these areas of metabolism that may be of much concern in future research.

2.3 Ozone

Consideration of the toxicity of O_3 is not complicated by its providing a source of an essential nutrient, as in the case with SO_2 and NO_x. We can therefore direct attention simply to the properties of the molecule itself.

At a high concentration, ozone causes a characteristic brown or white flecking to appear on the foliage of affected plants but there is a wide variation between species and between cultivars of the same species. This variation is particularly apparent between different cultivars of the tobacco plant, one of which, the variety Bel-W3, can be injured by hourly mean ozone concentrations above 40 ppb. Because of its high sensitivity, Bel-W3 is frequently used as a bio-monitor for O_3 to provide a cheap and convenient substitute for chemical monitoring methods, and it has been used in several studies in the UK.[29]

At lower concentrations, studies have shown that in the absence of visible damage, ozone may reduce the growth rate or physiological activity of plants. Again, however, there is considerable variation from species to species or between cultivars of the species. There are also considerable differences in toxicity

[27] R. G. Amundson and D. C. Maclean, in 'Air Pollution by Nitrogen Oxides,' ed. T. Schneider and L. Grant, Elsevier, Amsterdam, 1982, p. 501.
[28] A. R. Wellburn, J. Wilson, and P. H. Aldridge, *Environ. Pollut.*, 1980, **22**, 219.
[29] M. R. Ashmore, J. N. B. Bell, and C. L. Reily, *Nature*, 1987, **327**, 417.

in different environmental conditions. Initially this variation caused problems for researchers but more recently it has been looked upon as an opportunity to identify those characteristics of a plant, or the physiological states determined by the environment, that lead to enhanced susceptibility.[30] This approach can provide important clues to the mechanisms behind injury.

Cells inside the leaf, especially the photosynthetic mesophyll, have been identified as the primary sites of injury. Differences in the susceptibility of these cells does not, however, appear to be the most important cause of variation. It is the ability of the leaf to exclude O_3 from the sensitive sites that emerges as one critical point of control.

Like other gases in the atmosphere O_3 gains access into leaves by diffusing through stomatal pores. Figure 5 shows a surface view of a leaf of birch, *Betula pubescens*, as seen under a scanning electron microscope. The stomata can be clearly seen in the epidermal layer, and where this layer has been removed the underlying mesophyll can also be seen. This gives a good impression of the protection offered to the mesophyll by the epidermis and of the importance of stomata in allowing gaseous communication with the atmosphere.

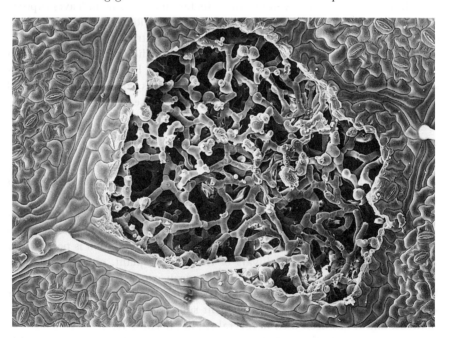

Figure 5 *Scanning electron micrograph of a frozen hydrated leaf of birch* (Betula pubescens). *The surface view of the epidermis shows the stomata, and where the epidermis has been removed the internal air spaces and mesophyll cells can be seen*
Provided by Dr. E. Neighbour and Dr. K. Oates, University of Lancaster.

---

[30] D. T. Tingey and G. E. Taylor, in 'Effects of Gaseous Air Pollution in Agriculture and Horticulture,' ed. M. H. Unsworth and D. P. Ormrod, Butterworth Scientific, London, 1982, p. 113.

There has been a great deal of research into the relationship between stomatal conductance and sensitivity to O_3 and there is general agreement that it is often the major determining factor. If the stomata are partially closed then some O_3 is excluded from the sensitive sites. Differences in stomatal opening can be inherent (*i.e.*, genetically determined) or can result from responses to environmental conditions. Stomata tend to close partially when the atmosphere is dry or when the water content of the soil is low. Differences in O_3 sensitivity are often well correlated with such changes.[30]

This correlation with stomatal conductance does not appear so clearly when we consider other pollutants, such as SO_2 and NO_x. The reason why it is specially important in relation to O_3 sensitivity has become clear quite recently as a result of the research of Mehlhorn and Wellburn.[31] They fumigated pea seedlings with 50–150 ppb O_3 for 7 h daily for their first 3 weeks of growth after germination. Even in 150 ppb O_3 (regarded as a high level of pollution) there was no evidence of any injury apart from a slight curling of the leaves.

On the other hand, if 3-week-old seedlings that had been grown in clean air were fumigated for just one day under the same conditions and in the same concentration of O_3, there was very severe damage to the leaves (Table 1).

Table 1   *Ethylene production and leaf injury in pea seedlings*

Treatment	(A) Three-week fumigation (C_2H_4 evolved nmol g^{-1} dry weight h^{-1})	(B) One-day fumigation (7 h) (C_2H_4 evolved nmol g^{-1} dry weight h^{-1})	Visible leaf injury (%)
Control	2.5 (0.1)*	2.6 (0.2)	none
50 ppbv O_3	1.9 (0.2)*	4.7 (0.2)	0–10
100 ppbv O_3	0.3 (0.1)*	7.0 (0.4)	10–35
150 ppbv O_3	0.2 (0.1)*	5.7 (0.5)**	>50

The data in column (A) are for plants grown for 3 weeks and exposed to O_3 for 7 hours daily. Those in column (B) are for plants grown in clean air for 3 weeks, then exposed to O_3 for 7 hours for 1 day only.
Results represent means (s.e.m. in parentheses) of at least 7 replicates. Treatments were compared using Student's t-test and the 100 and 150 ppbv treatments were highly significant at P < 0.01.
* No visible injury; ** also with raised rates of ethane evolution.
Reproduced from Mehlhorn and Wellburn[31] with permission from Macmillan Journals Ltd.

Subsequent studies showed that the two sets of seedlings differed markedly in the amounts of ethylene they were producing (Table 1). Ethylene is normally generated by plants and it is known to be able to regulate various aspects of growth and development. For this reason it is regarded as a hormone, and it is unique as the only known gaseous hormone in plants or animals.[32] The pathway

[31] H. Mehlhorn and A. R. Wellburn, *Nature*, 1987, **327**, 417.
[32] P. J. Davies, ed. 'Plant Hormones and their Role in Plant Growth and Development,' Martinus Nijhoff, Dordrecht.

for its biosynthesis from methionine is well established (Figure 6). The point of regulation may be at the level of synthesis of ACC (1-aminocyclopropane-1-carboxylic acid) and it is known that some of the other plant hormones such as the auxins can affect ACC production.

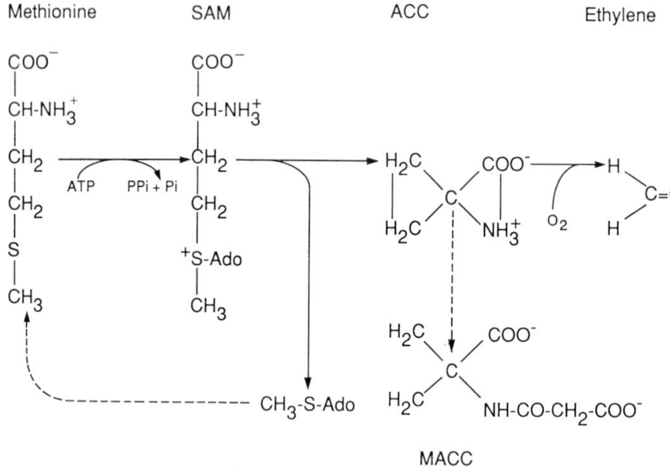

Methionine   SAM   ACC   Ethylene

Figure 6   *The pathway of ethylene biosynthesis from methionine*
SAM = *S-adenosylmethionine*
ACC = 1-*aminocyclopropane*-1-*carboxylic acid*
MACC = 1-*malonylaminocyclopropane*-1-*carboxylic acid*

Increased production of ethylene is associated with many events in plant development such as release from dormancy, stem and root growth, differentiation, abscission of leaves and fruits, flowering, and fruit ripening. In particular, ethylene production increases in response to environmental stresses of various kinds (*e.g.*, physical wounding, flooding).

In the experiments of Mehlhorn and Wellburn the rates of ethylene evolution were more than doubled in the 3-week-old plants fumigated for 7 h with O_3. The increased production of ethylene after O_3 fumigation was already well known from previous studies. The unique contribution by Mehlhorn and Wellburn was their proposal that the production of ethylene was a key factor in determining injury to O_3. They tested this hypothesis by treating plants with an inhibitor of ethylene biosynthesis, aminoethoxyvinylglycine (AVG), prior to exposing them to O_3. AVG inhibits the synthesis of ACC (Figure 6). The treatment with AVG reduced ethylene production by 85% and, also, almost completely prevented the damage normally caused to the 3-week-old plants by the short O_3 treatment.

Ozone can react with unsaturated hydrocarbons to produce water-soluble free radicals. Mehlhorn and Wellburn suggested that their formation in the intercellular spaces of the leaf around the mesophyll cells (see Figure 5) could trigger peroxidative processes inside the cells ultimately leading to damage. More recent studies have shown that there are free radicals present in leaf tissues, closely correlated with ethylene formation and ozone treatments as postulated (Mehlhorn and Wellburn, personal communication).

These important new findings explain why high stomatal conductance may be of special importance in connection with injury to leaves by O_3: only when the endogenous ethylene makes contact with the incoming O_3 does the critical formation of free radicals take place.

The parts of cells most vulnerable to free radicals in the intercellular air are likely to be the plasmalemma membranes. O_3 fumigation is known to cause damage to membranes so that the retention of solutes is reduced. The paper by Heath and Castillo[33] provides a valuable review of this topic.

2.4 Acid Rain and Acid Mist

The wet deposition of acidity from polluted air has probably increased the leaching of base cations from most soils in affected areas. This does, however, only lead to a change in acidity if a soil's reserves of base cations are low.

There is now quite good scientific evidence that the pH values of some soils in Scandinavia and Central Europe have decreased over the past few decades.[34,35] Increased base leaching has also been recognized in parts of North America but here there is little to suggest that soils have yet reached dangerously low pH values.

The input of acids (principally sulphuric and nitric) from the atmosphere is not the only way by which soil pH can be decreased. Various biotic processes and humus accumulation can also contribute.[36] There is still some uncertainty about the relative contributions of these different processes but, nevertheless, most soil scientists now accept that acidity in rainfall is playing an important part in the changes that are occurring. The direct and indirect effects on plants in the affected soils are much more open to dispute. In the Federal Republic of Germany the yellowing (chlorosis) of older needles of Norway spruce and silver fir has increased markedly on nutrient-poor soils. This phenomenon appears to be associated with deficiencies in cations, especially magnesium. The problem has been found over large geographical areas, many of which are long distances from the main sources of primary pollutants. It is, however, recognized that acidic deposition can occur far away from such sources.

The hypothesis that deterioration of the soil is a major factor in the decline in the health of trees has had much support but other ideas have also been extensively discussed. In particular, the increased frequency of episodes of O_3 pollution at the damaged sites has attracted attention and O_3 has been suggested to be either the predominant agent in causing damage, or at least an inciting stress. The injury to membranes caused by O_3 could, it is suggested, be the primary factor in the leakage of nutrients and their subsequent leaching from

[33] R. L. Heath and F. J. Castillo, in 'Air Pollution and Plant Metabolism,' ed. S. Schulte-Hostede, N. M. Darrall, L. W. Black, and A. R. Wellburn, Elsevier Applied Science, London, 1988, p. 55.
[34] E. Matzner and B. Ulrich, in 'Effects of Atmospheric Pollutants on Forests, Wetlands, and Agricultural Ecosystems,' ed. T. C. Hutchinson and K. M. Meema, Springer-Verlag, Berlin, 1987, p. 25.
[35] G. Abrahamsen, in 'Effects of Atmospheric Pollutants on Forests, Wetlands, and Agricultural Ecosystems,' ed. T. C. Hutchinson and K. M. Meema, Springer-Verlag, Berlin, 1987, p. 321.
[36] S. I. Nilsson, H. G. Miller, and J. D. Miller, *Oikos*, 1982, **39**, 40.

foliage by acid mists.[37] Another recent suggestion is that the deposition of nitrogen to forests from the increasing atmospheric concentrations of NO_x has caused an imbalance in the major nutrients required for growth (see section 2). Some authors have suggested that 'nitrogen saturation' is beginning to occur in forest ecosystems but as far as we know there is no clear definition of what 'saturation' means in this context.

The main reason for the slow progress in testing these and other hypotheses is the difficulty of conducting critical experiments on trees and a poor general understanding of the physiology of trees compared with herbaceous plants. The practical possibilities of conducting meaningful experiments on mature trees are now being actively explored but it will be some time before much progress can be made because many of the necessary experiments will last several years.

Most rainfall chemistry studies rely on the use of rain gauges for the collection of samples. Apart from other problems,[16] simple rain gauges can also under-estimate the amount of water from the atmosphere that is captured by vegetation in mist or clouds. This additional input of water from the atmosphere is known as 'occult' precipitation.[38] Not only is the volume of precipitation likely to be underestimated when plants are in cloud, but also, the amounts of dissolved materials that are deposited.

Collection of occult deposition separately from rain is difficult and special techniques have only recently been developed. On Great Dun Fell, in the northern Pennines in England, it has been estimated that the concentration of ions such as H^+, SO_4^{2-}, and NO_3^- were 2 to 4 times higher in cloudwater than in rain.[38]

There are occasions when cloudwater may simultaneously condense on foliage and evaporate from it and this can lead to a considerable increase in the concentration of ions in the liquid film or droplets on exposed surfaces.[39] This is thought usually to occur for short periods, for example, when the weather is improving after a dense cloud cover, but in thin cloud when some solar radiation can reach the surface it may be more prolonged.

Experimental studies of the effects of simulated rain on leaf surfaces have generally shown that there is little direct damage unless the pH is below 3.0, with the exception of a few cultivars of field crops that appear to be specially sensitive.[40] In some cases the growth of the plants was increased probably because of the fertilizing effects of additional nitrogen and sulphur. Such low pH values are of very infrequent occurrence in rain but it seems likely that they could occur much more frequently during occult deposition. For this reason, the effects of cloudwater on forests at high altitude are now being considered as a further contributory factor towards the damage to trees.

---

[37] R. A. Skeffington and T. M. Roberts, *Oecologia*, 1985, **65**, 201.
[38] M. H. Unsworth and D. Fowler, in 'Air Pollution and Ecosystems,' ed. P. Mathy, D. Reidel Publishing Co., Dordrecht, 1988, p. 68.
[39] M. H. Unsworth, *Nature*. 1984, **312**, 262.
[40] L. S. Evans, K. F. Lewis, E. A. Cunningham, and M. J. Patti, *New Phytol.*, 1982, **91**, 429.

3 INTERACTIONS BETWEEN POLLUTANTS

It has long been recognized that air pollutants rarely occur singly, yet much of the literature covers their separate effects rather than their joint action in realistic mixtures. There are two main reasons for this: mechanisms behind injury are necessarily studied in the simplest context, at least in the first instance; and the facilities available to researchers have until recently rarely allowed for the design of experiments involving different pollutants in factorial combinations, especially for long-term studies.

The situation that has emerged so far is complex. The effects of combinations of pollutants are sometimes greater and sometimes less than we would predict from the separate effects and only rarely are responses simply additive. In the space available here we shall concentrate on two specific pollutant combinations whose effects may be specially important.

3.1 Mixtures of SO_2 and NO_2

Combinations of these two primary pollutants have received attention because they do occur together in many situations. They are often produced simultaneously by the same sources, for example, during the combustion of S-containing fossil fuels. Joint action of SO_2 and NO_2 may be important in urban areas where there are regular episodes of high concentrations or in nearby rural areas where there may be longer exposures to both gases. Yearly mean concentrations of SO_2 in rural areas of Europe range from $<5\,\mu g\ m^{-3}$ close to the Atlantic seaboard to $30\,\mu g\ m^{-3}$ in rural areas near industrial regions. Although monitoring data are available for fewer rural sites the spatial distribution of NO_2 appears to be very similar to that of SO_2. Based on this assumption the mass ratio of $NO_2 : SO_2$ has been estimated to be about 0.8.[41]

Short-duration fumigations with SO_2 and NO_2 have often led to severe injury, far greater than that found with the pollutants separately. Tingey *et al.*[42] were the first to perform a detailed study; they found surprisingly high toxicity of some mixtures of SO_2 and NO_2 in 4 hour exposures, at concentrations around 10% of those required for injury by the individual gases. Since this early work, there has been a mixture of confirmatory and contradictory reports, but there have been more than enough observations of synergism between SO_2 and NO_2 to suggest that simultaneous exposure to the two pollutants could be a cause of unexpected damage in the field.

Studies of long-term effects of SO_2 and NO_2 on grasses and cereals have shown that there are much greater growth reductions during the winter than in spring and summer.[14,43] This suggests that slower rates of metabolism determined by environmental conditions may predispose plants to this form of injury.

[41] D. Fowler and J. N. Cape, 'Effects of Gaseous Air Pollutants in Agriculture and Horticulture,' ed. M. H. Unsworth and D. P. Ormrod, Butterworth Scientific, London, 1982, Chapter 1, p. 3.
[42] D. T. Tingey, R. A. Reinert, J. A. Dunning, and W. W. Heck, *Phytopathology*, 1971, **61**, 1506.
[43] P. C. Pande and T. A. Mansfield, *Environ. Pollut. Ser. A.*, 1985, **39**, 281.

As discussed above (Section 2), the enzyme nitrite reductase is considered to play a critical part in the metabolism of the solution products of NO_2 after its entry into cells. Yoneyama and Sasakawa[44] made use of $^{15}NO_2$ to demonstrate the appearance in cells of nitrite and nitrate ions and to show that after their reduction the ^{15}N was incorporated into amino acids. Analyses by Wellburn and his colleagues[45,46] of plastids extracted from leaves of grasses after exposure to SO_2, NO_2, or $SO_2 + NO_2$ revealed important differences in enzyme activities. Fumigation with NO_2, on its own, led to a statistically significant increase in the activity of nitrite reductase but no such increase was found when the NO_2 was accompanied by SO_2 (Figure 7). This obviously suggested that SO_2 prevented the induction of greater nitrite reductase activity, as normally occurs in the presence of NO_2. This means that the plants may have been unable to detoxify the solution products of NO_2. Thus there may be a fairly simple biochemical explanation of synergistic effects of SO_2 and NO_2 in damaging plants.

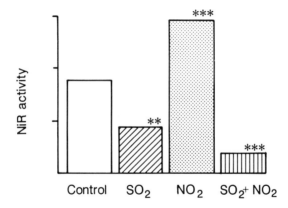

Figure 7   *Nitrite reductase activity in* Phleum pratense *(* Timothy grass*) exposed for 20 weeks to clean air or atmospheres containing 69 ppb SO_2 and NO_2 alone or in combination.* ** *and* *** *indicate significant differences at* $p < 0.01$ *and* $p < 0.001$, *respectively*
From Wellburn and used with permission of Butterworth Scientific.

3.2   Mixtures of SO_2 and O_3

Experimental studies of this combination of pollutants have produced a much more confusing picture. The earliest studies involving short-term exposures provided evidence of synergism not unlike that often found with SO_2 and NO_2.[47,48] The nature of foliar injury caused by SO_2 and by O_3 is quite different and it was noted in these and in some later experiments that the enhanced damage in combinations of the two resembled that caused by O_3 rather than by

[44] T. Yoneyama and H. Sasakawa, *Plant Cell Physiol.*, 1979, **20**, 263.
[45] A. R. Wellburn, in 'Effects of Gaseous Air Pollutants in Agriculture and Horticulture,' ed. M. H. Unsworth and D. P. Ormrod, Butterworth Scientific, London, 1982, p. 169.
[46] A. R. Wellburn, in 'Gaseous Air Pollutants and Plant Metabolism,' ed. M. J. Koziol and F. R. Whatley, Butterworth Scientific, London, 1984, p. 203.
[47] R. L. Engle and W. H. Gabelman, *Pro. Am. Soc. Hort. Sci.*, 1966, **89**, 423.
[48] H. A. Menser and H. E. Heggestad, *Science*, 1966, **153**, 424.

SO_2. This might be explained by the enhanced opening of stomata that can occur in SO_2-polluted leaves (see next section). Thus SO_2 could assist the access of O_3 into the leaf's interior.

In more recent studies no consistent picture of the combined effects of SO_2 and O_3 has emerged. In a review of the subject Kohut[49] reported evidence for additive, less-than-additive, and more-than-additive effects. The only consistency in the data from different sources was that most more-than-additive effects were found in short-term exposures. It seems unlikely that a simple description of the dose-response relationships for effects of these two pollutants on plants can be produced. This has serious implications for any modelling exercises aiming to predict the economic consequences of pollutants in the field.

The difficulties encountered by researchers in coming to a clear view of the combined effects of SO_2 and O_3 emphasize the complexity of the perturbations that can occur for plant life when we alter the atmospheric environment to even a small degree.

4 POLLUTANT/STRESS INTERACTIONS

The majority of experimental fumigations of plants with air pollutants have been conducted under conditions that are favourable for growth. There are good reasons for this: stress factors, biotic, and abiotic, are often difficult to maintain and control at a predetermined level, hence their inclusion in experiments that are already technically very difficult has been seen as an unmanageable complication. Yet the evidence from a few well-designed experiments, supported by many observations in the field, suggests that this may be a very important topic for future investigation.

4.1 Biotic Factors

There is now some good evidence that exposure to pollutants can predispose plants to attack by pathogens, *e.g.*, insects or fungi. There are, for example, often good correlations between insects feeding on plants and the presence of particular pollutants. Hughes *et al.*[50] found that the rate of development and the fecundity of the Mexican bean beetle on soybeans was increased after the plants had been exposed to SO_2. Ambient air in London, when compared with charcoal-filtered air, increased the growth rate of aphids on broad bean.[51] Aphids feed specifically on the sap of the phloem in plants, which is the main transport pathway for sugars, amino acids, *etc.* It is possible to raise some species of aphids on artificial diets providing the same nutritional factors as phloem. When such artificially maintained aphids are exposed to low concentrations of pollutants they are unaffected. Thus there is little doubt that it is changes in the physiology of the plant that are responsible for the differences in aphid activity in polluted air.

[49] R. Kohut, in 'Sulfur Dioxide and Vegetation,' ed. W. E. Winner, H. A. Mooney, and R. Goldstein, Stanford University Press, 1985.
[50] P. R. Hughes, J. E. Potter, and L. H. Weinstein, *Ent. Soc. Am.*, 1982, **11**, 173.
[51] G. P. Dohmen, S. McNeill, and J. N. B. Bell, *Nature*, 1984, **307**, 52.

Warrington[52] produced a dose-response curve showing how precisely the growth rate of aphids can be related to SO_2 concentration (Figure 8). There was a linear increase from zero up to about 110 ppb SO_2, after which there was a steep decline. This was thought to be the concentration at which the SO_2 began to be directly toxic to the aphids.

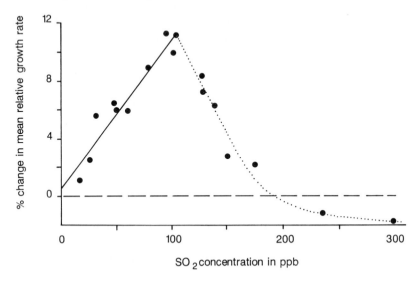

Figure 8 *Effects of atmospheric SO_2 concentration on the mean relative growth rate of pea aphids (Acyrthosiphon pisum). The aphids fed for 4 days on garden peas growing in the various SO_2 concentrations, to which the plants had also been exposed for about 20 days prior to the placement of the aphids. Percentage changes compared with aphids on plants in clean air are shown*
From Warrington, and used with the permission of Applied Science Publishers Ltd.

When plants are exposed to pollutants such as SO_2 and NO_2, they show increases in nutrients that are significant in the diet of aphids, *e.g.*, free amino acids.[53,54] It seems likely that this is the main cause of the increases in the growth rate of the aphids, increases which may occur when the plants are themselves showing negative growth responses to the pollutants. There are thus likely to be greater final impacts on growth when exposure to pollutants is accompanied by aphid infestation. The economic consequences cannot, however, be determined from our present knowledge.

The possibility that fungal pathogens may sometimes show increased activity on polluted plants has recently received attention because in areas of Central Europe where pollution levels are high, some cereal pathogens have recently

[52] S. Warrington, *Environ. Pollut.*, 1987, **43**, 155.
[53] S. Braun and W. Flückiger, *Environ. Pollut. Ser. A.*, 1985, **39**, 183.
[54] S. McNeill, J. N. B. Bell, M. Amino-Kano, and P. Mansfield, in 'How are the Effects of Air Pollutants on Agricultural Crops Influenced by Interactions with Other Limiting Factors?' report of Working Party III, Concerted Action on 'Effects of Air Pollution on Terrestrial and Aquatic Ecosystems,' p. 108, CEC, Brussels.

caused epidemics of economic significance. It has been speculated that pollutants have been responsible as a result of disturbed physiology of the plants but there is no clear evidence yet to support this.[55] At one time it was thought that the major effect of air pollution was an inhibitory one via the effect of SO_2 on fungal pathogens, *e.g.*, on *Diplocarpon rosae* causing blackspot of roses,[56] but it seems that there is a need for a reassessment of the situation now that the balance of pollutants has changed.

4.2 Abiotic or Environmental Factors

4.2.1 Low Temperatures. There is no simple relationship between temperature and individual physiological processes in plants. Different species are adapted to different environments and within a species those from different provenances can display distinctive behaviour. There are two main ways in which plants are normally damaged by low temperatures, *viz.* 'chilling' injury caused by low temperatures above 0 °C, and freezing injury. Some plants from warm climates cannot tolerate either chilling or freezing and so their geographical distribution is limited to areas where these stresses do not occur. Plants native to temperate regions can also be severely damaged or killed by chilling or freezing if they are not correctly acclimatized.[57] The process of acclimatization is usually called 'hardening' and it is achieved mainly by a period of prior exposure to temperature slightly above the threshold for injury. In some plants, woody perennials for example, hardening may also require or be enhanced by decreasing photoperiods in the autumn and it is therefore part of the dormancy cycle.

There has been an accumulation of circumstantial evidence for many years that air pollutants reduce the frost resistance of woody plants in the Nordic countries,[58] and recently some experimental evidence has begun to appear. Keller[15] was one of the first to perform controlled experiments under winter conditions and he found that exposure of dormant beech seedlings to SO_2 in winter led to increased concentrations of sulphur in the leaves that expanded in spring and death of some of the terminal buds of the type associated with frost injury. Keller[59] also found that Norway spruce suffered frost injury in late winter if exposed to SO_2 from October to April.

SO_2 can also enhance frost injury in herbaceous plants. Davison and Bailey[60] found that the ability of perennial ryegrass to undergo hardening against damage by sub-zero temperatures was reduced by prior exposure to 87 ppb SO_2

[55] H. Fehrmann, A. Von Tiedemann, L. W. Black, B. Glashagen, and T. Eisenman, in 'How are the Effects of Air Pollutants on Agricultural Crops Influenced by Interactions with Other Limiting Factors?' report of Working Party III, Concerted Action on 'Effects of Air Pollution on Terrestrial and Aquatic Ecosystems,' p. 98, CEC, Brussels.

[56] P. J. W. Saunders, *Ann. Appl. Biol.*, 1966, **58**, 103.

[57] J. Levitt, 'Responses of Plants to Environmental Stresses, Vol. 1, Chilling, Freezing and High Temperature Stresses,' Academic Press, New York, 1980.

[58] S. Huttunen, in 'Air Pollution and Plant Life,' ed. M. Treshow, John Wiley, Chichester, 1984, p. 321.

[59] T. Keller, *Gartenbauwissenschaft*, 1981, **46**, 170.

[60] A. W. Davison and I. F. Bailey, *Nature*, 1982, **297**, 400.

for 5 weeks, and Baker *et al.*[61] exposed winter wheat to SO_2 in the field and noticed that a natural frost ($-9\,°C$) in mid-January caused increased injury to the fumigated plants. Freer-Smith and Mansfield[62] found that there were small but consistent increases in frost injury on the needles of Sitka spruce cooled to -5 and $-10\,°C$ after exposure to 30 ppb SO_2 and 30 ppb NO_2 followed by a period of hardening in clean air.

The concentrations of SO_2 and NO_x in many locations where there is marked forest decline are very low throughout the year, but summertime O_3 concentrations can be high. There have been relatively few studies to determine whether O_3 can predispose plants to frost damage.

Although atmospheric pollution has been implicated as a likely contributing factor to forest declines in Europe and N. America, as yet no clear consensus regarding the primary causes responsible has emerged. Of the many recent hypotheses put forward as an explanation for tree injury many involve ozone as a major factor, either acting alone or in combination with other pollutants or environmental stresses.[63]

The concentrations of SO_2 and NO_x in many locations where there is marked forest decline are very low throughout the year and it is thought unlikely that these pollutants acting alone are responsible. The involvement of O_3 is based on evidence[64] that in some of the affected areas, atmospheric O_3 concentrations in the summer have been found to be in excess of 100 µg m^{-3}. This is especially true for high elevation sites (>800 m) where, due to meteorological and topographic factors, these areas experience far fewer seasonal and diurnal fluctuations in atmospheric O_3 concentrations than at lower elevations.[3,65] In addition, winter stresses, such as frosts, will also occur with a greater frequency at higher altitudes. Controlled fumigations with O_3 alone or in combination with acid mists and either at or in excess of 100 µg m^{-3} have failed to produce the symptoms of chlorosis or needle loss characteristic of the decline which suggests that a complex combination of physical and chemical stresses may be involved.[37,66]

Until recently there have been no studies on the effects of ozone in relation to frost injury, presumably because this pollutant is thought of predominantly as a summer phenomenon. Evidence is now accumulating, however, from controlled experiments suggesting that exposure to O_3 during the summer months may enhance the susceptibility of conifers to frost injury in the following autumn.[66,67,68] Although a clear effect of the photooxidant was observed in these experiments, the basic mechanism of acclimatization was not disrupted,

[61] C. K. Baker, M. H. Unsworth, and P. Greenwood, *Nature*, 1982, **299**, 149.

[62] P. H. Freer-Smith and T. A. Mansfield, *New Phytol.*, 1987, **106**, 237.

[63] S. B. McLaughlin, *J. Air Pollut. Contr. Ass.*, 1985, **35**, 512.

[64] B. Prinz, G. H. Krause, and H. Stratman, 'Forest Damage in the Federal Republic of Germany,' Land Institute of Pollution Control of the Land North-Rhine Westphalia, Wessen, West Germany, 1982.

[65] United Kingdom Photochemical Oxidant Review Group Interim Report, 'Ozone in the United Kingdom,' Department of the Environment, London, 1987.

[66] P. W. Lucas, D. A. Cottam, L. J. Sheppard, and B. J. Francis, *New Phytol.*, 1988, **108**, 495.

[67] K. A. Brown, T. M. Roberts, and L. W. Blank, *New Phytol.*, 1987, **105**, 149.

[68] J. D. Barnes and A. W. Davison, *New Phytol.*, 1988, **108**, 159.

because irrespective of treatment or species (Norway spruce and Sitka spruce) the majority of tree seedlings later developed full frost hardiness to temperatures below $-20\,^\circ\text{C}$.

A full understanding of the mechanisms responsible for acclimatization has yet to be established. It is generally agreed, however, that during the autumn perennial plants of temperate and Arctic latitudes acclimatize to low temperature by adopting one or a combination of strategies, such as the accumulation of soluble solutes (proline, arginine, and simple carbohydrates), changes in cell metabolism away from glycolysis towards the pentose phosphate pathway[69] or modifications to the cell membranes which involve changes in phospholipids, sterols, and glycolipids.[70] The way in which ozone or other pollutants may affect these processes is at present unknown but the hypothesis that alteration of membrane properties is involved appears to be plausible in view of their important physiological role during cold hardening. The recent findings of Mehlhorn and Wellburn with regard to the likely vulnerability of plasmalemma membranes to free radicals are relevant here (Section 3).

4.2.2 Water Deficits. After temperature, water availability is the main environmental factor governing the distribution of plants. Evolution has provided them with several features and mechanisms for the acquisition of water and to improve the economy of its usage. Some effects of air pollutants on plants have been identified which appear likely to disturb a plant's water economy.

Two pollutants, SO_2 and O_3, and mixtures of SO_2 and NO_2 have often been found to cause a change in the allocation of material between roots and shoots. There is reduced translocation of newly manufactured photosynthates from the leaves to the roots and the eventual effect on the plant can be a considerable increase in the shoot:root dry mass ratio. This change in the balance of growth could have important implications for plants growing in drying soil, where the increased penetration of roots is important for the maintenance of water supplies for the shoot.

Although the effects of pollutants on dry mass allocation have been well defined [reviewed by Mansfield *et al.*[71]], there has been little exploration so far of their significance for the water relations of plants. Heggestad *et al.*[72] found that soil moisture stress enhanced the deleterious effects of O_3 on yields of soybeans so long as the concentrations of O_3 were fairly low (below 80 ppb for 7 hours a day). At higher concentrations of O_3 opposite effects were found, *i.e.*, there was a reduced impact of water shortage on yield. At higher concentrations, O_3 is known to induce closure of stomata and this may have had the dominant effect by reducing water loss from the leaves.

Effects of pollutants on stomatal behaviour have also attracted a lot of attention. The opening of stomata ideally achieves an acceptable compromise

[69] C. J. Andrews and M. K. Pomeroy, *Plant Physiol.*, 1979, **64**, 120.

[70] B. Yoshida and M. Uemura, *Plant Physiol.*, 1984, **75**, 31.

[71] T. A. Mansfield, P. W. Lucas, and E. A. Wright, in 'Air Pollution and Ecosystems,' ed. P. Mathy, Reidel, Dordrecht, 1988, p. 123.

[72] H. E. Heggestad, T. J. Gish, E. H. Lee, J. H. Bennett, and L. W. Douglas, *Phytopathology*, 1985, **75**, 472.

between the plant's need to acquire atmospheric CO_2 for photosynthesis and the loss of water by transpiration. Any alteration in stomatal functioning can be serious because of interference with carbon gain or water usage.

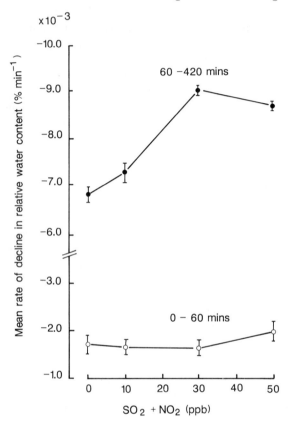

Figure 9   *The effects of exposure to SO_2 and NO_2 at the concentrations shown on the rate of water loss from detached leaves of* Phleum pratense. *Nine leaves were used for each treatment and the rates of water loss were estimated from the equation:*

$$\log R_t = \log R_o - kt$$

where R represents percentage relative water content, t is time in minutes, and k is a rate constant. The equation was fitted separately to two curves for each treatment over the period 0–60 mins and 60–420 mins

There is a considerable amount of evidence in the literature that SO_2, and mixtures of SO_2 and NO_2, can sometimes cause abnormal stomatal opening [see reviews by Mansfield and Freer-Smith and by Black[73,74]]. This is probably the result of damage to the cells surrounding the stomatal guard cells. The latter are

[73] T. A. Mansfield and P. H. Freer-Smith, In 'Gaseous Air Pollutants and Plant Metabolism,' ed. M. J. Koziol and F. R. Whatley, Butterworth Scientific, London, 1984, p. 131.
[74] V. J. Black, in 'Sulfur Dioxide and Vegetation,' eds. W. E. Winner, H. A. Mooney, and R. A. Goldstein, Stanford University Press, Stanford, California, 1985, p. 96.
[75] G. R. Squire and T. A. Mansfield, *New Phytol.*, 1972, **71**, 1033.

known to be more resilient than the other epidermal cells to adverse treatments.[75] The stomatal pore is opened as a result of turgor pressure generated in the guard cells, which push against the resistance offered by the turgor of surrounding cells. If the surrounding cells lose turgor then greater stomatal aperture occurs and there may not be enough pressure exerted on the guard cells to bring about closure of the pore. Under conditions of severe water stress the stomata are required to close quickly to protect the leaf. The simplest way to look for changes in the ability of the stomata to achieve this is to detach leaves and weigh them at intervals. If this is done with leaves previously exposed to SO_2 and NO_2 pollution, clear differences from those in clean air are revealed (Figure 9).

The full implications of changes in patterns of root growth and in the ability of stomata to close quickly in leaves under water stress, will have to be evaluated during the next phase of research into this important subject. Very small changes in these factors in plants in the field may assume great importance during occasional periods of extreme stress, for example, when there is an occasional summer drought in North West Europe.

Acknowlegement. P. W. Lucas thanks the Department of the Environment for their financial support.

CHAPTER 13

The Health Effects of Environmental Chemicals

H. A. WALDRON

1 INTRODUCTION

There has been, and continues to be, a good deal of public concern about the effects which the release of chemicals into the environment may have on health. Implicit in this concern, and frequently explicit, is the notion that any effects will necessarily be adverse although it is true to say that human life is not possible in the absence of chemicals, both organic and inorganic. Many elements are essential for normal biological function whether in macro or trace quantities, and very many organic molecules are also required by the human organism in order that it may go about its day to day business in what is usually regarded as a healthy way.

The inorganic chemicals which are biologically necessary are only available during adult life from food or drink and thus ultimately, from chemicals which are free in the environment. One of the characteristics of the essential elements is that they must be kept within certain rather narrow ranges—so-called concentration windows—otherwise signs or symptoms of excess or deficiency may result. We have, therefore, evolved special mechanisms whereby these concentration windows are generally maintained. Many of these mechanisms operate by limiting uptake from the gut which is the route through which chemicals are naturally presented and absorbed. One important change which has resulted from industrialization is that the concentration of many chemicals in the atmosphere has increased very considerably due to their release in factory chimney emissions. If these chemicals are in a form which is soluble in body fluids and if they are gaseous or their particle size is sufficiently small to allow them to penetrate into the deepest parts of the lung, then they are likely to be

261

absorbed since there are no mechanisms in the alveoli which will hinder this process to any substantial degree.

Two further factors need also be kept in mind when considering the effects of the release of chemicals into the environment. The first is, that despite the furore which is presently engaging the press, the contamination of food and drink by chemical (and biological) agents is considerably less than it was certainly until the first half of this century. During the nineteenth century the adulteration of food was a national scandal and must have resulted in a very great deal of morbidity. Thus, although the use of chemicals has increased to an enormous degree, it is likely that overall, exposure to many potentially toxic materials has actually decreased because the amounts in the diet have decreased. This is particularly the case with many of the toxic metals; I shall return to this point later in the chapter.

The second factor is that of dose. The concept that the toxicity of a material resides in the quantity to which an individual is exposed is generally attributed to Paracelsus:

> 'All substances are poisons [he wrote]; there is none which is not a poison. The right dose differentiates a poison and a remedy.'

Thus, the presence of a chemical or chemicals in the environment should not be taken as an indication *per se* that it is harmful to health; it may be in too low a concentration to pose any real threat to health and even if there is a small theoretical risk, this has to offset against any benefits which may accrue from the use of the material. Clearly if there are no benefits from its use, its use should be abandoned.

Different types of exposure to environmental chemicals (and hence the dose which is experienced) may be distinguished. There is catastrophic exposure which results from the massive release of material into the environment such as occurred at Seveso or Bhopal, for example. Endemic exposure may be said to occur when very large numbers of people are exposed, usually as the result of misuse of chemicals; the outbreaks of mercury poisoning in Iraq and elsewhere as the result of the use of seed grain treated with organic mercurials to make flour are examples of this kind of exposure. Finally, there is what might be called concomitant exposure which is that which is inescapable in a society which depends upon the use of chemicals to maintain its way of life.

2 CATASTROPHIC EXPOSURE

Some interesting comparisons can be drawn between the two best known examples of catastrophic exposure to chemicals, Seveso and Bhopal. On 10 July, 1976 a massive release of 2,3,7,8-tetrachlorodibenzo-*p*-dioxin occurred from a chemical plant in the vicinity of Milan in northern Italy which was manufacturing 2,4,5-trichlorophenol (2,4,5-TCP). This is only one of the many instances in which dioxin has been released from chemical plants when the exothermic reaction producing 2,4,5-TCP has got out of control. At Seveso a safety disc in a reaction vessel ruptured and a plume of chemicals containing 2,4,5-TCP blew 30 to 50 m above the factory. As it cooled the material in it was deposited over a cone shaped area down wind from the factory about 2 km long and 700 m wide. In all an area of 3–4 km^2 was contaminated and an estimated 3 to 16 kg of dioxin was released.[1] There were almost 28 000 people living in the vicinity of the factory. Those who lived in the immediate area downwind were evacuated 14 days after the explosion and the area was closed off. About 5 000 people in the most heavily contaminated area were allowed to stay in their homes but they were not allowed to cultivate or consume local vegetables or fruit nor to raise or keep poultry or other animals.

Dioxin is both extremely toxic and extremely stable and it is known to have teratogenic, fetotoxic, and porphyrinogenic effects, all of which are well documented. It is not used commercially but is found as a contaminant when 2,4,5-TCP is synthesized by the hydrolysis of tetrachlorobenzene at high temperatures. 2,4,5-TCP itself is used to make 2,4,5-T and 2,4-D which are used as herbicides; dioxin is often present in trace amounts in these compounds. The population at Seveso was screened shortly after the accident happened and a number of positive findings were noted. A few months after the accident, 176 individuals, mostly children, were found to have chloracne, 50 of whom came from the most contaminated area. This constituted about 7% of the population estimated to be at risk. A further round of medical screening in February, 1977 revealed a further 137 cases of chloracne but subsequent follow-ups showed that the incidence had decreased and that the individuals already affected had improved.

In addition to the chloracne, some neurological abnormalities were noted. These included polyneuropathy with some symptoms which were due to effects on the central nervous system. These observations were more common amongst the people who lived in the most heavily contaminated zone and the incidence of abnormal nerve conduction tests was significantly increased in those who had chloracne.[2]

Finally, there was evidence of liver enlargement in about 8% of the population and again this was most noticeable amongst the most heavily exposed individuals. Measurement of liver enzyme activity showed some abnormalities which had returned to normal about a year after the explosion. It is interesting and noteworthy that there was no evidence that the immune system had been affected, and there has been no evidence of chromosomal abnormalities or of any damage to the foetus. There were no deaths.[3,4]

The explosion at Seveso excited a great deal of public alarm particularly because dioxin was involved but the major harmful effects were directed towards the environment; many farm animals died and the site became a wasteland of dying plants and deserted homes. At Bhopal, the effects were directed almost entirely on the population living around the factory which was involved in the catastrophic release of methyl isocyanate (MIC).

[1] A. Giovanardi, in 'Proceedings of the Expert Meeting on the Problems Raised by TCDD Pollution,' Milan, 30 September and 1 October, 1976, p 49.
[2] G. Fillipini, B. Bordo, P. Crenna, N. Massetto, M. Musicco, and R. Boeri, *Scand. J. Work Environ. Health*, 1981, **7**, 257.
[3] F. Pocchiari, V. Silano, and A. Zampieri, *Ann. N.Y. Acad. Sci.*, 1979, **320**, 311.
[4] G. Reggiani, *J. Toxicol. Environ. Health*, 1980, **6**, 27.

The accident at Bhopal occurred on 3 December, 1984 at the Union Carbide factory which had been producing the insecticide carbaryl for about eighteen years; MIC was one of the main ingredients. The MIC itself was produced from monomethylamine (MMA) and phosgene, phosgene being produced on site by reacting chlorine and carbon monoxide. The MMA and chlorine were brought by tanker from other plants in India, stored and used when required; chloroform was used as a solvent throughout the process. There were in this plant, then, a variety of extremely toxic materials in use. On the night of the accident it seems that some water inadvertently got into a tank where 41 tonnes of MIC were being stored causing a runaway chemical reaction. The heat of the reaction, possibly augmented by reactions with other materials present in the tank as contaminants, produced vaporization of such momentum that it could not be contained by the safety systems which may, in any event, have been defective or of inadequate capacity. The safety valve on the tank remained open for about two hours allowing MIC in liquid and vapour form to escape into the surroundings. The prevailing wind carried the cloud towards the north of the plant and then towards the west and affected approximately 100 000 people living in the vicinity. It is ironic that none of the workers on night duty at the plant was harmed. There were at least 2 000 deaths. The most frequent symptoms in those who survived were burning of the eyes, coughing, watering of the eyes, and vomiting.[5]

The fate of the victims of Bhopal has been intensified by the inability of the Indian or the state governments adequately to be able to compensate them; and the great procrastination which seems inevitably to follow the entry of any complicated law case into the American courts.

3 ENDEMIC EXPOSURE

There have been many outbreaks of endemic exposure to toxic chemicals in the environment but a rather small number of chemicals has actually been involved. These have included mercury, hexachlorobenzene, cadmium, TOCP, PCBs and PBBs, dioxin, and, in historical times, lead. I am considering here only exposure to 'pure' chemicals and only exposures which can be considered to be unnatural. In terms of overall morbidity and mortality, probably nothing can surpass the damage which has been caused by exposure to the by-products of the burning of coal. Not only have there been remarkable episodes in which sudden increases in the number of deaths have been noted to follow exceptionally high pollution, for example, in the Meuse Valley in 1930, in Donora in 1948, and in London in 1952, but careful epidemiological studies have shown how closely both morbidity and mortality were related to the atmospheric levels of sulphur dioxide and smoke.[6] It should be noted in passing that there are instances where harmful effects on health can follow upon exposure to naturally occurring high levels of toxic chemicals; the most obvious example is the endemic fluorosis which is found in India, Iran, Taiwan, and in some parts of Texas.

[5] N. Andersson, M. Kerr Muir, V. Mehra, and A. G. Salmon, *Brit. J. Industr. Med.*, 1988, **45**, 469.
[6] World Health Organisation, 'Environmental Health Criteria 8. Sulfur Oxides and Suspended Particulate Matter,' WHO, Geneva, 1979.

3.1 **Mercury**

Mercury in its organic form has accounted for many of the episodes of endemic environmental exposure. Probably the best known of these is Minimata Bay disease.[7] This was first noted at the end of 1953 when an unusual neurological disorder began to affect the villagers who lived on Minimata Bay on the south west coast of Kyushu, the most southerly of the main islands of Japan. It was commonly referred to as *kibyo*, that is, the mystery illness. Both sexes and all ages were affected and those who were affected presented with a mixture of signs relating to the peripheral and central nervous systems. The prognosis of the condition was poor, many patients became disabled and bedridden and about 40% died. The disorder was associated with the consumption of fish caught in the bay but mercury poisoning was not considered in the early investigations of the condition because many of what were then considered the classic symptoms of mercury poisoning were not present and it was only when it was realized that the signs of the disease were similar to those described in a man who had died after being poisoned during the manufacture of alkyl mercury fungicides that the possibility was raised.

The source of the mercury was effluent released into the bay from a factory which was manufacturing vinyl chloride using mercuric chloride as a catalyst. It is claimed that *inorganic* mercury was released by the factory and that this was methylated by micro-organisms living in the sediments in the bay. However, the rate of conversion is extremely slow, much too slow to have accounted for the large amounts of methyl mercury which are calculated to have accumulated in the waters of the bay. It is therefore much more likely that the mercury was actually released in the organic form; at the time there were no regulations forbidding this in Japan.

About 700 people were affected at Minimata Bay and there was a second outbreak of methyl mercury poisoning in Japan in 1964 in Niigata following pollution of the Agano River by industrial effluent. Some 500 individuals were affected in this outbreak. It should be noted that people who depend upon fish as the staple part of their diet may still be at risk of excessive exposure to mercury, even if not on the scale experienced at Minimata and Niigata. For example, blood methyl mercury levels have been found to be almost ten times higher in a Peruvian population who ate on average 10.1 kg of fish per family of (on average) 6.2 persons compared with a control population whose fish intake was considerably more modest. (The mean in the high fish eating population was 82 ng ml^{-1} and in the control population, 9.9 ng ml^{-1}.) Moreover, 29.5% of the heavily exposed population had signs of a sensory neuropathy.[8] In a fish eating population in New Guinea, hair mercury concentrations were between two and three times that of a control group (6.4 µg g^{-1} compared with 2.4 µg g^{-1}); there were no demonstrable ill effects in this group.[9]

[7] M. Katsuna, 'Minimata Disease,' Kumamoto University, Japan, 1968.
[8] M. D. Turner, D. O. Marsh, J. C. Smith, J. B. Inglis, T. W. Clarkson, C. E. Rubio, J. Chiriboga, and C. C. Chiriboga, *Arch. Environ. Health*, 1980, **35**, 367.
[9] J. H. Kyle and N. Ghani, *Arch. Environ. Health*, 1982, **37**, 266.

Large scale outbreaks of organic mercury poisoning have also occurred from eating bread made from seed grain which had been treated with mercurial fungicides. Outbreaks have been reported in Pakistan and Guatemala but the most severe have been in Iraq. The largest episode occurred in 1971–72 when the Iraqi government imported a large consignment of treated seed grain which was then distributed to the largely illiterate rural population. The distribution was accompanied by warnings that the seed was for sowing not for eating and the sacks were marked with warning labels—in English or Spanish! The seed had been treated with a red dye to distinguish it from edible grain but the farmers found that they could remove the dye by washing and equated this with the removal of the poison. The grain was first used to make bread in November, 1971 and the first cases of poisoning appeared in December. By the end of March, 1972 there had been 6 530 admissions to hospital and 459 (7%) of these had died.[10] These represented only the most severe cases and the true extent of this outbreak will probably never be known although it has been suggested that the incidence of the disease may have been as high as 73 per 10^3.[11]

3.2 Hexachlorobenzene

An episode of hexachlorobenzene poisoning occurred in Turkey in 1956 which was again due to the ingestion of treated seed grain. Several surveys of the affected population found about 3 000 individuals with porphyria cutanea tarda.[12] New cases continued to appear until 1961 although hexachlorobenzene was withdrawn from the market in 1959. There was a 10% mortality but mortality was as high as 95% amongst infants born to mothers who had ingested the compound.

3.3 Cadmium

Environmental exposure to cadmium was considered to be the cause of a disease which was first reported in 1955 as occurring in a localized area downstream from a mine on the Juntsu River in the Toyama Prefecture in Japan. It was called *itai-itai* disease meaning 'it hurts'. The condition was almost entirely confined to elderly women who had borne several children and was characterized by severe bone pain, waddling gait, severe osteomalacia, pathological fractures, and some signs of renal impairment. The water which was used to irrigate crops was frequently contaminated by outpourings from the mine which contained zinc, lead, and cadmium. Levels of cadmium in rice samples were shown to contain about ten times the amount normally present and the view was gradually formed that it was the cadmium which was responsible for the

---

[10] F. Bakir, S. F. Damluji, L. Amin-Zaki, M. Murtadha, A. Khalidi, N. Y. al-Rawi, S. Tikriti, H. I. Dhahir, T. W. Clarkson, J. C. Smithy, and R. A. Doherty, *Science*, 1973, **181**, 230.
[11] G. Kazantzis, A. W. al-Mufti, A. al-Jawad, Y. al-Shawani, M. A. Majid, R. M. Mahmoud, M. Soufi, K. Tawfiq, M. A. Ibrahim, and H. Dabagh, *Bull. WHO*, 1976, **53**, Suppl 37.
[12] R. Schmid, *New Eng. J. Med.*, 1960, **263**, 297.

disease.[13] There are some dissenting voices raised against this proposition, however. The disease has been found in women who were certainly not exposed to cadmium and it tended to appear in those women who, for religious reasons, covered themselves and kept themselves indoors and out of the sun. It seems likely that deficiencies of calcium and vitamin D were at least partially to blame and that cadmium may have been acting only as one factor in what was a multi-factorial aetiology for this disease.[14]

3.4 Triorthocresyl Phosphate (TOCP)

Endemic TOCP poisoning has resulted from the accidental or deliberate adulteration of food and drink with this compound. The first recorded cases were described in the United States during prohibition. During this period, all sorts of curious drinks were imported to meet the apparently insatiable demand for alcohol. One of these drinks was made from an extract of Jamaican ginger and some of the brew was fortified with TOCP. Large numbers of cases of paralysis followed and were colloquially known as 'ginger paralysis' or 'jake leg.' The US Bureau of Prohibition estimated that there were 20 000 cases of paralysis in 1930 as the result of drinking contaminated Jamaican ginger. Some of the victims recovered fairly promptly but others remained affected for life.[15]

Another massive outbreak of TOCP poisoning occurred in Morocco in 1959. A mixture of olive oil and aircraft lubricating oil was sold as edible oil and about 10 000 people were involved in the epidemic which followed.[16] The perpetrators of this deliberate fraud were subsequently tried and sentenced to death. Other episodes of TOCP poisoning have arisen from the accidental contamination of food during shipping or storage[17] or from the substitution of TOCP for cooking oil.[18]

3.5 Toxic Oil Syndrome

Cooking oil was at the heart of a more recent endemic of poisoning in Spain, although in this case TOCP was not to blame and, indeed, the toxic agent has not been identified satisfactorily to this day. The syndrome manifested itself in May, 1981 in an eight year old boy who died with acute respiratory insufficiency. He was one of a family of eight living in Madrid of whom six eventually became ill. By June, 2 000 patients with the condition had been admitted to hospitals in Madrid and a further 600 to hospitals in the provinces. By the end of August, 13 000 people had been treated in hospital and 100 had died. The final toll was about 20 000 persons affected with 400 deaths, a case fatality rate of 2%.

[13] L. Friberg, T. Kjellstrom, G. Nordberg, and M. Piscator, 'Cadmium in the Environment, III,' US Environmental Protection Agency, Washington, 1975.
[14] K. Tsuchiya, *Fed. Proc.*, 1976, **35**, 2412.
[15] W. J. Hayes, 'Pesticides Studied in Man,' Williams and Wilkins, Baltimore, 1982, p. 320.
[16] H. Geoffroy, A. Slomic, M. Benebadji, and P. Pascal, *World Neurol.*, 1960, **1**, 244.
[17] M. Sorokin, *Med. J. Aust.*, 1969, **1**, 506.
[18] A. W. Jordi, *J. Aviation Med.*, 1952, **23**, 623.

The illness began with a fever which was followed by severe respiratory symptoms and a variety of skin rashes which led some of the victims to be diagnosed as having measles or german measles. Many of the patients developed signs of cerebral oedema and many had cardiological abnormalities.

The cause of the disease was traced to adulterated cooking oil which was fraudulently sold to the public as pure olive oil. The oil was sold by door to door salesmen in five litre plastic bottles with no labels; since olive oil is an expensive commodity in Spain it was the poorer families in the working class suburbs of Madrid who were visited by the salesmen and it was they who almost entirely bore the brunt of the disease. The composition of the oil varied but rapeseed oil accounted for up to 90%; there were varying amounts of soya oil, castor-oil, olive oil, and animal fats. The oil also contained between 1 and 50 ppm of aniline and between 1 500 and 2 000 ppm of acetanilide. It is illegal to import edible rapeseed oil into Spain, only oil which has been denatured with 2% aniline may be admitted.

It seems that those who perpetrated the fraud tried to refine out the aniline and in doing so produced a number of chemical species, one of which was acetanilide which had reacted with fatty acids in the oil to produce oleoanilide which was originally presumed to be the toxic agent.[19] Later work, however, showed the presence of a number of other anilides, the main component of which was a fatty acid diester of 1,2-propanediol-3-aminophenyl. The other aniline compounds were considered to be hydrolysis products of diacyl 1,2-propanediol-3-aminophenyl or positional isomers. It has been suggested that these compounds may have been responsible for the symptoms produced by the toxic oil.[20]

3.6 Polychlorinated Biphenyls (PCBs)

Episodes of human poisoning with PCBs occurred in Japan (which seems to have had rather more than its share of environmental disasters) and in Taiwan. The disease first made its appearance in Japan in 1968 in the western part, when a number of families were noted to have developed chloracne, the skin condition which affected the victims at Seveso. Chloracne is more severe than the type which occurs in adolescents and it has a rather different distribution on the body. Characteristically, cysts containing a straw coloured liquid are found in chloracne but not in the teenage variety. Epidemiological studies brought other cases to light and it was found that the factor which the cases had in common was exposure to a particular batch of one brand of rice oil. Chemical analysis showed that the oil was contaminated with PCBs.

The PCBs were shown to have leaked into the oil from equipment which had been used to process the oil, the PCBs having been used as a heat-transfer fluid as was commonly the case in industry. By the end of 1977, 1 665 individuals were considered to have met the diagnostic criteria for what has come to be known as Yusho disease.[21]

[19] J. M. Tabuenca, *Lancet*, 1981, **2**, 567.
[20] A. V. Roncero, C. J. del Valle, R. M. Duran, and E. G. Constante, *Lancet*, 1983, **2**, 1025.
[21] H. Urabe, K. Koda, and M. Asahi, *Ann. N.Y. Acad. Sci.*, 1979, **320**, 273.

In addition to chloracne, the patients with Yusho disease had a number of systemic complaints, including loss of appetite, lassitude, nausea and vomiting, weakness, and loss of sensation in the extremities. Some also had hyperpigmentation of the face and nails.

The patients were followed up from 1969 to 1975 and in 64% of cases, the skin lesions improved. A number of non-specific symptoms persisted, however, including a feeling of fatigue, headache, abdominal pain, cough with sputum, numbness and pain in the extremities, and in the females, changes in menstruation. Objective findings included a sensory neuropathy, retarded growth in children, and abnormal development of the teeth. Children who had been exposed *in utero* were small for dates and were hyperpigmented.

Some patients were found to be anaemic and some had other abnormalities but the most striking observation was a marked increase in serum triglyceride levels. The mean in the patients was 134 ± 60 mg/100 ml compared with a mean of 74 ± 29 mg/100 ml in normal controls. When the PCB concentrations in the serum were measured by gas chromatography it was found that the patients with Yusho had an isomeric pattern which was different from that seen in controls whose only exposure had been from the general environment.

Yusho disease appeared in Taiwan in the spring and summer of 1979 in two prefectures in the middle part of the country.[22] The signs and symptoms were indistinguishable from the Japanese outbreak and by the end of 1980, more than 1 800 people had been affected. The source of the PCBs was again contaminated rice oil. The blood of those with the disease ranged from 54 to 135 ppb. Later studies showed that polychlorinated dibenzofurans (PCDFs) and polychlorinated quaterphenyls (PCQs) were also present in the blood.[23]

To what extent the symptoms seen in the patients with Yusho disease were *entirely* due to PCBs is difficult to say since it is known that they were also exposed to PCDFs and to PCQs. These compounds are formed when PCBs are heated. In animal models PCDFs and PCQs are more toxic than PCBs and it seems probable that there may have been some synergistic effects between the different compounds in the oil. The fact that the PCB isomers in the oil were different from those in the general environment may also be of importance since they may have been more toxic than those to which the population at large was exposed.

3.7 Polybrominated Biphenyl (PBB)

PBB is used mainly in plastics as a fire retardant but in May and June of 1973 some ten to twenty bags were sent in error to a grain elevator instead of a livestock food additive. The chemical company which made the PBB normally supplied magnesium oxide to go into the cattle feed but both products were packed in the same colour bag and although the PBB was labelled 'Firemaster'

---

[22] P. H. Chen, J. M. Gaw, C. K. Wong, and C. J. Chen, 'Levels and gas chromatographic patterns of Polychlorinated Biphenyls in the blood of patients after PCB poisoning in Taiwan,' *Bull. Environ. Contam. Toxicol.*, 1980, **25**, 325–329.

[23] T. Kashimoto, H. Miyata, S. Kunita, T-C. Tung, S-T. Hsu, K-J. Chang, S-Y. Tang, G. Ohi, J. Nakagawa, and S-I. Yamamoto, *Arch. Environ. Health*, 1981, **36**, 321.

rather than 'Nutrimaster', and although this difference was actually noted by the staff at the grain elevator, it was nevertheless incorporated into the feed and distributed throughout the state to be fed to the unsuspecting cows.

Reports of sick cows began to surface in August, 1973 and towards the end of the year it was realized that the feed was to blame. Despite this, the contamination continued both because there was cross-contamination of otherwise normal feed from the grain elevator and because the tainted feed was resold at a discount after it had been returned to the suppliers. Not until PBB was identified in the feed in May, 1974 was any attempt made to limit the contamination.

From the time that the feed had become contaminated dairy products containing PBB had been sold throughout Michigan and cows and other livestock which had been given it had been slaughtered for meat. A representative sample of 2 000 individuals was surveyed and more than half had a concentration of PBB in excess of 10 parts per billion (ppb) in their body fat.[24] At that time, most members of the general population would have had no detectable level of PBB in their fatty tissues. Farmers and others who consumed produce directly from contaminated farms had the highest levels of PBB.

An initial study of 217 farmers concluded that the exposure had caused no deleterious effects to their health but the study was criticized on the grounds that the control group had also been exposed to PBB.[25] A second study was then carried out in which of over 1 000 farmers were compared with unexposed farmers from Wisconsin. This second study did find some adverse effects; acne, dry skin, hyperpigmentation, and discoloration of the nails were all more common in the exposed group and they also complained more of headaches, nausea, depression, and a number of other non-specific symptoms. Serum levels of hepatic enzymes were higher in the Michigan farmers than their neighbouring controls.[5] Individuals with symptoms were also more likely to have elevated enzyme levels and it was also shown that there were some changes in the immune system[26,27] and that some individuals had enlargement of the liver and a sensory neuropathy.[28] In follow up studies, the PBB levels in the serum were found to decrease but it was interesting that elevated PCB levels were also found and these were actually higher than the PBB levels. There was no relationship between abnormal liver function tests and serum PBB concentrations but there was a slight (and not statistically significant) negative correlation with serum PBB levels and some tests of thyroid function.[29]

It is of interest that in none of these studies did the subjective or the objective findings correlate with serum or fat PBB concentrations. This may have been

[24] I. J. Selikoff, 'A Survey of the General Population of Michigan for Health Effects of PBB Exposure,' Final Report Submitted to the Michigan Department of Public Health, 1979.
[25] H.E.B. Humphrey and N. S. Hayner, 'Polybrominated Biphenyls: An Agricultural Incident and its Consequences. An Epidemiological Investigation of Human Exposure,' Michigan Department of Health, Michigan, 1975.
[26] J. G. Bekesi, J. F. Holland, and H. A. Anderson, 'Lymphocyte function of Michigan dairyfarmers exposed to Polybrominated Biphenyls,' *Science*, 1978, **199**, 1207–1210.
[27] P. J. Landrigan, K. R. Wilcox, and J. Silva, *Ann. N.Y. Acad. Sci.* 1979, **320**, 284.
[28] J. E. Stross, R. K. Nixon, and M. D. Anderson, *Ann. N.Y. Acad. Sci.*, 1979, **320**, 368.
[29] K. Kreiss, C. Roberts, and H. H. Humphrey, *Arch. Environ. Health*, 1982, **37**, 141.

because there was another toxic contaminant present which was acting independently of the PBB, or that the levels of PBB had fallen in the interval between ingestion and the beginning of the studies. It may also be the case that levels of PBB in blood and fat are not good indicators of levels in target organs.

The humans who were exposed in this episode did not suffer severely at the time; the same could not be said for the animals. More than a quarter of a million cows, 1.6 million chickens, and thousands of pigs either died or were slaughtered.

3.8 Dioxin

Endemic exposure to dioxin occurred in Missouri in the early part of 1971 when waste products from a plant which produced hexachlorophene and 2,4,5-T were mixed with waste oils and sprayed on various sites throughout the state for dust control. Dioxin was one of the contaminants in the oil and levels of up to 33 ppm were reported in some areas of the state, but levels in residential areas were generally considerably below this. There were no reported cases of acute poisoning although a child who played in a riding area which was contaminated at the highest level developed haemorrhagic cystitis.[30] Two follow up studies were carried out on people living in a mobile home park where soil dioxin levels were up to 2.2 ppm (the highest level found in a residential area). The first[31,32] reported no adverse findings other than some depression of the immune response on skin testing. This observation was not confirmed by a later study of the same population by Evans *et al.*[33]

3.9 Lead

The outbreaks of endemic lead poisoning in classical and historical times are too well known to go into in great detail here. Suffice it to say, that the adulteration of food and drink with lead has been a significant contributor to morbidity and perhaps to mortality in the past. The most serious form of endemic lead poisoning arose from the habit of adulterating wine with lead to improve a poor vintage and make it more saleable. During the eighteenth century there were a number of famous outbreaks of lead poisoning, the Devonshire colic being perhaps the most well known. In this case, the adulteration of cider with lead arose accidentally due to the presence of lead in the pounds and presses used to make the cider.

The widespread use of lead in cooking utensils, in glazes, and in pewter, together with deliberate adulteration, resulted in a degree of exposure during the eighteenth and nineteenth centuries which was probably 2–3 times greater than

[30] C. D. Carter, R. D. Kimbrough, and J. A. Liddle, *Science*, 1975, **188**, 738.

[31] P. A. Stehr, G. Stein, H. Falk, E. Sampson, S. J. Smith, K. Steinberg, K. Webb, S. Ayres, W. Schramm, H. D. Donnell, and W. B. Gedney, *Arch. Environ. Health*, 1986, **41**, 16.

[32] R. E. Hoffman, P. A. Stehr-Green, and K. B. Webb, *JAMA*, 1986, **255**, 2031.

[33] R. G. Evans, K. B. Webb, A. P. Knutsen, S. T. Roodman, D. W. Roberts, J. R. Bagby, W. A. Garrett, and J. S. Andres, *Arch. Environ. Health*, 1988, 43, 273.

it is today, notwithstanding the increase in air lead concentrations which have resulted from industrialization and the use of lead in petrol.[34] The present generation is probably not much more exposed to lead than were those who were living in Britain before the Roman invasion and this has some relevance when we come to discuss concomitant exposure.

4 CONCOMITANT EXPOSURE

Exposure to chemicals in the environment is an inescapable part of life; this has always been the case and indeed, it is essential for life. When our ancestors followed a hunter–gatherer existence, exposure was to what one might call natural environmental levels and these increased once this way of life was abandoned, more so, when mining and other industrial processes were developed. Two remarkable changes in exposure can be noted subsequently; the first followed the industrial revolution, particularly during its consolidation in the nineteenth century. The quantity of chemicals which were thrown into the environment increased enormously and there was a rapid worsening in pollution; that due to the burning of fossil fuels was especially awful. The nineteenth century was a time of the most widespread pollution and there can scarcely have been any food which was not adulterated to some degree; and until Chadwick managed to find the means by which to separate sewage from drinking water towards the end of the nineteenth century, that which was not contaminated by chemicals was likely to present a bacteriological or viral hazard.

Despite the frequent claims to the contrary and despite present scares about *Salmonella* and *Listeria*, our food and drink are less adulterated and safer to eat than they have probably ever been. The remarkable change which has been brought about in the last half century is the increase in the numbers of different chemicals and in the amounts which are made; the production of organic chemicals has increased during that time, for example, by well over two orders of magnitude. Since there is no means of preventing the escape of chemicals into the surroundings in which they are made and used, there can be no doubt at all that we are exposed to a much richer variety of chemicals than ever before.

The fact that chemicals can be detected in all compartments of the environment, albeit sometimes in trace amounts, has engendered a great deal of both informed and uninformed concern.

Many chemical substances have been held culpable for the major or minor ills to which we are prone; pesticides and herbicides, preservatives, food additives, fertilizers, and several metals including mercury, cadmium, lead, and, most recently, aluminium all have (or have had) their detractors. Certainly there is evidence to suggest that some diseases are likely to have made their appearance following some environmental change; for example, rheumatoid arthritis and Parkinson's disease both seem to be 'modern' diseases in the sense that there are no reliable descriptions of them before the nineteenth century. What the agent is which may have caused these diseases to appear is not at all clear although in the

---

[34] T. Waldron, in 'Diet and Crafts in Towns,' ed. D. Serjeantson and T. Waldron, BAR, Oxford, 1989, p 55.

case of Parkinson's disease, interest in the role of environmental chemicals has been greatly stimulated by the production of the disease by MPTP in some drug addicts. This compound was inadvertently generated during the home production of opiate derivatives and it rapidly induced lesions in the brain identical with those seen in patients with the naturally occurring form of the disease; it can also produce these lesions in animal models. There is, thus, every justification for regarding the presence of chemicals in the environment with some circumspection. Our attitudes, however, must be guided by hard evidence rather than plausible speculation. In my view, the alarm which has been engendered about the presence of lead in petrol falls into the latter rather than the former category.

The notion that lead in petrol is harmful to the health of the general population, and of the health of children in particular, seems to have arisen from the knowledge that tetra-ethyl lead (TEL) is extremely toxic. TEL is one of the principal anti-knock agents added to petrol to raise the octane rating. There is no dispute about its toxicity and the most stringent precautions are taken by those who are required to handle it and on this account there have been very few cases of industrial TEL poisoning. The supposition therefore arose that organo-lead compounds must therefore be harmful to the general population since lead is emitted in the exhaust gases of petrol engines, although not in the organic form.

It was soon noted that the concentration of lead in the air of cities could be correlated with the density of motor vehicles and no one would disagree with the proposition that most of the lead in the air of modern cities comes from this source; typically air lead concentrations in cities are of the order of $1-2 \mu g \, m^{-3}$. The next link in the chain of reasoning is to suppose that this represents an *increase* in exposure, and that this would be reflected in higher blood lead concentrations in those who were exposed to this source of environmental lead. Finally, since lead may cause adverse effects on the central nervous system, it may be supposed that children with elevated blood lead concentrations (derived from lead in petrol) would show some psychological abnormalities compared with those with lower concentrations.

The notion that lead in petrol is actually damaging the central nervous system of children is now so firmly established in the public mind that it is not likely to be dislodged even though the evidence to support most of the arguments in the chain of reasoning is tenuous.

It would be expected that children living in areas of high traffic density would have higher blood lead concentrations than those living in the country but this is not the case. Thus Elwood et al.[35] found that blood lead levels in Ebbw Vale were similar to those in the island of Tory, where there is no traffic, and those on the island of Sark where motor cars are prohibited were the same as those on Jersey.

There has also been a general trend downwards in blood lead concentrations as shown by the results of the two EEC blood surveys conducted in the UK in

[35] P. C. Ellwood, R. Blaney, R. C. Robb, A. J. Essex-Cater, B. E. Davies, and C. Toothill, *J. Epidemiol. Community Health*, 1985, **39**, 256.

1979 and 1981[36]; on average the levels seem to be falling by about 5%/year. This is frequently attributed to the reduction in lead in petrol, but this reduction is counterbalanced by the increase in petrol consumption.

Part of the rationale for supposing that environmental levels of lead might have adverse effects on children was based on the observations of Byers and Lord in 1943[37] who first noted the intellectual and behavioural sequelae of chronic lead poisoning. It is of some interest, however, that they also stated in their paper that 'the presence or absence of evidence of involvement of the nervous system bore no relation to the eventual intellectual development of the children; indeed the only truly successful member of the group had the most severe encephalopathy encountered amongst the 20 children.'

There have been very many studies carried out to examine the relationship between lead levels (usually as measured by blood or tooth concentrations) and intellectual ability (usually in the form of an IQ). Greatly to simplify; the results from studies in the United States have tended to show a decrement in verbal but not performance IQ whereas studies in this country have tended to show little if any deficit.

One of the most influential studies was that of Needleman and his colleagues[38] in which tooth lead concentrations were compared with performance and it was found that the group with the highest tooth leads performed less well than the group with the lowest lead concentrations. Subsequently Needleman's data were re-examined[39] and it was stated that the findings could not be accepted as valid because of sampling and statistical errors. Nevertheless, Needleman's data were confirmed by a study carried out in Britain by Smith *et al.*[40] who reported significant differences in groups of children with different tooth lead levels with IQ scores favouring the low lead group. When these data were subjected to further analysis, a number of factors were found to be associated with the IQ of the children, the most important being the IQ of the mother, followed by family size and social class, and the quality of inter-family relationships. When these factors were taken into account, the effect of tooth lead concentration disappeared.[41]

The conclusions which British investigators have drawn from their studies have been either that lead has no effect, or that it is marginal. In a study in Birmingham the amount of variance in IQ which could be accounted for by the blood concentration was less than 1%[42]; it was between $2\frac{1}{2}$ and $4\frac{1}{2}$ in a study in London.[43] More recently the London study was replicated and this concluded

[36] M. J. Quinn, *Int. J. Epidemiol*, 1985, **14**, 420.
[37] R. K. Byers and E. E. Lord, *Am. J. Dis. Child.*, 1943, **66**, 471.
[38] H. L. Needleman, C. Gunnoe, A. Leviton, R. Reed, H. Peresie, C. Maher, and P. Barrett, *N. Eng. J. Med.*, 1979, **300**, 689.
[39] E. Marshall, *Science*, 1983, **222**, 906.
[40] M. Smith, T. Delves, R. Lansdowne, B. Clayton, and P. Graham, *Dev. Med. Child. Neurol.*, 1983, **25**, 1
[41] S. J. Pocock, D. Ashby, and M. Smith, *Int. J. Epidemiol.*, 1987, **16**, 57.
[42] P. J. Harvey, M. W. Hamlin, and R. Kumar, *Sci. Total Environ.*, 1984, **40**, 45.
[43] W. Yule, R. Lansdowne, I. B. Millar, and M. A. Urbanowicz, *Dev. Med. Child. Neurol.* 1981, **23**, 567.

that there was no evidence for any significant effect of lead on IQ.[44]

The evidence, at least of those studies which have been conducted in this country, suggests that the effect which concomitant amounts of lead in the environment may have on children's performance is marginal at most and certainly not nearly as important as social factors. (This is not the case with children who have suffered from an attack of lead poisoning with encephalopathy when there is a serious danger of permanent sequelae unless treatment is prompt and effective. In the majority of cases of childhood lead poisoning, the source of the lead is old paint.) In spite of this it is impossible to shake the conviction which very many people have that the contrary is, in fact, the case. One reason may be the need which Dunster[45] has identified in another context. That is, 'the need to attribute natural adversity to an identifed cause, [to] some form of universal culprit. If we cannot positively identify the cause, we look for any plausible candidate.' This need to externalize the cause of illness or misfortune is a well recognized human failing but it does lay the field open for what may be called paratoxicology and all too often the plausible candidate causing the natural adversity is some chemical in the environment.

5 CONCLUSIONS

Catastrophic and endemic exposure to chemicals in the environment undoubtedly cause much human suffering and may be associated with a considerable mortality and morbidity. Accidents which lead to catastrophic exposure are by their nature unpredictable but steps can be taken to minimize the likelihood of their occurrence. Much depends upon the installation of adequate safety systems and their proper maintenance and monitoring and on informing and educating the workforce.[46]

Endemic exposure is the most serious of all the forms of environmental exposure and it is not easy to see how it could easily be controlled since much of it arises from deliberate acts rather than as the result of accidents and the means have not been found whereby the more undesirable features of human nature can be changed or contained. The control of emissions into rivers or other ground waters, for example, can only be achieved if there is a body which can effectively police them and if the penalties which can be imposed upon polluters are such that they are a true deterrent. It is a sad reflection on western industrialists that they will 'export' processes which are considered too hazardous to be carried out in their own countries to developing countries where technological growth may be given a higher priority than safety.

Concomitant exposure, the lowest in the ranking order, seems to evoke most alarm in the general population and may give rise to fear about consequences which are insignificant compared with other effects. To consider the effects of lead in petrol on children's behaviour, for example, without also wishing to

[44] R. Lansdowne, W. Yule, M. A. Urbanowicz, and J. Hunter, *Int. Arch. Occup. Environ. Health*, 1986, **57**, 225.

[45] H. J. Dunster, *Nuclear Engineer*, 1985, **26**, 35.

[46] International Labour Office, 'Major Hazard Control: A Practical Manual,' ILO, Geneva, 1989.

alleviate those social factors—inadequate housing, poor diet, poor caring, or poor schooling—which are incomparably more important seems not to be throwing out the baby with the bath water, but not to be giving him a bath at all.

Proper public education about relative risks so that appropriate choices can be made may be a role for governments, both nationally and internationally. Recent history indicates that the industrial lobby is untrustworthy in this respect and that legislation without adequate means of enforcement and without truly deterrent penalties for the law breakers is insufficient to protect the public from potential sources of harm.

CHAPTER 14

The Legal Control of Pollution

R. MACRORY

1 FUNCTIONS AND CHARACTERISTICS OF POLLUTION LAW

In a modern industrial society, the control and management of pollution is underpinned by the law. Pollution laws are both complex and dynamic, and the present decade, in particular, has seen British law in this field being subjected to a series of important changes.* But while the detailed substance of the law will never be entirely static, the underlying functions of different legal provisions tend to remain constant, and a valuable, if not essential, starting point of any analysis of the law is therefore to identify first the main purpose of the provisions in question. The first section of this chapter identifies these main functions, while section 2 considers the various sources of pollution law in this country. Section 3 discusses typical characteristics of British pollution law, and these are illustrated in the following three sections which examine in more detail the principal legal controls in the fields of waste, water, and industrial air pollution respectively. The concluding section deals with a number of significant policy themes which are currently affecting the discussion and design of effective controls.

Legal principles can, for a start, provide remedies for those whose interests are damaged or threatened by pollution. Such legal actions originally formed the mainstay of British pollution law, and continue today to have some significance.[1] But in themselves private remedies are unlikely to provide a wholly satisfactory basis for regulation. Individuals may have neither the resources nor motivation to enforce their rights. Equally importantly, many aspects of the environment

[1] D. Hughes, 'Environmental Law,' Butterworths, London, 1986, Chapter 2 discusses the main common law remedies available.

* The examples of pollution laws and administrative set-ups discussed in this chapter are those relating to England and Wales only, though similar arrangements are applicable to other parts of the United Kingdom. The law is stated as at December, 1988, though a number of the more significant changes in prospect are indicated.

may not be explicitly owned or protected by private interests but are subject to common enjoyment—water, the marine environment, wild flora and fauna are all examples of such 'common' property. Legal controls over pollution are therefore now dominated by forms of 'public' law which provide various mechanisms of control over sources of pollution, irrespectively of possible damage to private interests, and with task of enforcement assigned to some official agency or public body.[2]

The typical modern pollution law which performs this public regulatory function will contain, within its provisions, a number of key elements. These include, first and foremost, the root mechanism of legal control in the field covered by the law in question.[3] This could take the form of straightforward prohibition of a specified activity, backed by criminal sanctions.* More common, though, is some form of licensing or consent system, again ultimately enforced by criminal penalties, but allowing public authorities far greater flexibility to permit discharges or emissions into the environment to take place but under controlled conditions.† The second critical part of a pollution law will be those provisions concerned with the administration of the controls. These may identify the particular authority or body responsible for implementation and enforcement, and will often deal in considerable detail with matters of process— for example, the procedural steps required when considering an application for a consent; powers available to enter and search premises and take samples; rights of the general public to be involved in consent procedures or to have access to information in the hands of the public authority. All these elements, though sometimes resulting in dense legal provisions of some complexity, are in practice vitally important for the effective operation of the controls. Finally, the pollution law may contain provisions concerning the overall goals of the controls. Such provisions may be drafted in the broadest and most generalized terms or, at the other extreme, may take the form of more specific requirements, expressed in numerate and quantitative terms, and perhaps containing time-limits for attaining targets.

The first two types of provision—those concerning the basic control mechanism and the administration of the control—are necessary elements of any pollution law. It is less essential, however, for the third type to be present. The basic aims or goals of the controls can be matters that are left to be determined at the discretion of those responsible for administering them rather than stated in the law, and this has in fact long been the case with many British pollution laws.

The main functions of different legal provisions can thus be distinguished. But

---

[2] Lord Nathan, 'Fencing Our Eden,' 1987 Garner Environmental Law Lecture, United Kingdom Environmental Law Association, London, 1987, pp. 16–18.

[3] A. Ogus, 'The Regulation of Pollution,' in G. Richardson, A. Ogus, and P. Burrows, 'Policing Pollution: A Study of Regulation and Enforcement,' Clarendon Press, Oxford, 1982, especially at pp. 40–48 where various forms of pollution offences are analysed.

* *e.g.*, Clean Act, 1956, 'Subject to the provisions of this Act, dark smoke shall not be emitted from a chimney of any building.'

† *e.g.*, Food and Environment Protection Act, 1985, 'Subject to the following provisions of this Part of this Act, a licence under this part of the Act is needed . . . for the deposit of articles within United Kingdom waters, either in the sea or under the sea-bed . . .'

individual pollution laws also tend to reflect underlying policy approaches towards pollution. Broadly, we can distinguish (i) regulation founded on technology-based standards and concentrating on what the originator of the pollution can achieve by way of minimization and (ii) regulation based on ambient environmental quality standards, where the regulator looks to the receiving environment as the starting point. Provisions in the law may expressly identify the dominant approach being followed, but often this has to be determined more from what is being implicitly stated. British water pollution law, for example, has been largely based on the ambient quality approach, while key air pollution controls are founded on a technology-based approach. Current trends in British pollution laws suggest a trend to combine elements of both approaches.

2 SOURCES OF LAW

Determining the applicable law to any aspect of pollution is not necessarily straightforward. There is no authoritative codification of law in the United Kingdom, nor is there a single, comprehensive law on pollution, and it is therefore necessary to draw from a variety of sources. Nearly all relevant legal principles in this field are now contained in Acts of Parliament, passed by Parliament and generally drafted and promoted by Government. These contain at the least the key control mechanisms and administrative arrangements. The modern tendency, though, is for provisions in the primary legislation to give power to a Government minister to make regulations or other forms of subsidiary legislation at a later date. Details which would otherwise overload the body of the statute can then be flushed out and, since the process of changing and updating this subsidiary legislation is much simpler than for primary Acts of Parliament, there are obvious attractions in this approach.*

Statutes and their associated subsidiary legislation are the primary source of law in the field of pollution law. A secondary, and quite distinct source, derives from the principles developed by the courts when handling individual cases. Case-law has traditionally provided an important and dynamic input in the British legal system. For a start, the ultimate authority on the interpretation of statutory provisions (which, however well drafted, often contain ambiguities or unanticipated problem areas) rests with the courts. The courts also play a critical role in developing principles of law which are independent from those contained in actual statutory provisions, though again always in the context of deciding particular cases before them. In the present context, such case-law principles are especially significant in determining the grounds on which private actions may be sought to compensate for damage suffered by pollution or prevention of further harm. Case-law alone provides the main source of relevant law in this area, with statutory provisions contributing little.

A third, and final significant source of law derives from international legal obligations, in the form of Treaties and similar agreements made between

---

* Attractive, that is, from the administrator's point of view. The extent to which the details of some modern pollution controls have been left to the discretion of Government regulations has caused concern in other circles.

countries on a bilateral or multilateral basis. In contrast to the position in some countries, British courts have never recognized that provisions of international treaties can by themselves form part of our national law which can be directly enforced in the courts. They are essentially treated as political obligations which must be converted by Governments into legislation if they are to be given true legal status. As such, they should, perhaps, be regarded primarily as an important lever on shaping national pollution laws.

The Treaty of Rome, which forms the basis of the European Economic Community, is, in one sense, an international legal obligation of the type described above. But it also represents a distinct and powerful source of law, with growing importance in most areas of pollution law. There are two main reasons for this. The Treaty established new permanent European institutions including a body responsible for developing and securing implementation of new policies (the European Commission based in Brussels) and an international court providing final, authoritative legal interpretation on the implication and meaning of EEC policies (the European Court based in Luxembourg). Accession to the Treaty implies an acceptance by member states of the role and authority of these institutions.[4] Individual policies, though developed by the Commission, cannot become part of EEC law unless agreed by the Governments of member countries (acting in the European Council of Ministers), though in certain fields countries have agreed that decisions may be taken by a form of majority voting.

The second distinguishing feature of EEC law is that in certain instances, provisions of the Treaty itself and of subsidiary EEC policies may have direct effect within the national legal system in that they may be enforced or relied upon in national courts, despite the absence of national implementing legislation. This aspect of EEC law, quite distinct from other forms of international law, is reflected in both the Treaty itself, and the British Statute upon which accession was based, though its precise scope and application to individual situations is still being developed.[5]

The original EEC Treaty made no express reference to pollution or the environment, but amendments made in 1987 now provide an explicit basis for policy development in this field.[6] The absence of specific Treaty provisions did not, however, inhibit the development of EEC policies, and since 1973 the Commission and other European institutions have been highly active, producing a series of Four-Year Action Programmes, and over 200 individual EEC laws in the environmental field, around 60 of which directly relate to pollution. These laws have generally taken the form of 'Directives,' meaning that strictly they represent obligations on member countries to achieve the aims of each Directive usually within a specified time-limit, but leaving considerable discretion to each

---

[4] A. Bradley, in the 'The Changing Constitution,' ed. J. Jowell and D. Oliver, Clarendon Press, Oxford, 1985, pp. 32–40.

[5] European Communities Act, 1972, section 2(1).

[6] Title VII 1957 Treaty Establishing the European Economic Community as added by Single European Act, 1986.

country as to the means chosen to implement the Directive.* EEC laws now represent a considerable influence on many areas of pollution policy in this country, notably in the fields of air and water pollution.[7]

3 UNDERLYING CHARACTERISTICS OF BRITISH POLLUTION LAWS

Before examining in more detail the relevant law in some key areas, it is worth highlighting a number of prevailing characteristics. These are not necessarily common to every aspect of pollution law but are sufficiently consistent to mark a strongly etched pattern.

First, statutory provisions (both Acts of Parliament and subsidiary legislation) now provide the major source of formal pollution law in this country and have tended to be developed on a specialized basis, dealing with different aspects of pollution, normally on a medium by medium basis (air, water, land, *etc.*). Inconsistencies in approach between different laws can be readily identified, and these are often the result of historical or sectoral influences operating on the particular law in question rather than distinctive factors relevant to the areas of pollution themselves. There is now, though, a more apparent concern to achieve greater coherence and to secure a more integrated approach towards pollution control.[8]

Underlying the design of much of the legislation has been an approach towards pollution control which resists the concept of absolute standards relating solely to the properties of the substances being discharged. Instead, its inter-action with the receiving environment and the targets to be protected (whether human, other living organisms, or physical features) must be taken into account, with allowance being made of environmental heterogeneity and the resilience of the physical environment to handle and absorb material.[9] This approach, reinforced by the variability of the country's physical characteristics, has resulted in a tendency for the legal provisions to delegate the implementation and enforcement of controls to a local or regional level, and for the law itself to prescribe little in the way of precise standards applicable across the country as a whole. Central Government plays little part in the direct enforcement of the laws, apart from some areas (such as radioactive waste disposal or certain sources of industrial pollution) where for reasons of strategic importance, technical complexity, or political sensitivity it has been considered appropriate to retain control. This tradition of localized standards and localized enforcement still

---

[7] N. Haigh, 'EEC Environmental Policy and Britain,' 2nd Edition, Longman, London, 1987. Haigh's book provides a lucid account of each EEC Directive and its implementation in Britain.
[8] UK Department of the Environment, 'Integrated Pollution Control,' Consultation Paper, DoE, London, 1988.
[9] M. Holdgate, in 'Britain, Europe, and the Environment,' ed. R. Macrory, Imperial College Centre for Environmental Technology, London, 1983, p. 12.

---

* Should a Member State fail to implement a Directive, the European Commission may commence legal proceedings, eventually bringing the case before the European Court of Justice. This is a regular occurrence in the environmental field, though to date action against Britain has been rare.

dominates fields such as waste disposal or neighbourhood noise, but it is one that is increasingly subject to challenge in certain areas. EEC laws in particular have tended to contain express standards applicable across the country with limited room for variation.

Finally, it is worth emphasizing the non-legalistic and non-confrontational approach which has long been adopted by many British authorities, whether at local or national level.[10] Nearly all statutory pollution controls ultimately rest on criminal offences of various types, enforceable by the regulatory authorities through the criminal courts, and these offences are normally framed in 'strict' terms, meaning that the mere act of non-compliance is sufficient to secure conviction; it is not necessary, as with most types of crime, to prove that the defendant actually had criminal intentions. Despite this backdrop, most authorities, faced with non-compliance, have treated prosecution very much as a last resort, to be pursued only against the recalcitrant or gross violator. The reluctance to go to court can be explained in part by the unpredictability of the legal process and a dissatisfaction with what appears to be low levels of fines often awarded in the courts.[11] But at least as important is the fact that the majority of those who actually apply and implement the legal controls in practice have a scientific or technical background. In contrast to countries such as the United States lawyers do not play the dominant role in the process, and many regulatory authorities would regard their role as primarily concerned with effective environmental management rather than simply one of law enforcement. One of the consequences of this non-litigious stance is that the British courts have had little opportunity to provide much of an input into the interpretation of statutory pollution laws.[12] Cases which are brought by authorities tend to be those where the defendant will clearly plead guilty rather than dispute points of law, and as a result there remain many examples of ambiguous terminology employed in pollution laws where the higher courts have yet to provide guidance on their meaning or scope.

4 WASTE DISPOSAL ON LAND

4.1 The Control Framework

Nearly all solid wastes arising in the United Kingdom, both domestic and commerical, are disposed of on engineered land-filled sites. Until the 1970's, the main legal controls over such disposal operations consisted of a combination of Town and Country Planning laws, requiring prior planning permission for proposed new use of land for waste disposal, and various statutory nuisance laws which provided powers to deal with specific localized problems but generally

[10] D. Vogel, 'Environmental Policy in Great Britain and the United States,' Cornell University Press, Ithaca, 1986, especially pp. 83–90 for an illuminating American view of British practice.
[11] R. Macrory and S. Withers, 'Application of Administrative and Criminal Punishments concerning Hazardous Wastes in England,' Report for European Commission, Imperial College Centre for Environmental Technology, London, 1985.
[12] R. Macrory, 'Environmental Case-Law in the United Kingdom' in 'La Giurisprudenza Ambientale Europea,' ed. A. Postiglione, Dott. A. Giuffre, Milan, 1988.

only after pollution had occurred. Part I of the Control of Pollution Act, 1974 introduced an additional, and more specialized set of controls, based on the individual licensing of facilities and special monitoring of movements of particularly dangerous types of waste.

The main controls in Part I of the Act apply to 'controlled waste,' defined in section 30(1) to mean household, industrial, or commercial waste, and the meaning of these terms is further elaborated in both the Act and regulations made by the Secretary of State where particular descriptions of waste are defined as falling within the categories.[13] Two important general points should be noted. The controls do not apply to all types of wastes. In particular, wastes from mines and quarries, and from agricultural operations are expressly excluded, and are still largely regulated by planning controls under the Town and Country Planning legislation; similarly, the disposal of radioactive waste is excluded, being dealt with under specialized laws. Secondly, the legislation applies to types of waste categorized by their source, rather than their toxicity or similar qualities, and the same basic sets of controls cover wastes whether they can be considered to be hazardous or non-hazardous. This can be contrasted to the position in some countries where quite distinct regimes apply to pre-determined categories of hazardous and non-hazardous waste.

4.2 Site-Licensing

Under section 3 of the Control of Pollution Act a person may not deposit controlled waste on land unless the site in question is occupied by the holder of a licence, and the disposal is in accordance with any conditions attached to the licence. The use of plant and equipment for disposal or treatment of waste by various means, including incineration, fuel production, and composting, is similarly prohibited unless in accordance with a disposal licence.

As with many examples of licence regimes, the Act permits the Secretary of State to exempt particular classes of disposal from the scope of the Controls in order to prevent the system being overloaded by innocuous or minor examples of disposals. Examples of exemptions under current regulations include dredging waste deposited alongside river banks, sewage sludge deposited as a fertilizer, and the temporary storage of restricted quantities of waste by its producer pending disposal elsewhere; none of these exemptions, however, will apply where the disposal in question could give risk damage to humans, animals, or to water supplies (collectively defined as an 'environmental hazard' in the Act).[14]

The central regulatory device in the Act is therefore a licensing system backed by criminal penalties, and many of the provisions in the Act are concerned with administrative and procedural matters surrounding the controls. Licences are issued by 'Disposal Authorities' who are also responsible for enforcement of the controls. The Disposal Authorities are designated local government bodies in every area, though the administrative structure has been made all the more

---

[13] Control of Pollution Act, 1974, section 30; Collection and Disposal of Waste Regulations, 1988, S.I. 1988, No. 819.

[14] Collection and Disposal of Waste Regulations, 1988, op. cit.

complex in recent years as a result of general reorganization of local government, and at present there are over 200 Disposal Authorities in the country, of varying sizes and operating at different levels of local government structure.

In dealing with a disposal licence application, the Authority may only refuse to grant the licence where this would prevent damage to humans or water supplies, but the Act gives them a broad discretion to attach conditions to a licence, and, in any event, the site in question must also have any necessary planning permission under the Town and Country Planning Acts. More importantly, they have a wide power to vary and modify conditions at a later date, underlining the fact that the controls are intended to provide a continuous and flexible form of regulation. The applicant or holder of a licence has the right to appeal against the decision of the Disposal Authority to the Secretary of State for the Environment, and these appellate functions represent almost the only example of Central Government's direct legal involvement in the system. In other respects, it has mainly confined its role to one of providing both policy and technical guidance, leaving the actual enforcement of the controls and their primary application to the more locally based Disposal Authorities.

4.3   Tracking System for Movements of Waste

The disposal licence system is concerned with regulating operations at the disposal site or facility itself. Regulations made under Section 17 of the Control of Pollution Act contain an additional control mechanism, aimed at monitoring the movements of particularly dangerous types of waste. The system applies to a limited number of types of wastes, known legally as 'special waste,' mainly defined by a combination of listed families of substances and a toxicity test.[15] In essence, the regulations require producers of such wastes to provide the Disposal Authority of the area in which the wastes are to be disposed with a minimum of three days notice of consignments. The 'home' waste Disposal Authority is also notified, and during the movement of wastes until its eventual disposal, those involved in the chain of operations must similarly provide notification to the Authorities. Clearly, the system (colloquially known as 'cradle to grave' tracking) is bureaucratically demanding, though where operated properly, it should provide valuable warning signals of misplaced consignments failing to reach their intended disposal destination.

4.4   Imports and Exports of Wastes

The main legal controls are currently contained in the Transfrontier Shipment of Hazardous Waste Regulations 1988.[16] These are intended to implement an EEC Directive on the subject and introduce a system of consignment notes and documentation for international movements of waste rather more comprehensive

---

[15] Control of Pollution (Special Waste) Regulations, 1980, S.I. 1980, No. 1709.
[16] Transfrontier Shipment of Hazardous Waste Regulations, 1988, S.I. 1988, No. 1562.

than that already existing under the Special Waste regulations.[17] The requirements are limited to wastes containing specified substances (almost coextensive
with Special Wastes), and consignees of waste into the United Kingdom must
not accept shipments unless accompanied by required consignment notes. In
addition, the Disposal Authority must be given advance notice of imports, and
has one month in which either to acknowledge the consignment note (only then
may the holder of the waste proceed with shipment) or to make a written
objection. The grounds for objection are contained in the EEC Directive, and
imply that the Disposal Authority may only object if the proposed shipment
would be illegal in some way (*e.g.*, the intended disposal site was not licensed to
accept the types of waste notified); the fact that the import might be politically
controversial would not in itself be grounds for objection by the Disposal
Authority.

The 1988 regulations therefore build upon the Special Waste consignment
note system, and are primarily intended to give the opportunity for improved
monitoring of international movements of hazardous wastes, rather than to
restrict or prohibit such trade. The controls apply only to certain types of waste,
and there is no specific waste legislation relating to the import of non-hazardous
and municipal waste, though various laws concerning the entry of potentially
harmful material into the country may be relevant. But as a result of growing
public and political concern over the whole question of imports, future legislation in this field is likely to contain more explicit powers which would enable the
import or export of any type of waste to be prohibited or restricted in certain
circumstances.[18]

4.5 Enforcement of Consistent Standards

A common criticism made of the UK controls in recent years has been that while
the basic legal framework is reasonably sound its implementation in practice has
been extremely variable across the country.[19] This is in part the consequence of
a system involving a large number of Disposal Authorities, which differ considerably in size and expertise, together with the strong degree of discretion inherent
in the legal controls. The waste disposal industry, with the private sector
dominating many areas of waste disposal, itself is highly diverse, and in the
future the legal controls are likely to have to pay increasing attention to the
regulation of individual disposal operators as well as the disposal sites and
facilities. Ensuring an effective administrative structure will be equally important in order to secure both higher standards and improved consistency. One
particular area of concern is the fact that under current arrangements, local
authorities responsible for enforcing site licensing controls also operate their own
sites where they must regulate themselves. Self-regulation by pollution control
authorities of their own operations giving rise to potential pollution has long

[17] Council Directive 84/631/EEC on the supervision and control within the European Community of
the transfrontier shipment of hazardous waste.
[18] UK Department of the Environment, 'Waste Disposal Law Amendments: Follow Up Consultation Joint Paper,' DoE/Welsh Office, London, 1988.
[19] Royal Commission on Environmental Pollution, 'Managing Waste: The Duty of Care,' Eleventh
Report, Cmnd. 9675, HMSO, London, 1985.

been a structural weakness of British administration. This particular problem is already being tackled in the field of water pollution as a result of the proposed privatization of the water industry (see Section 6 below), and some equivalent response must be expected in this area.

5 INDUSTRIAL AIR POLLUTION

5.1 Prescribed Classes of Works

A critical distinction in British air pollution law is made between listed and non-listed categories of industrial process. The listed categories, prescribed in regulations[20] and updated from time to time, cover much of heavy manufacturing industries, the chemical industry, and those types of industry considered to give rise to potentially serious air pollution problems, and the legislation and administrative arrangements that apply to prescribed classes of processes are quite distinct from those applicable to all other industrial sources.*

Under section 5 of the Health and Safety at Work *etc.* Act 1974, a person with control of prescribed processes is under a general duty to use the best practicable means (i) 'for preventing the emission into the atmosphere from the premises of noxious or offensive substances' and (ii) for rendering harmless and inoffensive such substances as may be so emitted. The duty is supervised by an arm of central government, Her Majesty's Inspectorate of Pollution, who possess considerable powers of entry, taking samples and various methods of enforcement, including criminal prosecution for non-compliance with the duty. The legal approach involving central government control over a limited number of special cases, and having as its core a broad-based, continuous duty which avoids reference to numerate standards or emission limits has a long history in the United Kingdom, stretching back well over 100 years.[21]

The core central duty is expressed in terms of the 'best practical means'. What is involved in this concept may change over time in the light of scientific developments, may vary from industry to industry, and requires economic implications to be balanced against technical possibilities. The duty is imposed directly on the industry itself, and in theory the courts would be the final arbiters of what is the best practicable means at any particular time. But in practice, the Inspectorate, and its predecessors, in effect provide authoritative guidance on the standards currently expected of particular industries, and advice on best current practice is both given by individual Inspectors in respect to plants falling within their jurisdiction, and provided in a series of published Notes relating to

---

[20] Health and Safety (Emissions into the Atmosphere) Regulations, 1983, S.I. 1983, No. 943, contains the present list.

[21] E. Ashby and M. Anderson, 'The Politics of Clean Air,' Clarendon Press, Oxford, 1981.

* Examples include thermal power stations, oil refineries, and cement production plants. In December, 1986, there were in England, Wales, and Scotland 3 370 scheduled processes at 2 290 individual works coming under these controls.

classes of industries.[22] These notes contain details of emission standards which are presumed to be BPM at the time, together with other technical guidance on matters considered to fall within the broad duty, such as maintenance of plant and training of operators, and are regularly updated, after extensive discussion and negotiation with the industries and relevant trade associations. In addition to the general duty under section 5, plants must be annually registered with the Inspectorate.

The non-legalistic way in which these controls have been interpreted and applied is one of their dominant characteristics; even the core phrase, 'best practicable means,' as used in the air pollution legislation, has never been subject to litigation in the higher courts, despite the fact that the concept is one that is inherently subject to conflicting interpretations. The relationship between the Inspectorate and the industries within their jurisdictions has, as a consequence, been somewhat of a mystery to those outside the system, not helped by the fact that for many years information concerning the actual standards being required of any particular plant was regarded as confidential. But both practice and the legal base has been subject to considerable change in recent years. More extensive information is now made available by the Inspectorate, explicit EEC standards are applicable in some cases, and a more formalized consent process has recently been proposed (see below section 5.3).

5.2 The Clean Air Acts

The Clean Air Acts of 1956 and 1968 provide the main legal controls over air pollution both from non-prescribed industrial processes and domestic premises, with enforcement resting with the environmental health departments of district and London Borough Councils rather than Central Government. The legal provisions contain no generalized duty equivalent to that contained in section 5 of the Health and Safety at Work *etc.* Act, nor is there a formalized consent system for atmospheric emissions. Instead, the legislation contains a number of quite specialized provisions focused on combustion processes, and concerned mainly with smoke, dust and grit emissions. The legal methods of control adopted cover a broad spectrum, and include specific offences of emitting 'dark' smoke subject to limited exceptions; powers to remedy smoke problems giving rise to a nuisance; the need to obtain approval for the installation of new industrial furnaces above a specified size; similar provisions to ensure that new chimneys carrying atmospheric emissions are sufficiently high to prevent the emissions becoming prejudicial to health or a nuisance; and unusually, power for central government to lay down national emission standards in respect to grit and dust from certain furnaces.

The most innovative type of legal control introduced by the Clean Air Act, 1956 is the power of local authorities to establish 'smoke control areas' (colloquially known as 'smokeless zones') covering all or part of their district. Once an

---

[22] Her Majesty's Inspectorate of Pollution, 'Best Practicable Means: General Principles and Practice,' BPM Notes I/88, Department of the Environment, London, 1988. A further twenty-five Notes relating to particular classes of industry have been published in the latest series.

area has been established, the burning of fuels giving rise to smoke through chimneys is prohibited from classes of premises named in the order, unless types of fuel authorized in regulations are used. Smoke control areas have been primarily directed at domestic premises, and the legislation provides for the making of financial grants for owners or occupiers of older private dwellings who incur capital expenditure such as the alteration of old cookers, in order to comply with the requirements. The combination of what amounts to an absolute ban on smoke emissions (without the need to prove the existence of 'dark smoke,' or a nuisance) together with the provision of financial incentives to ensure compliance has made these provisions remarkably effective at speeding up the social changes needed to reduce urban smoke levels that were once so prevalent in Britain. Despite this, it is apparent that the present legal controls available to local authorities in this field are in need of reform. The preoccupation with smoke and combustion processes reflect earlier policy priorities, and in the future the development of more adaptable and comprehensive methods of regulation, mirroring those relating to the prescribed classes of industrial processes can be expected.

5.3 EEC Legislation and Future Directions

The controls outlined above are now applied in the light of agreed EEC policies in this field, and indeed will often provide the means to achieve them.[23] Member States are obliged to ensure that the quality of air does not exceed ambient concentrations specified in Directives concerning smoke and sulphur dioxide (80/779/EEC), (85/203), and lead (82/884) and nitrogen dioxide (85/203). The framework Directive (84/360) requires that there exists an effective authorization procedure for new or altered industrial plants of classes specified in the Directive. Before granting an authorization, the appropriate authority must be satisfied that appropriate preventive measures are being taken including the use of the best available technology not entailing excessive cost (BATNEEC), and that significant air pollution will not be caused by the emission of any of eight specified substances, including sulphur dioxide, asbestos, and fluorine. These provisions by and large reflect the existing approach of the controls relating to prescribed classes of works, though the Directive has required the introduction of more formalized and open authorizing documentation than existed before.

Where the Directive does mark a significant shift in British practice is its provision that subsequent Directives may prescribe emission limits relating to specific classes of plants, though such limits must also take account of BATNEEC.* At the end of 1988 the first Directive to contain such limits was agreed. Council Directive of 24 November, 1988 (88/609/EEC) lays down numerate

[23] N. Haigh (1987), op. cit.; R. Macrory, 'Industrial Air Pollution: Implementing the European Framework,' Proceedings of the National Society of Clean Air Conference, Llandudno, 1988.

* Though the terminology very much in current vogue, 'BATNEEC' of course, involves concepts as rich in qualitative criteria as 'best practicable means.' The European Commission, in conjunction with national officials is now working on the production of European guidance notes on its application to various industries.

emission limits for sulphur dioxide, nitrogen oxides, and dust from new combustion plants with a rated thermal input equal to or greater than 50 MW. The Directive also introduces another type of policy instrument dealing with national air pollution problems by requiring each member state to seek a progressive reduction of its total annual emissions of sulphur dioxide and nitrogen oxides from existing plants, based on 1980 as a base point. Target dates and percentage reductions are specified, though significantly these are not precisely the same for each country, a common feature of many existing international agreements. Instead, some degree of variation is permitted since, according to the Directive's preamble, 'due account has been taken of the need for comparable effort, whilst making allowances for the specific situations of member states'.*

The need to implement the Framework Directive in the United Kingdom, and the continuing development of EEC policies on industrial air pollution has stimulated a fundamental review of air pollution controls, with a major consultation paper issued by the Government in 1986, with follow up papers in 1988.[24] At the time of writing, no new legislation had yet been published, but the basic framework for future changes was reasonably clear. An expanded two-part list of scheduled processes would replace the current list, with Her Majesty's Inspectorate of Pollution still controlling those in Part A, but with local authorities regulating any in Part B. Both categories would be subject to authorization by the relevant authority, and these authorizations would contain a combination of explicit conditions of a technical nature, and a general residual duty to use the best practicable means to prevent and reduce damaging emissions. Breach of these conditions would be a criminal offence, bringing the overall approach more in line with the consent systems for waste or effluent disposal.

6 WATER POLLUTION

6.1 Consent Based System

Legislation concerning water pollution is particularly complex.[25] Different statutes apply to discharges depending on whether they are made into public sewers (Public Health Acts, 1936 and 1961 and Public Health (Drainage of Premises) Act, 1937), the marine environment (Food and Environment Protection Act, 1985), and inland, estuarine, and tidal waters (Part II of the Control of Pollution Act to be replaced by the Water Act, 1989). But, despite distinct administrative arrangements and legal provisions, there is a common form of central control mechanism—a consent or licensing system operated by public

---

[24] UK Department of the Environment, 'Air Pollution Control in Great Britain: Review and Proposals,' Consultation Paper, DoE, London, 1986; UK Department of the Environment, 'Air Pollution Control in Great Britain,' Follow-Up Consultation Paper, DoE, London, 1988; UK Department of the Environment, 'Air Pollution Control in Great Britain: Works Proposed to be Scheduled for Authorization,' Consultation Paper, DoE, London, 1988.

[25] W. Howarth, 'Water Pollution Law,' Shaw and Sons, London, 1988 for a comprehensive, recent account.

* Britain, for example, must secure a 40% reduction of SO_2 emissions by 1998, Netherlands 60%, and Spain 24%.

authorities, and underpinned by criminal offences, often drafted in sweeping terms.

The control of discharges into inland waters provides a typical example. Sections 32 and 33 of the Control of Pollution Act, 1974 (which will be replaced by single section written in similar terms under section 107 of the Water Act, 1989) make it an offence to 'cause or knowingly permit' the discharge into inland waters of (i) 'noxious, poisonous, or polluting matter' and (ii) any trade or sewage effluent (*i.e.*, whether or not it is potentially harmful). Various defences are applicable, the main being that the discharge in question was in accordance with a consent granted by a water authority. The leading decision of the House of Lords in Alphacell Ltd. *vs.* Woodward (1972) 3 All ER 475 interpreting the meaning of the word 'cause' in a similarly drafted previous water pollution offence emphasized the strictness of the provisions: provided there was a causal connection between the defendant's acts and the discharge in question, the offence could be committed even though there was no actual intention or negligence present. This approach was confirmed in Wrothwell Ltd. *vs.* Yorkshire Water Authority (1984) Crim. L.R. 43 where a company director who poured herbicide down a drain connecting to a river was convicted of the offence causing effluent to enter waters, even though he genuinely believed the drain led only to the public sewer.

Consents, which provide the main defence to these offences, are therefore the main regulatory tool for controlling polluting discharges, and currently in England and Wales there are around 30 000 consents relating to discharges of industrial effluent with a further 4 355 authorizing discharges of sewage effluent from sewage treatment works.[26] Such consents are granted by regional water authorities in England and Wales for industrial discharges and the Department of the Environment for effluent from water authorities' own sewage works (but after proposed privatization of the water industry a single new public agency, the National Rivers Authority will be responsible for issuing consents other than those for discharges to sewers). The legislation gives extensive powers to attach and revise conditions, and as might be expected, many of the provisions are concerned with matters of procedure—one of the innovatory features of the 1974 Act was to introduce an element of public consultation and participation into the consent setting process. The legislation also reflected a traditional style of British pollution law by remaining largely silent on questions of the quality of receiving water. Until recently, the substantive goals of the system and the criteria on which decisions concerning consents should be based have rested on a non-statutory and often quite informal basis, though since 1978, more explicit but again non-statutory water quality objectives for different stretches of rivers and other waters have been developed.[27] The advent of EEC legislation in this field (discussed in the subsequent section) together with the proposed privatization of water authorities has set in train a substantial reappraisal of this long-established approach—ambient environmental quality will remain the basis of most of the

---

[26] Figures quoted in W. Howarth, op. cit., pp. 36 and 37.

[27] National Water Council, 'River Water Quality. The Next Stage: Review of Discharge Consent Conditions,' National Water Council, London, 1978.

controls, but central government will possess the power to establish water quality objectives, including a time-limit for achievement, on a statutory basis.[28]

6.2 EEC Directives

The development of a number of important Directives concerning various aspects of water pollution was a central aim of the early years of the Community environmental action programmes. The culminative effect of the Directives agreed to date has brought about a radical change to the climate of legal regulation in this country by introducing into the system a layer of quantitative numerate standards in respect of water quality. Under EEC law, these are legally binding on the UK Government and initially have been transposed into the national controls by non-statutory means: Government administrative circulars requiring control authorities to ensure their regulatory powers are used to achieve the aims of the Directives. In the future, however, these EEC standards are likely increasingly to be reflected explicitly in national legislation.

Many of the EEC Directives relating to water pollution contain a variety of quality standards for various categories of water depending on its use (*e.g.*, bathing waters, waters supporting shellfish) but one of the most critical pieces of EEC law is EC Directive 76/464 concerning the discharge of Dangerous Substances into the Aquatic Environment. It is sometimes known as a 'mother' or framework Directive since it essentially lays down a set of general principles concerning the application of controls over the discharge of effluents, leaving numerate standards for individual substances to be prescribed in subsequent, 'daughter' Directives. The framework Directive specifies a number of broad families of chemical substances, dividing them into two Lists: List I, containing those considered most toxic (the 'black list'), and List II (the 'grey list'), containing those that are less dangerous. The stated aims of the Directive are to 'eliminate' pollution caused by List I substances, and to 'reduce' pollution caused by List II.

The Directive envisages two possible control regimes for List substances. Subsequent daughter Directives relating to the individual substances must specify both a minimum emission standard relating to the actual discharge point (known as a 'Limit Value') or a numerate standard concerning the quality of the receiving water (described as a 'Quality Objective'), and member states have the option to choose either approach. The United Kingdom, with its long-standing emphasis on the need to take account of heterogeneous receiving environments in determining individual pollution consents, pressed strongly for the inclusion of the Quality Objective approach, and to date has been the only country to choose that option for List I substances. A small number of daughter Directives concerning List I substances have been agreed since 1976, but progress has been extremely slow, particularly in the light of the European Commission's identification in 1982 of 129 priority substances falling within List I families.[29] The

---

[28] Section 105, Water Act, 1989.

[29] Council Directives on mercury from chlor-alkali industry (82/176/EEC); mercury from other sources (84/156/EEC); cadmium (83/513/EEC); hexachlorocyclohexane (84/491/EEC); DDT, carbon tetrachloride, and pentachlorophenol (86/280/EEC).

production of List I Directives should, however, speed up as a result of the agreement in 1986 to the Directive on Limit Values and Quality Objectives for Discharges of Certain Substances (86/280). This should considerably simplify procedures in the future by introducing a set of provisions intended to be common to all subsequent List I substances, and thereby allowing future Limit Values and Quality Objectives to be added simply by means of annexes to this Directive.* A further policy development concerns the basic dichotomy of the Limit Value and Quality Objective options contained in the 1976 Directive, where increasingly it is being recognized that both approaches have strengths and weaknesses, and that a combination of the best elements of both may prove a more effective policy base.[30] Specific EEC legislation has yet to emerge to meet this argument, but can be expected in the future.

6.3   Future Administrative Arrangements

The prospective privatization of the water industry, as set out in the Water Act, 1989, will provide a major shift in the administration of legal controls. The functions of the regional public water authorities in England and Wales, set up in 1974, stretched across the hydrological cycle and included the regulation of water abstraction, provision of water supplies, sewage treatment, and pollution control. Whatever the economic and technical advantages of this integrated approach, it was recognized that environmental protection was a matter that could not rest solely in the private sector once the industry was privatized, and hence a new public body, the National Rivers Authority, has been established.[31] The Authority's main functions will be to set discharge consents, and monitor and enforce the relevant pollution controls.

The legal provisions in Part II of the Control of Pollution Act are replaced by new measures in the Water Act, but for the most part these are simply a restatement, and the controls will still rest on a consent system combined with criminal offences. One important change, though, is the proposal to give the Secretary of State the power to establish statutory water quality objectives, which will provide a legal goal of the controls; there will be a general duty on the National Rivers Authority to secure achievement of these objectives.[32] The Secretary of State will also have the power to establish Water Protection Zones, within which he may prohibit or restrict activities—whether occurring on land or water—which may result in water pollution.[33] This is a distinct form of

[30] UK House of Lords Select Committee on the European Communities. Session 1984–85, 15th Report 'Dangerous Substances,' HMSO, London, 1985; see also UK Department of the Environment, 'Inputs of Dangerous Substances to Water: Proposals for a Unified System of Control,' DoE/Welsh Office, London, 1988.

[31] UK Department of the Environment, 'The National Rivers Authority: The Government's Proposals for a Public Regulatory Body in a Privatized Water Industry,' Consultation Paper, DoE/Welsh Office/Ministry of Agriculture, Fisheries, and Food, London, 1987. Section 1, Water Act, 1989.

[32] Section 106, Water Act, 1989.

[33] Section 111, Water Act, 1989.

* The process is now exemplified with the addition of a further seven substances under Council Directive of 16 June, 1988 amending Annex II of the 1986 Directive (88/347/EEC)

legal regime aimed at minimizing pollution problems resulting from activities which a consent based system for discharges does not effectively regulate. The Secretary of State will have extensive discretion as to what may or may not be restrained with such Zones.

7 RECENT TRENDS IN POLLUTION LEGISLATION

7.1 Public Involvement

The extensive controls concerning land-use planning contained in the Town and Country Planning legislation have long contained measures providing for various forms of involvement by the general public in decision-making, ranging from public registers of planning applications, public consultation, up to the holding of formal public inquiries. Until quite recently, such provisions have scarcely been mirrored in pollution legislation, mainly on the grounds that pollution control decisions often involved complex technical issues on which public comment has little to offer, and that commercial secrets might be exposed by greater public access.

The water control provisions in Part II of the Control of Pollution Act, 1974 marked a major break with traditional assumptions, by introducing rights of legal access to information concerning discharge consents and sample results, and a degree of public consultation where new applications for consents were being considered by water authorities. The scheme includes protective procedures for industries who feel that their commercial secrets would be prejudiced by public access. The precedent of the Control of Pollution Act has tended to undermine arguments based on fundamental principle that this form of participation should not be extended to other areas of pollution control, and the independent Royal Commission on Environmental Pollution has consistently made the case for a presumption of openness in this field.[34] The Government has in principle agreed with the philosophy, though as yet little in the way of positive legal provisions have emerged.[35] Nevertheless, legal rights concerning public involvement of some form in pollution control can be expected to form an important element of future amendments and reforms.

7.2 Land-use Planning and Pollution Control

Effective pollution control legislation aims to prevent problems occurring before they cause damage. Where new land-use developments are involved, it is clear that the system of planning controls under the Town and Country Planning legislation has an important role to play, not so much in the enforcement of detailed operational standards but in ensuring that poor locational choices are

---

[34] Royal Commission on Environmental Pollution, Tenth Report, 'Tackling Pollution—Experience and Prospects,' Cmnd. 9149, HMSO, London, 1984, paras. 2.33–2.78.
[35] UK Department of the Environment, Pollution Paper No. 22, 'Controlling Pollution: Principles and Prospects,' HMSO, London, 1984; UK Department of the Environment, Pollution Paper No. 23, 'Public Access to Environmental Information,' HMSO, London, 1986.

not made which would otherwise exacerbate subsequent pollution.[36] Potential pollution loads from a proposed development are clearly a relevant issue for planning authorities to consider, though not necessarily an overriding one. The goals of the planning controls are not always those of pollution minimization, hence the occurrence of environmentally sub-standard planning decisions. Nevertheless, successful intermeshing of land-use planning and pollution legislation is an important aspect of a preventive control system, though not one easy to achieve, and made more complex by the differing decison-making bodies that may be involved.

Until recently, the interlinking of these two fields of law was secured largely by administrative practice rather than positive legal measures. In 1988, however, regulations were made, introducing formal environmental assessment procedures into the land-use planning system.[37] These regulations, brought about by the need to implement the EEC Directive on the subject, require certain procedural steps, including the gathering together of information concerning the likely environmental effects of a proposed development, to be taken before a decision on a planning application is made. The requirements do not apply to all types of development, but are limited to two lists of classes of development. Those in List I (which include power stations and waste disposal facilities) must always be subject to environmental assessment procedures, whilst those in List II will only be so where planning authorities consider they are likely to have a 'significant effect' on the environment. The legal rules are complex, and no doubt will lead to litigation in the future, but they shy away from requiring that the assessment information in respect of a proposed development must be given an overriding priority by a planning authority—it must simply 'be taken into account' before the decision is made. To that extent, the new procedures probably add little more to what should have already been good planning practice. But the formal legal linking of land-use planning controls and pollution regulation may well prove one of the more significant and long-lasting effects of the assessment requirements.

7.3   Integrated Pollution Control

As will have been apparent, existing pollution laws have been developed on a medium by medium basis, dealing with air, water, and waste under separate control systems. In recent years, there has been concern that such arrangements, often involving different administrative bodies, may result in decisions which fail to reflect interconnections between pollution streams, and therefore lead to sub-optimal results. The term, 'best practicable environmental option' (BPEO), has been coined to describe a new policy approach which would ensure that the environmental consequences of particular strategies for controlling pollution are

---

[36] C. Wood, 'Town Planning and Pollution Control,' Manchester University Press, Manchester, 1976, especially 188–193; see also C. Miller and C. Wood, 'Planning and Pollution,' Clarendon Press, Oxford, 1983, which includes detailed case-studies.

[37] Town and Country Planning (Assessment of Environmental Effects) Regulations, 1988, S.I. 1988, No. 1199.

fully weighed against each other—'The BPEO procedure establishes, for a given set of objectives, the option that provides the most benefit or least damage to the environment as a whole, at acceptable cost, in the long term as well as in the short term.'[38] Though a powerful concept, which is supported in principle by Government, its full implications and practical effects still remain somewhat obscure.

As yet, few provisions in British pollution laws reflect this more rounded approach, with one exception in the controls concerning licensing for the disposal of waste at sea.[39] Instead, various informal consultative and liaison arrangements between existing regulatory bodies have been developed, but a more structured response was initiated in 1987 with the formation within the Department of the Environment of a new body, Her Majesty's Inspectorate of Pollution (HMIP). This Inspectorate encompassed the former Industrial Air Pollution Inspectorate, the Radiochemical Inspectorate, the Hazardous Waste Inspectorate, and various water pollution functions, and its initial prime function was to provide expert technical advice on multi-media pollution issues in addition to continuing the former specialist expertise of the constituent components. But its legal powers of direct regulation over industrial pollution are limited under present laws to air pollution (for scheduled processes) and radioactive discharges. The Government have therefore proposed that HMIP should have new regulatory powers, giving it clear responsibility for controlling discharges to all media from certain classes of processes or industries (including those giving rise to specific types of discharges or producing particularly hazardous wastes).[40]

These latest proposals in integrated pollution control would require new primary legislation, and would represent a significant step in bringing a more coherent and sophisticated structure to present legal arrangements. Problem areas remain, not least concerning the relationship of HMIP and the new National Rivers Authority. At the same time, the whole concept of the boundaries of pollution control is likely to change, with the notion of anticipatory prevention, alongside an integrated approach, continuing to grow as a powerful policy tool. But the practical implementation of such policies raises intriguing institutional and legal issues. To what extent, for example, should a pollution control authority now be legitimately concerned with minimizing production of potentially harmful material from an industry rather than confining its attention to controlling emissions or deposits? Should those regulating atmospheric pollution now address more forcibly questions of energy conservation? How far should legal standards ignore current technical consensus in favour of a more cautious but possibly unfounded approach? These are, perhaps, questions more of policy rather than law, but at the very least the structure of existing legislation will affect the way they are addressed. British pollution laws have for over a

---

[38] Royal Commission on Environmental Pollution, 12th Report, 'Best Practicable Environmental Option,' Cm. 310, HMSO, London, 1988, para 2.1.

[39] Food and Environment Protection Act, 1985, section 8(2).

[40] UK Department of the Environment, 'Integrated Pollution Control,' A Consultation Paper, DoE, London, 1988.

century remained fairly static in their underlying structure. But as will have been apparent from this chapter, recent years have seen a rare series of fundamental reappraisals, emerging from both national and international pressures, which have challenged well-established and familiar approaches. A new legal framework is emerging as a consequence—one that will provide a more sophisticated and transparent basis of control for modern conditions—but whose real test will be its ability to cope with pollution issues as yet unforeseen.

CHAPTER 15

The Control of Industrial Pollution

W. L. LINTON

1 WHAT IS POLLUTION?

Pollution is a judgement, not necessarily in the sense of a Judgement laid upon us by the Almighty although clergymen have recently drawn parallels between the concepts of spiritual purity and 'ecological purity' with the implied conclusion that, unless we achieve suitable standards in both, we are in deep trouble. This elevation of environmental concern to a theological level accords with a popular view that pollution is a departure from some natural and intended order of things, a change brought about by man or, more usually by industry (*i.e.*, by others); if man would stop interfering with nature and exercise proper steward-ship an unpolluted world would be regained.

However, as total de-industrialization of the developed countries is improb-able, a more practical view of pollution is required, one which allows sustainable development of healthy and vigorous industrial activity to provide the goods and services required by modern civilization. This concept of sustainable develop-ment, with industrial activity and environmental protection both interdepen-dent, was one theme of the report by the Bruntland Commission which considered the environmental consequences of the industrialization of presently under-developed countries. Concern at the level of residual pesticides in food-stuffs is legitimate in the prosperous countries, but the wholesale banning of pesticides, including exports, is unlikely to help the starving in the Sahele. Differing priorities will lead to differing perceptions of pollution, and *vice versa*.

Pollution is, perhaps surprisingly, not defined in the Control of Pollution Act, 1974. As chemical analysis becomes ever more sensitive it is found that the occurrence of 'pure' substances or environments becomes ever more elusive— 'there is no such number as zero, only a limit of detection.' In order to distinguish between this insignificant contamination and pollution, the European Commission has defined pollution of aquatic systems as:

'The discharge by man of substances or energy into the aquatic environment, the results of which are such as to cause hazards to human health, harm to living resources and to the aquatic ecosystem, damage to amenities or interference with other legitimate uses of water.'

Thus hazard to health and/or other detriments are required, with the implication that impurities below the level of detriment constitute mere contamination rather than pollution. A similar distinction is implied by the Mahler criteria* (named after a former Chief Alkali Inspector) which underscore the UK's concept of best practicable means for controlling emissions to the atmosphere. Unfortunately this useful distinction between contamination and pollution simply transfers the subjectivity to related questions: how big a hazard to human health, how much harm to living resources, how severe the damage to amenities, before pollution can be proved?

A subsidiary question is who should judge whether the detriment is acceptable, and the different responses from environmentalists and industrialists illustrates the different world views of the protagonists. Industrialists, particularly those in the process industries, emphasize the need for scientific evidence of harm before pollution is admitted, whereas some environmentalists profess little faith in science. Industry, not surprisingly, gives importance to wealth creation, to tangibles and other material values; environmentalists, while acknowledging material benefits, advocate that the non-material 'quality of life' should have higher priority and for some the expression 'sustainable development' is self-contradictory. The argument is, at least partly, about onus of proof and the ordering of priorities for improvement (we *all* want a better future). From this, it appears that no single, generally acceptable distinction can be made between contamination and pollution—the latter is a value judgement made by the observer rather than a property of the observed environment.

It is for politicians to synthesize from these opposing views the legal and regulatory mechanisms which control industrial activity and modern life in general; it is by this means that the degree and extent of acceptable detriment to the environment is balanced against the manifest benefits derived from industrial development. In the UK, as elsewhere in Europe, all political parties are becoming 'greener,' partly in response to public concern about specific topics such as deterioration of the ozone layer, acid rain, and the greenhouse effect, partly in response to an anti-science view which opposes 'synthetics' and mistrusts expert opinion, and partly due to an anti-industry view. EEC Directives on environmental matters, many arising from 'Green' pressures in Germany, Holland, and Denmark, will increasingly dictate the regulatory scene in the UK. A Canute-like stand by industry against a tide of increasingly 'green' legislation would be contrary to public opinion and probably counter-productive, but industry, adopting a positive attitude to environmental, health,

* see section 2, below

and safety improvements, must seek to ensure that regulations are soundly based scientifically, are not beyond the limits of practicable technology, and do not have unforeseen and unintended adverse consequences. In pollution terms, some degree of disposal of industrial wastes must be permitted, but this must be limited such that the natural absorptive capacity of the environment, and degradation processes, prevent a cumulative reduction in environmental quality. In the following sections some aspects of the UK control procedures are outlined.

2 AIR POLLUTION

The legal framework for controlling and limiting air pollution in the UK is described in Chapter 14 where the long-standing concept of 'best practicable means' to limit emissions is explained. For scheduled processes, subject to control by Her Majesty's Inspectorate of Pollution (HMIP), the following criteria (by Mahler) apply to the emission of potential pollutants to the atmosphere:

1 No emission can be tolerated which constitutes a demonstrable health hazard, either short or long term.
2 Emissions, in terms of both concentration and mass, must be reduced to the lowest practicable amount.
3 Having secured the minimum practicable emission, the height of the discharge chimney must be arranged so that the residual emission is rendered harmless and inoffensive.

For heavy metals, vapours, and gases the ground level concentration to achieve these objectives is sometimes taken as one-fortieth of the occupational exposure level (OEL). This OEL is the concentration of a substance to which most people *at work* may be exposed for a normal working week, year upon year, without adverse effects on health; the divisor of forty is applied to compensate for longer exposure by the public and to give a safety factor for infants, the sick, and the elderly. Protection of public health is thus a first requirement, but the criteria also provide protection for buildings, animals, vegetation, amenity, *etc.*, where these are the critical targets. It will be apparent that emissions must be reduced where new scientific information reveals hitherto unknown effects or when technology develops new methods of reducing emissions; in the latter case a timetable for reductions by existing plants is agreed with HMIP.

The UK is no longer fully independent in prescribing environmental regulations, and 'best practicable means' will be replaced by 'best available technology not entailing excessive cost' (BATNEEC) as required by the EEC Directive on Emissions from Industrial Plant 84/360/EEC. The effects of this change cannot, with any certainty, be predicted because various EEC member countries are asserting that their present diverse systems are, in effect, BATNEEC, and that it is the *other* countries who must change their systems. The Commission intend that all member countries should operate the *same* system, and it is clear that some members will, to achieve this aim, have to modify their approach. The two main views of BATNEEC at present are:

1 A *substance-based* approach in which standard emission limits are set for a wide range of listed substances, to be applied regardless of the activity on the premises, with only minor relaxations permitted. This approach is codified in the TA Luft system used in the FRG.

2 A *process-based* approach be adopted similar to 'bpm,' which specifies the emission control system, operating requirements, *etc.*, as well as emission limits, on a case-by-case basis. By being more specific about, for example, materials handling, this approach can control fugitive emissions from stockpiles, roof louvres, *etc.*, sources which can be missed by a 'stack-only' approach.

A resolution of the meaning of BATNEEC in a European, rather than national, context is awaited, but it seems probable that for new plants the emphasis will be on 'BAT' rather than 'NEEC.' This would accord with the preventive principle, strongly advocated in the FRG, that all emissions should cease or be reduced to the technically feasible minimum regardless of whether environmental damage is being caused. A similar principle, but with more allowance for practicability (*i.e.*, cost factors *inter alia*) and with greater flexibility, is contained in bpm.

3 LIQUID EFFLUENT CONTROL

EEC Directives aimed at controlling pollution of rivers, estuaries, and coastal waters permit member states to operate either of two approaches: environmental quality objectives (EQO) or limit values (sometimes referred to as fixed emission standards). In the EQO approach the aim is to ensure that the *receiving water*, after mixing with discharges, is of a quality suitable for its designated use. Such uses include abstraction for potable water, abstraction for food processing or for agriculture, protection of fish, bathing and water contact sports, protection of shellfish, *etc.* Suitability for the designated use is determined by reference to environmental (water) quality standards which specify maximum limits for a wide range of contaminants such as cadmium, nickel, copper, arsenic, ammonia, pH, *etc.* Industrial discharges to the river are limited to ensure that the EQS values in the river are not exceeded; this is done on a case-by-case basis unlike the limit value approach which applies the same concentration limit to *all* discharges regardless of whether the receiving water is a sewer, river, or estuary. The UK is the only EEC member country to opt for control by the EQO system for all discharges, all other members adopting the limit value approach for the so-called 'black list' of the most dangerous pollutants (including cadmium, mercury, carcinogens, and some pesticides). For substances listed in Annex II of the Directive (the 'grey list'), including most of the heavy metals such as zinc, arsenic, copper, lead, and chromium, the parent Directive (Dangerous Substances in the Aquatic Environment) requires all countries to use the EQO approach, although it seems probable that the range of EQO values permitted by the Commission will be so restricted as to preclude the flexibility which characterizes the present system in the UK.

A significant advantage of the EQO approach is its ability to allow for inputs to rivers from diffuse sources, such as nitrates in agricultural run-off, whereas limit values apply only to point sources such as effluent pipes. Limit values, with the same permitted concentration applying to all discharge pipes regardless of whether discharging to a small stream or to a well-mixed estuary, ignore the capacity of different receiving waters to accommodate different loads without causing pollution. On the other hand, EQOs are generally derived from data in the literature on the toxicity of a pollutant to the relevant fish species, and where the data are deficient or contradictory the validity of the EQO is questionable. Limit values, usually technology based and increasingly including BATNEEC, have the advantage of simplicity both in specifying the emission limits and in minimizing the subsequent monitoring, but by ignoring the quality of the receiving water they do not allow for diffuse sources. Clearly a combination of the two approaches is required, and the UK has anticipated other EEC countries by announcing that, to coincide with the legislation associated with the privatization of the water industry, EQO values for UK rivers will be given statutory authority. Some pollutants which are considered to be toxic, bio-accumulative, and non-biodegradable, will comprise a 'red list' of substances (cadmium, mercury, several pesticides), and discharges containing these substances will be controlled by limit values based on BATNEEC while the quality of receiving waters will continue to be required to comply with the relevant EQOs.

4 SOLID WASTE DISPOSAL

A recent opinion poll shows unprecedented levels of public concern about man's effect on the environment, and worries about the disposal of chemical waste now rank higher than worries about the disposal of nuclear waste. The justification for such worry is debatable and may derive in part from the media habit of prefixing the word 'waste' with 'hazardous' or 'toxic,' and the assumption that all waste disposal operations are themselves hazardous to neighbours, but the belated realization by the control authorities that 'rubbish dumps' generate methane has done little to reassure the public that waste disposal is fully understood and adequately controlled. The perception of such dumps as litter-strewn and rat-infested, although generally outdated, still remains. To achieve public acceptance, or at least toleration, operating standards in the handling and final disposal of waste must be raised to the level already achieved by the best practitioners in the UK.

Waste disposal on land, whether to landfill or incineration, is regulated by the conditions specified in a site licence issued by the Waste Disposal Authority; conditions typically will specify the type and quantity of the waste permitted, site operating conditions such as access, fencing, depth of fill, and monitoring requirements such as leachate sampling. Detailed guidance on site licensing is given in the revised DoE Waste Management Paper No 4. Similarly, disposal of waste at sea is subject to licensing by the Ministry of Agriculture, Fisheries, and Food (MAFF), the licence specifying the type and quantity of waste and the location of the dumping grounds or the incineration area. Disposal at sea is now licensed by MAFF only if there is no practicable land-based alternative, and incineration at sea, already a costly option used mainly for noxious substances or those producing acid fumes on combustion, is to be phased out by the end of 1994. The ending of this option for disposal will increase the pressure for land-based alternatives, and a number of waste disposal contractors are planning to construct new low-polluting incinerators. The logic behind the ban on incineration at sea is unclear but is based, at least partly, on a view that disposal by dilution and dispersion will contribute to future environmental problems; a mistrust of disposal which occurs away from public view may also play a part. The arbitrary removal of disposal routes without evidence of need will reduce the value of the 'best practicable environmental option' philosophy.*

Organizations representing industrial waste producers are developing the idea of a 'duty of care' to raise waste disposal standards. Under common law a person owes a duty of care to 'persons who are so closely and directly affected by my act that I ought reasonably to have them in contemplation.' It is now proposed that a duty of care be included in statute law, accompanied by a code of practice which lists the specific actions a waste producer must take to demonstrate that a duty of care has been effected. One example of this concept, too detailed to be applied in all cases by a statutory duty, is that already being used by some members of the chemical and pharmaceutical industry. It includes some of the following:

—demonstration that a suitable disposal route and disposal contractor have been selected after a proper evaluation of the alternatives (*i.e.*, the waste producer has *not* simply accepted the cheapest offer and disclaimed all responsibility for what happens to his waste after it leaves his premises);

—giving adequate information to the transporter and disposal contractor to enable them to carry out their operations safely;

—making proper contractual arrangements for disposal of the waste (*i.e.*, it is *not* a 'no questions asked' arrangement);

—auditing the contractor's operations by visiting the disposal site to see if it is a well-run operation with which the producer is happy to be associated, checking on records of waste disposal, compliance with site licensing conditions, relationships with neighbours, *etc.*

Thus, although *legal ownership* of the waste may transfer from waste producer to the disposal contractor at the factory gate, the producer accepts a continuing responsibility for how his waste is handled. This is merely enlightened self-interest, as failure to improve disposal standards generally will make the creation of new disposal sites in future increasingly difficult in the face of an antagonistic and fearful public opinion.

5 INTEGRATED POLLUTION CONTROL

Regulatory control in the UK has traditionally segregated the above three topics of air pollution, liquid effluents and solid waste disposal, with little recognition at

* See section 5

the administrative level that they are interdependent: cleaning a gas by wet scrubbing converts a potential for air pollution into one for water pollution, and precipitation of dissolved substances converts it further into a solid waste disposal problem. The realization that these constitute alternative disposal options, each with its own potential for creating environmental problems, led the Royal Commission on Environmental Pollution to advocate, in their 5th Report (1976), a central inspectorate to 'be responsible for dealing with all pollution problems arising from industrial processes.' This inspectorate would, in the RCEP's view, aim to achieve the 'best practicable environment option' (BPEO) taking account of all waste streams from a process and the technical possibilities for dealing with it. The concept of BPEO was further developed by the RCEP in their 12th Report (1988).

A central inspectorate, HMIP, was formed in 1987 by amalgamating the former Industrial Air Pollution Inspectorate, the Hazardous Waste Inspectorate, and the Radiochemical Inspectorate. Its mandate concentrates on the control of pollution from *industrial sources* only and only from scheduled processes; it does not have direct responsibility for the control of pollution from diffuse sources, such as agriculture, nor does it have *direct* responsibility, other than auditing and advisory functions, for solid waste disposal which remains under the control of Waste Disposal Authorities. HMIP contains a new water pollution inspectorate whose functions are at present unclear but they will control discharges from scheduled processes, perhaps by specifying BATNEEC, and have some functions in the control of Red List substances discharged to sewers, rivers and estuaries. The main responsibility for control of discharges to water will lie with the newly created National Rivers Authority (NRA) who will control all non-Red List discharges to rivers, including discharges from sewage treatment works operated by the privatized water services companies (the present regional water authorities). The NRA will also be responsible for ensuring that river waters comply with the statutory water quality objectives which are yet to be published.

The way in which HMIP will seek to apply the BPEO principle is still unclear. The RCEP, in their 12th Report, indicate that the BPEO principle establishes the option that provides the most benefit (or least damage) to the environment as a whole, at acceptable cost, and argues that it is not enough to consider the BPEO for the disposal of a waste stream without also examining the production process to see whether the waste can be avoided, reduced, or its nature modified. Industry anticipates that HMIP will, for scheduled processes, require BAT-NEEC to apply to all waste disposal routes on new plants, with existing plants being updated on an agreed timetable. The application of BPEO, however, would suggest that applying BATNEEC to each *separate* disposal route will not, in itself, be considered sufficient to minimize environmental damage, and that a review of overall environmental effects may show that *all* waste from a process should be disposed of by one, or perhaps two, environmentally preferred routes. Establishing the BPEO will be, in many cases and particularly for scheduled processes, a joint activity between HMIP and the company concerned, and industry will welcome any advice that HMIP can offer on process modifications to reduce or avoid waste production. However, industry cannot be expected to welcome any dictation on the choice of process being used, as process choice is based on many interrelated factors such as price and availability of raw materials and energy, capital cost of plant, location of plant, product quality, *etc.*, with environmental factors but one in a complex of other considerations. Fortunately, the history of relationships between HMIP's predecessors and the process industry suggests that discussion and resolution of problems, rather than imposition of a decision, is likely to be the method used.

The prevention of pollution from diffuse sources seems likely to be controlled by restricting the use, or perhaps the production, of the polluting material. Thus the production of CFCs is controlled under an international agreement (the Montreal Protocol) because of the effect of some CFCs on stratospheric ozone, and paints containing tributyltin compounds have been banned for use on small craft because of the toxic effects of organotin compounds on shellfish. Restrictions on the use of nitrate fertilizers may be introduced to protect river waters which are abstracted for potable water. There is, as yet, no suggestion that BPEO be formally extended to study such problems as the use of nitrate fertilizer, where the benefits of such use might be compared with the environmental consequences, alternatives to the use of nitrates would be considered, and options for the removal of nitrate from water compared with the case for restriction of use. It is perhaps in these major decision areas, such as balancing the merits and disadvantages of nuclear power *versus* fossil fuel power stations, that BPEO concept could most usefully be applied. For many process plants the BPEO for waste disposal will be self-evident, and the concept will be devalued if the phrase is used for trivial applications. Limitation of manpower resources within HMIP will also make it impractical to apply BPEO rigorously to every scheduled plant.

6 EMPLOYEE PROTECTION

Although public perceptions of pollution, and media attention, focus naturally on the external environment, for industry the protection of workers against risk to health at work is equally important. The Health and Safety at Work Act, 1974 lays a general duty on employers to ensure, so far as is reasonably practicable, the health, safety, and welfare of all his employees. This duty has been developed and refined by regulations, sometimes applying to specific industries and substances (The Control of Lead at Work Regulations, 1980), and by Guidance Notes and Codes of Practice issued by the Health and Safety Executive. Typical of such detailed guidance is the HSE Guidance Note EH40, published each year, which lists occupational exposure limits for about 750 gases, vapours, and airborne dusts.

These industry- and substance-specific guides and regulations have been of particular use in the process industries, particularly the chemical manufacturing industry. The general duty in the 1974 Act, however, applies to *all* employers and workers, and it has been elaborated in the Control of Substances Hazardous to Health Regulations (COSHH) which were introduced in October, 1988. These Regulations aim to protect workers against risk to health from substances handled at work, such substances including gases, vapours, liquids, solids, dusts, and micro-organisms, and the Regulations apply to *all* workplaces, including garages, workshops, shops, offices, laboratories, *etc.*, as well as other service and all manufacturing industry. The Regulations require:

—a *suitable* and *sufficient* assessment of the risks to the health of employees, based on a review of the substances to which workers may be exposed, the degree of exposure and the toxicity of the substances;

—that employers prevent the exposure of employees to substances hazardous to health, or, if this is not reasonably practicable, that exposure be adequately controlled;

—that the measures used to control exposure are checked regularly to ensure their continuing effectiveness;

—that, where relevant, employee exposure be monitored by measuring the airborne concentrations of hazardous substances;

—that, where relevant, employees are subject to health surveillance;

—that records of personal exposure, and the corresponding medical records, be kept for 30 years (for future epidemiological research into the relationship between illness and occupational exposure to hazardous substances).

The Regulations *prohibit* any work which is liable to expose any employee to a substance hazardous to health unless the employer has carried out the assessment referred to above. There can be little doubt that, properly implemented and enforced, the COSHH Regulations will control the incidence of occupationally-induced ill health. The establishing of adequate in-plant control, however, depends on adequate assessment of the toxicity of substances, and the setting of proper control limits, by the regulatory authorities. These are particularly problematical in the case of carcinogens, where some groups believe there is no such thing as a safe limit ('one fibre of asbestos can kill you').

7   CARCINOGENS

The COSHH Regulations do not explicitly include carcinogens, but a supplementary Code of Practice on the Control of Carcinogenic Substances indicates that the COSHH philosophy applies: assessment, control of exposure, monitoring of exposure, health surveillance. The particular difficulties and limitations of health surveillance for carcinogens are recognized, where a latent period of perhaps 20 years can occur between exposure and the development of clinical symptoms, and employees may be advised of the need for continuing health surveillance after exposure has ceased.

The methods of identifying the carcinogenic potential of chemicals are:

—Epidemiology, where the incidence of disease in an exposed group is compared with that in a similar non-exposed group, with allowance made for the inevitable minor differences between the groups. This technique is useful in confirming suspicions, and may identify a process as hazardous even though a specific chemical cause is not identified. It is, of course, a historical review, not predictive.

—Animal experiments, usually on rats and mice, is a predictive technique and does not involve exposure of humans. Results may be misleading because of differences in the metabolic pathways between humans and test animals, and due to the incidence of spontaneous tumours in test animals there is a high statistical 'noise' against which the effect must be sought. This means that many tests are carried out at such high doses as to make the results of doubtful relevance to man.

—*In vitro* tests, which assume a link between cancer and genetic damage, look at changes occurring in cultured cells, yeasts, or bacteria (the 'Ames test'). These tests are comparatively inexpensive, do not involve animals, but are not infallible: for example, they fail to detect asbestos and give a false-positive for vitamin C.

—Chemical structure, particularly similarity to a known carcinogen, is a useful preliminary screen. For example, nitrosamines as a group are suspect carcinogens.

The most widely regarded list of carcinogens is that based on the reviews by the International Agency for Research on Cancer (IARC), part of the World Health Organization. Of 734 separate substances reviewed by IARC, 30 substances or groups of substances are judged to be associated with cancer in humans, including arsenic, asbestos, benzene, chromium, and some of its compounds, and vinyl chloride. A further 68 substances or groups, including cadmium and nickel and some of their compounds, are considered as probable human carcinogens, and a further 214 have produced 'sufficient evidence' of carcinogenicity in test animals.

The IARC classification focuses only on hazard, an intrinsic property of a substance, and it is IARC policy to avoid quantifying the risk of cancer arising from exposure to listed carcinogens. This, and other, qualifying aspects of the IARC list may be overlooked by regulatory authorities who tend to the administratively simple view that if a substance, mixture, or article contains a listed carcinogen it must itself be a carcinogen. If the 'ionic theory' of carcinogenicity of metal salts prevails (if one compound of a metal is a carcinogen then all other compounds must also be considered as carcinogens), the number of listed carcinogens will increase dramatically. Nickel now has generic classification as a carcinogen in the USA although only two compounds, the sub-sulphide and carbonyl, found only in the smelting industry, are known carcinogens. Alloys are treated as though they are simple mixtures, so that alloys which are imported to the USA and which contain nickel must bear a 'may cause cancer' label. This logic, which might lead to the stainless steel from which many everyday items such as cutlery are made being labelled as possible carcinogens, is surely mistaken. Warning labels based on 'right to know' will lose their psychological impact if applied on the basis of hazard rather than on risk.

8 ACTIONS BY THE PROCESS INDUSTRIES

Faced with this apparent deluge of restriction and regulation, what should be the attitude of manufacturing industry? Industry recognizes that it exists only by consent, such consent applying to products as well as to production processes, but it is hard for the science-based process industries to accept that a belief in the ability of science to provide all with a better future is on the wane. Public perceptions may be inconsistent—a 'medicine' that cures us invokes gratitude to our doctor, but if it has an unpleasant side effect this is the fault of the 'drug' manufacturer. Nevertheless most people will wish to be on the side of the angels in environmental matters, and the need to ensure that future products are 'green' or 'environmentally friendly' is predictable. 'Biodegradable' may become more than a marketing ploy, rather a real requirement where practicable. Consideration may have to be given to how consumer products are to be disposed of at the end of their useful life: a small industry is presently forming to recover CFCs from refrigerators, freezers, and air conditioners to prevent release of the CFCs from waste tips, and the original manufacturers of CFCs are arranging the recycling or destruction of this recovered material. This concept of 'product stewardship' is growing within the chemical industry and is likely to become more widely recognized.

Regardless of the views of any UK government on deregulation of industry, new regulations will continue to flow from the European Commission. Industry must ensure that its views on new legislation are adequately represented at a sufficiently early stage, usually through trade associations informing the relevant UK government department which form the official channels to the EC Directorates (usually the DoE, although DTI should be encouraged to play a more significant role). Industry groups whose trade associations are not alert to developments in Brussels are likely to suffer by being inadvertently overlooked when the effects of legislation are being considered. Direct contact with MPs on matters of concern to individual companies may produce more response than a letter to a departmental minister.

In the control of pollution from existing manufacturing operations industry in the UK will seek a system which:

—is based on criteria which are scientifically sound;
—allows consultation between the control authorities and the controlled to ensure a workable scheme which, when implemented, achieves the intended result;
—puts emphasis on field inspection and enforcement by an adequately resourced and technically capable staff (a legalistic procedure which emphasizes permits but underplays monitoring, as with some aspects of the USA system, with decisions determined by a slow process through the courts, would not be acceptable);
—treats the environment as an integrated whole, where necessary applying the 'best practicable environmental option' to the disposal of wastes and emissions which inevitably arise.

The control system must also be sufficiently open, intelligible, and accessible to the public on whose behalf it is being conducted. HMIP must have adequate resources and operate with sufficient visibility to maintain credibility and public confidence in the control system. Without such confidence self-appointed groups of concerned people will clamour for immediate action over perceived and imagined sources of pollution, with government action based on populism rather than reason.

CHAPTER 16

Organometallic Compounds in the Environment

J. R. ASHBY and P. J. CRAIG

1 INTRODUCTION

1.1 Scope

Organometallic compounds usually enter the environment following use as commodities, although there are instances of the formation of these compounds in the environment (*e.g.*, mercury methylation). Some organometallic products are applied directly to the environment as biocides, in anti-fouling paints, or in petrol. Others reach the wider environment indirectly (*e.g.*, leaching from organotin-based PVC stabilizers). In general it is necessary to consider not only the direct toxicity of the compound but the toxicities of possible metabolites at points other than those of initial application. Formation of organometallic compounds in the environment is important because the organic derivatives usually are of greater toxicity than their parent inorganic metals or ions. Hence the complete cycling or transport of the original organometallic compound in the environment must be considered.

The scope of this chapter includes only compounds with metal–carbon bonds. Metal organic complexes and inorganic reactions are covered elsewhere. Some metalloidal elements which form methyl derivatives in the environment are included (*e.g.*, arsenic and some aspects of selenium chemistry). Legal aspects and organometallic compounds in the workplace are also covered elsewhere. Certain veterinary uses of organometallic compounds do have environmental impact and this will be noted. Theoretical aspects of environmental methylation (biomethylation) will be mentioned only where directly relevant to the environmental situation. A recent discussion of this topic is available.[1]

[1] P. J. Craig, ed., 'Organometallic Compounds in the Environment,' Longman, 1986.

With regard to environmental methylation an important difficulty has concerned the low levels at which some of these compounds are formed. Methyl mercury is generated at the part per billion level $(ng\,g^{-1})$, and the methyl tin derivatives have been observed at the part per trillion level $(pg\,g^{-1})$. These observations are significant; continuous formation of organometallic compounds in the environment even at low levels may result in food chain effects leading to much higher concentrations in organisms. The main analytical problem seems to be the adaptation of techniques that work well on standard test materials for operation in a mixed environmental matrix. Analytical problems are noted as they arise.

1.2 Source Material

Tin, lead, and mercury, whose organometallic compounds are of most environmental concern, have been reviewed in detail in monographs. Environmental concentrations, commercial uses, toxicity, persistence and methylation of the metal and compounds are discussed.[2-4] The cycling of metals and organometals in the environment has been reviewed recently,[5] and is also covered in a single volume work.[6] Biological methylation is discussed in a number of series.[7] Organometallic compounds in the environment was the subject of an American Chemical Society (ACS) monograph.[8] The US National Academy of Science and the ACS have published information on lead and mercury in the environment.[9,10] The World Health Organization has produced a series of monographs on Environmental Health Criteria, and mercury, lead, and tin organometallic compounds have been covered.[11-13] Organometallic compounds used for medicinal or veterinary use are noted in the respective Pharmacopoeiae.[14-16] Other

[2] 'The Biogeochemistry of Lead in the Environment,' (vols 1 and 2), ed. J. O. Nriagu, Elsevier North Holland, 1978.
[3] 'The Biochemistry of Mercury in the Environment,' ed. J. O. Nriagu, Elsevier North Holland, 1978.
[4] 'Organotin Compounds: New Chemistry and Applications,' ed. J. J. Zuckerman, ACS Advances in Chemistry Series, 157, ACS Washington, DC, 1976.
[5] P. J. Craig, in 'Handbook of Environmental Chemistry,' vol. 1, part 4, ed. O. Hutzinger, Springer Verlag, Berlin, and Heidelberg, 1980, p. 169.
[6] 'Metal Pollution in the Aquatic Environment,' ed. U. Forstner and C. T. W. Wittman, Springer Verlag, Berlin, Heidelberg, and New York, 1979.
[7] 'The Biological Alkylation of Heavy Elements,' ed. P. J. Craig and F. Glockling, Special Publication No. 66, The Royal Society of Chemistry, London, 1988.
[8] 'Organometals and Organometalloids—Occurrence and Fate in the Environment,' ed. F. E. Brinckman and J. M. Bellama, ACS Symposium Series No. 82, ACS Washington, DC, 1978.
[9] 'Trace Elements in the Environment,' ed. E. L. Kothny, ACS Adv. Chem. Ser. No. 123, ACS Washington, DC, 1973.
[10] 'An Assessment of Mercury in the Environment,' National Academy of Science, Washington, DC, 1977.
[11] 'Environmental Health Criteria 1, Mercury,' World Health Organization, Geneva, 1978.
[12] 'Environmental Health Criteria 3, Lead,' World Health Organization, Geneva, 1977.
[13] 'Environmental Health Criteria 15, Tin,' World Health Organization, Geneva, 1980.
[14] 'British Pharmacopoeia' (Vols. I and II), HMSO, London, 1980.
[15] Martindale, 'The Extra Pharmacopoeia,' The Pharmaceutical Press, 27th edn., London, 1979.
[16] 'British Pharmacopoeia,' (Veterinary), HMSO, London, 1977.

organizations have reported on mercury in the environment.[17-18] There is a recent monograph on lead in the environment.[19] The degradation properties of organotin compounds have also been reviewed recently.[20,21]

1.3 Units and Abbreviations

SI units are used where concentrations are quoted. These are related to 'trivial' units, as follows.

SI Unit	*Trivial Unit*
$\mu g\,g^{-1}$ ($mg\,kg^{-1}$)	ppm
$ng\,g^{-1}$ ($\mu g\,kg^{-1}$)	ppb
$pg\,g^{-1}$	ppt
$\mu g\,dm^{-3}$	$\mu g\,L^{-1}$
$\mu g\,m^{-3}$	

Concentrations unless otherwise noted refer to 'dry weight' of the matrix (*e.g.*, sediments or fish).

Where an element or simple compound is being referred to, the full terminology is used (*e.g.*, lead or methyl mercury). Complex materials are abbreviated (*e.g.*, Me_3Sn^+ and Me_4Pb). Methyl mercury refers to monomethyl mercury ($MeHg^+$) and dimethyl mercury appears in full. $MeCoB_{12}$ is methyl cobalamin; PVC is poly(vinyl)chloride. Alkyl groups are abbreviated conventionally. Analytical techniques are listed in the normal abbreviated form (*e.g.*, atomic absorption is written AA).

2 ORGANOTIN COMPOUNDS

2.1 Use and Toxicity

The chief uses of organotin compounds are as stabilizers for PVC and as biocides and both led to the introduction of these compounds to the environment. As a result their toxicity and their breakdown pattern are of importance in any consideration of their environmental role. Dialkyl tin compounds are used for polymer stabilization and triorganotin species have biocidal properties. Production of organotin compounds is more than 40 000 tonnes annually. A number of reviews on the organotin compounds exist.[20-24]

[17] 'Mercury and the Environment—Studies on Mercury Use and Emission, Biological Impact and Control, Organization for Economic Co-operation and Development,' Paris, 1974.
[18] 'Mercury Contamination in Man and his Environment,' International Atomic Energy Agency, Vienna, 1972.
[19] R. M. Harrison and D. P. Laxen, 'Lead Pollution, Causes, and Control,' Chapman and Hall, 1981.
[20] S. J. Blunden, L. A. Hobbs, and P. J. Smith, 'Environmental Chemistry,' ed. H. J. M. Bowen (Specialist Periodical Reports), The Royal Society of Chemistry, London, Vol. 3, 1984.
[21] S. J. Blunden and A. H. Chapman, *Environ. Tech. Lett.*, 1982, **3**, 267.
[22] 'Organotin compounds' (Vols 1–3), ed. A. W. Sawyer, M. Dekker, New York, 1971–72.
[23] R. C. Poller, 'The Chemistry of Organotin Compounds,' Logos Press, London, 1970.
[24] P. J. Craig, *Environ. Tech. Lett.* 1980, **1**, 225.

2.1.1 Triorganotin Compounds as Biocides. The more important feature is the nature of the organic group, rather than the anionic group. The main biocidal applications are summarized in Table 1, and account for about 8 000 tonnes of organotin production.

Table 1 *Some uses of triorganotin compounds*

Ph_3SnOAc	Fungicide, anti-fouling paint
Ph_3SnOH	Fungicide, anti-fouling paint
Ph_3SnCl	Anti-fouling paint
Ph_3SnF	Anti-fouling paint
$Ph_3SnSCSNMe_2$	Anti-fouling paint
$Ph_3SnOCOCH_2Cl$	Anti-fouling paint
$Ph_3SnOCOC_5H_4N_3$	Anti-fouling paint
$Ph_3SnOCOCH_2CBr_2COOSnPh_3$	Anti-fouling paint
Bu_3SnOAc	Anti-fouling paint
$Bu_3SnOCOPh$	Disinfectant
Bu_3SnCl	Rodent repellant
Bu_3SnF	Anti-fouling paint
$(Bu_3Sn)_2O$ (TBTO)	Fungicide: bacteriostatic (wood or stone preservative)
	Anti-fouling paint
Bu_3Sn adipate	Anti-fouling paint
Bu_3Sn benzoate	Germicide
Bu_3Sn methacrylate (copolymer)	Anti-fouling paint
$Bu_3SnOCOCH_2(CBr_2COOSnBu_3)$	Anti-fouling paint
Bu_3Sn (naphthenate)	Bacteriostatic (wood preservative)
$(Bu_3Sn)_3PO_4$	Bacteriostatic (wood preservative)
$(c\text{-}C_6H_{11})_3SnOH$	Insecticide (orchards)
$(c\text{-}C_6H_{11})_3Sn-N{\overset{N=C}{\underset{C=N}{\Big|}}}$	Insecticide
$[(PhC(Me)_2CH_2)_3Sn]_2O$	Insecticide, acaricide
$Et_3Sn(p\text{-}OC_6H_4Br)$	Nematocide
$(CH_2{=}CH)_3SnCl$	Herbicide
Me_6Sn_2	Insecticide

It can be seen from Table 1 that triorganotin compounds are introduced direct into the environment. Hence knowledge of their toxicity and degradation properties in the environment has been an important research goal in recent years. Toxic effects for organotin compounds are at a maximum when three organic groups are present (*i.e.*, R_3SnX). Usually maximum toxicity occurs for R = methyl to butyl; higher trialkyl tin species are of little toxicity. The basic cause of acute triorganotin toxicity arises from disruption of ion transport, and inhibition of ATP synthesis. Choice of alkyl group in the triorganotin biocides represents a balance between the excessive phytotoxicity of the lower alkyls and insufficient toxicity for the higher alkyl compounds.

Decay in the environment is most likely to occur through absorption of u.v. light, biological action, or chemical means. Essentially, decay occurs by stepwise loss of the organic group and the final product has been postulated as the non-toxic dioxide (cassiterite). This decay to progressively less toxic materials is an attractive environmental feature of organotin products. It does appear however, that the picture is not quite as simple as this; for example, many studies suggest that by no means all of the introduced tin is finally converted to the oxide. *In vitro* studies suggest that decay does occur to the oxide, but when biological or light induced decay studies are carried out in a realistic environmental model, decay is often incomplete and apparently stable organotin products may remain. This tends to be confirmed by the recent detections of methyl and butyl tin species at low concentrations (ng dm^{-3}) in environmentally dispersed situations, *e.g.* rivers, rainwater, *etc.* This suggests a greater stability in the environment than some laboratory studies on pure compounds have implied. That pure triphenyl-, dibutyl-, and dialkyl-tin compounds undergo stepwise decay under u.v. light to inorganic tin under abiotic conditions should be assessed from this point of view.[20,21]

Biological metabolism of trialkyltins has been modelled by the rat liver microsomal monooxygenase system and occurs mainly by hydroxylation of carbon atoms on the organotin.[25-27] The hydroxy metabolites decay to dialkyltin compounds by tin–carbon bond cleavage with further carbon hydroxylation and cleavage to tin(IV) oxide. In the environment similar biological oxidations are believed to decay other trialkyltins to inorganic tin via successive loss of hydroxyalkyl groups, although the mechanism may be more complex than that originally suggested.[28] Triphenyltin is metabolized but not by a hydroxylation route.[29] Aqueous triphenyl tin compounds are degraded by homolytic cleavage of the tin–carbon bonds to Ph_2SnO when exposed to light. Neither Ph_4Sn, monophenyl tin compounds nor inorganic tin are found as products, but a water-soluble organotin polymer is formed $(PhSnO_xH_y)_n$. A similar distribution of products is observed for the decomposition of aqueous di- and mono-phenyltin species.[29] It has previously been concluded that Ph_3SnOAc degrades through the di- and mono-compounds to inorganic tin[30] by simple sequential loss of phenyl groups to tin(IV) oxide.

In another study half an application of $(Bu_3Sn)_2O$ disappeared from unsterilized silt and sand loams in approximately 15 and 20 weeks. Small amounts of dibutyl tin derivatives were formed, and carbon dioxide from the butyl groups was evolved. Unextractable tin-containing residues were also formed in the soil. This was ascribed to irreversible adsorption to soil constituents but may have been an organotin polymer, or it could also be tin oxide. The mechanism

[25] R. H. Fish, E. C. Kimmel, and J. E. Casida, in Ref. 4, p. 197.
[26] E. H. Fish, J. E. Casida, and E. C. Kimmel, in Ref. 8, p. 82.
[27] E. C. Kimmel, J. E. Casida, and R. H. Fish, *J. Agric. Food Chem.*, 1980, **28**, 117.
[28] E. H. Blair, *Environ. Qual. Saf. Suppl.*, 1975, **1**, 406.
[29] C. J. Soderquist and D. G. Crosby, *J. Agric. Food Chem.*, 1980, **28**, 111.
[30] C. J. Evans, 'Tin and Its Uses,' No. 100, 1974, 3.

suggested was hydroxylation.[31] Similar decay to the dibutyl stage in spent anti-fouling coatings has also been reported.[32]

It has also been shown that residues of $(c\text{-}C_6H_{11})_3SnOH$ on apples and pears decline by 50% in about three weeks due to photodegradation. A 20–50% reduction may be achieved by washing.

Most studies show that triorganotin compounds are strongly adsorbed onto soil. Over a longer period however studies have not tended to show a build up of tin in the soil to beyond average natural levels in soils. It is likely that environmental leaching over periods of years removes more tin from the soil than is suggested by laboratory studies existing over periods of weeks.

At the present time the most important cause of organotin pollution to the aqueous environment arises from the tributyl tin used in marine anti-fouling materials.[33] The reported effects on commercial oysters (*Crassostrae gigas*) was first attributed to the proximity of boat yards and marinas in 1980,[34] and more recent work[35,36] has confirmed that TBTO can cause shell thickening deformities and abnormally weak larva in these oysters. The accumulating evidence[37-40] that TBTO, at very low levels ($20\ ng\ dm^{-3}$ or lower) is toxic to commercial oysters and a wide range of other aquatic organisms, has led to the banning of this compound in England for use in anti-fouling paints used on boats of less than 25 metres offered for retail sale (May, 1987). Other countries have also banned TBT based anti-fouling paints for use on small boats, *e.g.*, France in 1982, Ireland in 1987, and legislation is under way in America, and the Scandinavian countries. For information on TBT containing anti-fouling paints, their effects, and environmental analysis see Oceans 86, 87, 88[41] and Organotin compounds in the Aquatic Environment.[42] Here release is direct from the ship's hull to the surrounding water and is more environmentally significant in harbours than the open sea. The organotin species adhere to the sediment materials rather than remain in solution in water and this matrix appears to be a

[31] D. Barug and J. W. Vonk, *Pestic. Sci.* 1980, **11**, 77.

[32] W. R. Blair, J. A. Jackson, G. J. Olsen, F. E. Brinckman, and W. P. Iverson, in 'Abstracts Intern. Conf. Heavy Metals Environ.,' Amsterdam, Sept., 1981.

[33] C. J. Evans and P. J. Smith, *J. Oil Col. Chem. Assoc.*, 1975, **58**, 160.

[34] C. Alzieu, Y. Thibaud, M. Heral, and B. Bouter, *Rev. Trav. Inst. Pech. Marit.*, 1980, **44**, (4), 301–349.

[35] M. J. Waldock and J. E. Thain, *Aquaculture*, 1985, **44**, 133.

[36] C. Alzieu and M. Heral, 'In Ecotoxicological testup for the marine environment,' ed. G. Persoone, E. Jaspers, and C. Claus, State Univ. Chent and Inst. Mar. Scient. Res., Bredene, Belgium Vol 2, 187.

[37] G. W. Bryan, P. E. Gibbs, L. G. Hummerstone, and G. R. Burt, *J. Mar. Biol. Ass. UK*, 1986, **66**, 611.

[38] B. D. Bayne, *Ophelia*, 1965, **2**, 1.

[39] M. J. Waldock, J. E. Thain, and M. E. Waite, *Appl. Organometall. Chem.*, 1987, **1**, 287.

[40] G. E. Walsh, L. L. McLaughlin, E. M. Lores, M. K. Louie, and C. H. Deans, *Chemosphere*, 1985, **14**, 383.

[41] 'Oceans 86, 87, and 88,' (Organotin Symposium) available from IEEE Science Center, 445 Hoes Lane, Piscataway, NJ 08854, USA.

[42] 'Organotin compounds in the Aquatic Environment. Scientific criteria for assessing the effects on environmental quality,' J. A. J. Thompson, M. G. Sheffer, R. C. Pierce, Y. K. Chau, J. J. Cooney, W. R. Cullen, and R. J. Maguire. NRCC, Report No. 22494, NRC Canada, Ottawa, Canada, 1985.

potential source of any methylation of tin occurring in the environment. Significant methyl tin compounds have been analysed from harbour waters and sediments.

It can be concluded that the net result of chemical and biological degradation in the environment for alkyltins is a stepwise loss of alkyl groups leading eventually to tin(IV) oxide. Half-lives range from seconds to days and longer in soil. What is uncertain is whether conversion to inorganic tin is always complete, what the timescale for complete conversion is, and whether or not organotins are in effect persistent final decay products. However studies with $(Bu_3Sn)_2O$ at the $100\,\mu g\,g^{-1}$ level on soil show no adverse effects on micro-organisms or fertility.

Table 2   *Some uses of diorganotin compounds*

$[n\text{-}Oct_2Sn(C_4H_2O_4)]_n$ (maleate polymer)	PVC stabilization (including food contact)
$[n\text{-}Oct_2Sn(SCH_2COOi\text{-}Oct)]_2$	PVC stabilization (including food contact)
$Bu_2Sn(OAC)_2$	Polyurethane catalysts; cold curing of silicones
$Bu_2Sn(OCOi\text{-}Oct)_2$	Polyurethane catalysts; cold curing of silicones
$Bu_2Sn(OCOC_{11}H_{23})_2$ (dilaurate)	PVC stabilization; de-worming of chickens, catalysis
$Bu_2Sn(SCH_2COOi\text{-}Oct)_2$	PVC stabilization
$[n\text{-}Bu_2Sn(C_4H_2O_4)])_n$	PVC stabilization
$Bu_2Sn(OCOCH{=}CHCOOOct)_2$	PVC stabilization
$Bu_2Sn(SC_{12}H_{25})_2$	PVC stabilization
$Bu_2Sn(OCOC_{12}H_{25})_2$	Catalysis
$[Bu_2SnO]_n$	Catalysis
Bu_2SnCl_2	Glass strengthening; precursor for SnO_2
Me_2SnCl_2	Glass strengthening; precursor for SnO_2
$Me_2Sn(SCH_2COOi\text{-}Oct)_2$	Heat stabilizer and food contact (PVC)
$(BuOCOCH_2CH_2)_2Sn(SCH_2COOi\text{-}Oct)_2$	PVC stabilization
Me_6Sn_2	Insecticide; antifeedant

Table 3   *Some uses of monoorganotin compounds*

Compound	Use
$MeSn(SCH_2COOi\text{-}Oct)_3$	PVC stabilization
$MeSnCl_3$	Glass strengthening, SnO_2 precursor
$Bu(SCH_2COOi\text{-}Oct)_3$	PVC stabilization
$BuSnCl_3$	Glass strengthening
$(BuSnS_{1.5})_4$	PVC stabilization
$BuSn(OH)_2Cl$	Catalysts for transesterification
$[BuSn(O)OH]_n$	Catalysis
$OctSn(SCH_2COOi\text{-}Oct)_3$	PVC stabilization
$BuOCOCH_2CH_2Sn(SCH_2COOi\text{-}Oct)_3$	PVC stabilization

2.1.2 Mono- and Diorganotin Compounds as Catalysts or Stabilizers. About 40 000 tonnes of mono- and diorganotins are used as heat and light stabilizers for rigid PVC; 2 000 tonnes are used as homogeneous catalysts for silicone or polyurethane manufacture and transesterification reactions (see Tables 2 and 3). Here the toxicities of the organotin compounds are incidental to the desired use. The mechanisms of the tin compounds for these uses have been discussed.[1,43] The main routes by which the mono- or diorganotin compounds reach the environment are by leaching from PVC through weathering or solvent action, transport from land burial, or transport to atmosphere from incineration of waste products. Leaching from PVC is of most significance where the plastics are used in food contact or as potable water conduits. (For the latter use about 3 000 tonnes of methyltins were used in the USA in 1981. This low toxicity of methyl derivatives is unusual.)

The less toxic mono- and diorganotin compounds degrade along the same pathway as the triorgano products. The concentrations of PVC stabilizers used are at the 0.5–2.0% level. Leaching rates are low, making tin-stabilized PVC acceptable for food contact use, although there have been adverse tissue reactions in medical applications.[21] It is likely that direct biocidal use of triorganotin compounds, although smaller in tonnage terms, leads to a greater introduction of organotin compounds into the natural environment by virtue of the mode of application.

The toxic effects of the lower dialkyl tin compounds are due to their reaction with sulphydryl groups and interference with α-ketoacid oxidation.[44] The higher alkyls are of little toxicity.

Calculations of the amounts of organotin compounds reaching the environment after use have been made for the USA for 1976.[45] A number of detailed compilations of LD_{50} and LC_{50} doses for organotin compounds have been collected particularly by the International Tin Research Institute, London.[46]

2.2 Detection and Transformation of Organotin Compounds in the Environment

A number of organotin compounds have been detected in the environment both close to and remote from the points of application (Table 4). Detection of methyltin derivatives raises the question of their formation. Owing to the use of methyltin compounds in PVC water pipes, the environmental methyltins need not be the product of a natural methylation process in the environment (biomethylation). The levels reported are in the $ng\,dm^{-3}$ range in air to $\mu g\,dm^{-3}$ in urine[47]; $ng\,dm^{-3}$ levels in water samples were also reported.[48] Two other groups in North America have recently reported the presence of methyl tin

[43] B. Sugavanam, 'Tin and Its Uses,' No. 126, 1980, 4.
[44] P. J. Smith and L. Smith, *Chem. Br.*, 1975, **11**, 208.
[45] J. J. Zuckerman, R. P. Reisdorf, H. V. Ellis, III, and R. R. Wilkinson in Ref. 8, p. 397.
[46] P. J. Smith, 'Toxicological Data on Organotin Compounds,' Intern. Tin Res. Inst., Pub. No. 538.
[47] P. S. Braman and M. A. Tompkins, *Anal. Chem.*, 1979, **51**, 12.
[48] V. F. Hodge, S. L. Seidel, and E. D. Goldberg, *Anal. Chem.*, 1979, **51**, 1256.

compounds in waters, one group at the $\mu g\,dm^{-3}$ levels.[49,50] At least two reports[51,52] have shown mixed methyl-butyl tin species present in environmental samples. The possibility exists that butyl tin compounds from anti-fouling paint have been biomethylated but it is also possible that mixed methyl–butyl tin compounds were present as impurities in the original paint. Mixed methyl–phenyl or methyl–octyl tin compounds have not been detected as might be expected if biomethylation of tin was occurring. Possibly complete degradation to inorganic tin has to occur first or even replacement of the other organic groups by methyl.

Table 4   *Organotin species detected in the environment**

Me_4Sn	$BuSn^{3+}$	Bu_3Sn^+
Me_3Sn^+	$BuSnH_3$	$BuSnMe_3$
Me_2Sn^{2+}	Me_3SnH	Bu_2SnMe_2
$MeSn^{3+}$	Me_2SnH_2	Bu_3SnMe
Bu_2Sn^{2+}	$MeSnH_3$	

* Detected at other than the points of application of commercial organotin compounds.

There is now a substantial body of results reporting the presence of tributyl tin and its degradation products in water, see for example 41, 42, 53–55. Less is known about the levels of butyl tins in sediments. Some levels found include $2\,340\,ng\,dm^{-3}$ of TBT in water[56] taken from Port Hope, Ontario, Canada, $680\,ng\,dm^{-3}$ from the River Yealm, $230\,ng\,dm^{-3}$ in the Dart, and $270\,ng\,dm^{-3}$ in the Avon[57] all of which are in the South of England. Sediment levels of 10.78, 9.32, and $7.35\,\mu g\,g^{-1}$ from various sites in Vancouver Harbour, Canada, have also been reported.[58]

A number of modern analytical methods have been used to detect low levels of organotin compounds in the environment. Borohydride has been used to convert the organotin species to volatile organotin hydrides. These are trapped and examined later by atomic emission, absorption, mass spectroscopic, or GC flame photometric techniques. The use of Grignard agents to ethylate, or pentylate, after extraction and concentration is used extensively,[51,52,57–59] and recent

[49] Y. K. Chau, P. T. S. Wong, and G. A. Bengert, *Anal. Chem.*, 1982, **54**, 246.
[50] J. A. Jackson, W. R. Blair, F. E. Brinckman, and W. P. Iverson, *Environ. Sci. Technol.*, 1982, **16**, 110.
[51] R. J. Maguire, *Environ. Sci. Technol.*, 1984, **18**, 291.
[52] R. J. Maguire, R. J. Tkacz, Y. K. Chau, G. A. Bengert, and P. T. S. Wong, *Chemosphere*, 1986, **15**, 253.
[53] M. Stallard, V. Hodge, and E. D. Goldberg, *Environ. Monit. and Asses.*, 1987, **9**, 195.
[54] D. V. Hall, M. J. Lenkevich, S. W. Hall, A. E. Pinkney, and S. J. Bushong, *Marine. Poll. Bull.*, 1987, **18**, 2, 78.
[55] P. W. Balls, *Aquaculture*, 1987, **65**, 227.
[56] R. J. Maguire and R. J. Tkacz, *J. Chrom.*, 1983, **286**, 99.
[57] J. R. Ashby and P. J. Craig, *Sci. Total Env.*, 1989, **78**, 219.
[58] R. J. Maguire, Y. K. Chau, G. A. Bengent, and E. J. Hale, *Environ. Sci. Technol.*, 1982, **16**, 698.
[59] M. D. Muller, *Anal. Chem.*, 1987, **59**, 617.

developments using the derivatization agent sodium tetra-ethyl borate have allowed an easier ethylation procedure to be used.[57] Some workers use hydride-generation followed by cryogenic trapping.[60,61] A review of analytical procedures for both water and sediment is available.[62]

Bioaccumulation of organotins has been concerned mainly with tributyl tin and its degradation products. Bioconcentration factors for TBT of 2 600X (sheepshead minnow), 3 000 to 10 000X for oysters (*C. gigas*), and 2 000X for the oyster *Ostrea edulis* have been noted. A concentration factor of 1.9×10^3 was found for algae from one site in San Diego Bay.[41] Tin concentrations in foods after treatment with organotins are about 0.4 to 2.0 μg g^{-1} (surface, apples, and pears). Levels on crops do not often exceed 0.5 μg g^{-1} though degradation rates vary. Cows fed with sugar beet containing Ph_3SnOAc at 1 μg g^{-1} produced milk containing 4.0 ng g^{-1} of the acetate. In some countries (*e.g.*, Canada, USA) alkyl tin levels of up to 1 μg g^{-1} in food are allowed (from PVC wrapping). There seems little migration of organotins from PVC bottles into liquid foods inside.

There is evidence of bioconcentration by water plants downstream of factories using tin compounds (upstream and downstream concentration ratios were between 170 and 240) and sediments in the region showed increased tin levels downstream of the tin emission.[63]

There are now a number of reports in existence of laboratory studies of biomethylation of tin. Some of these are abiotic organometallic studies designed to test the feasibility of possible environmental methylating agents (*e.g.*, methyl cobalamin ($MeCoB_{12}$) or iodomethane). These will not be discussed in detail but are referred to here.[64,65] In general they have tended not clearly to confirm or deny the possibility of environmental tin methylation. $MeCoB_{12}$ is reported to react with divalent tin to give $MeSnCl_3$.[65] Iodomethane produces Me_4Sn with tin powder or with tin(II) salts in the presence of reducing agents.[64]

There are no reports of the incubation of tin(0) or tin(II) or inorganic tin(IV) compounds in sediments to produce Me_4Sn. However, a series of incubations leading to methyl tin species has been reported.[66] Various tin(II) and tin(IV) salts were incubated with sediment from Plastic Lake, Ontario, Canada, and the methyl tin products were analysed by conversion to volatile species, followed by GC–AA dectection. The trimethyl tin species were found at ng levels.

Incubation of organotin compounds in this system produced Me_4Sn only from Me_3SnCl (at approximately a $6 \times 10^{-4}\%$ conversion over 14 days). This production of Me_4Sn could arise by surface catalysed redistribution, sulphide-mediated dismutation (*vide infra*), or biological methylation.

[60] M. Chamsaz, M. I. Khasawreh, and J. D. Winefordner, *Talanta*, 1988, **35**, No. 7, 519.

[61] P. W. Balls, *Anal. Chim. Acta.*, 1987 **197**, 309.

[62] J. R. Ashby, S. Clark, and P. J. Craig, in 'The Biological Alkylation of Heavy Elements,' ed. P. J. Craig and F. Glockling, Special Publication No. 66, The Royal Society of Chemistry, London, 1988, p. 263.

[63] J. J. Zuckerman, R. P. Reisdorf, H. V. Ellis, III, and R. R. Wilkinson, in Ref. 9, p. 388.

[64] P. J. Craig and S. Rapsomanikis, *Env. Sci. Tech.*, 1982, 114.

[65] Y. T. Fanchiang and J. M. Wood, *J. Am. Chem. Soc.*, 1981, 5100.

[66] Y. K. Chau, P. T. S. Wong, O. Kramer, and G. A. Bengert, 'Abstracts Intern. Conf. Heavy Metals Environ.,' Amsterdam, Sept. 1981.

Monomethyl tin was produced from the incubation of the yeast *Saccharomyces cerevisiae* with inorganic tin(II) compounds,[67] and also from the reaction of methyl-cobalamin with tin(II) chloride,[68] under environmental conditions of pH and [Cl$^-$]. Analysis was by hydride generation followed by flame ionization detection or mass spectroscopy.

Inorganic tin(II) was also shown to be methylated when added to sediments or yeasts and incubated in the presence of carbanion donors.[69,70] Butyl and phenyl tin compounds when incubated produce small amounts of trimethyl tin; this would be of larger significance if confirmed at greater yields as it would represent the methylation of the decomposition products of commercial tin compounds to a toxic metabolite. So far the reported yields (about $1 \times 10^{-3}\%$) do not justify concern. Observation of methyl group attachment to inorganic tin compounds in these experiments is important evidence of environmental methylation, though negative results with sediments have been reported.

The first evidence for methylation of tin(IV) was presented in 1974.[71] Hydrated tin(II) chloride was incubated with a tin-resistant *Pseudomonas* strain and gave a species having the same fluorescence spectrum as a dimethyl tin control. It was also demonstrated that this strain produced methyl mercury when mercury was added, but much methyl mercury was produced when tin was also present. A transmethylation from tin to mercury was suggested. This work has been extended,[72,73] and the results suggest that the organism produces Me_4Sn and methyl stannanes (*viz.* Me_2SnH_2 and Me_3SnH) from incubations with hydrated tin(IV) chloride ($SnCl_4.5H_2O$). No volatile organotin products were obtained with tin(II) salts or in the absence of tin. The tin(IV) results were explained by biomethylation and reduction. Analysis was by selected ion monitoring with a GC–MS system for the major fragment ions together with calibration chromatograms from standards.

Micro-organisms from Chesapeake Bay, USA, have also been shown to transform $SnCl_4.5H_2O$ to methyl tin compounds.[74] Two methylated tin compounds were observed after borohydride treatment of the culture (*viz.* Me_2SnH_2 and Me_3SnH). Me_4Sn was not detectable in this experiment as it would, if present, have been removed by a purging step before analysis. Identification of products was carried out by GC retention times and mass spectroscopic techniques.

It has been shown that both biologically active and sterile sediments will convert trimethyltin hydroxide to Me_4Sn.[75] Production was greatest in biologically active sediments to which sodium sulphide had been added, but production still occurred at lesser yield in sterile sediments without sulphide. Maximum

[67] J. R. Ashby, and P. J. Craig, *Appl. Organometall. Chem.*, 1987, **1**, 275.
[68] J. R. Ashby, and P. J. Craig, *Heavy Metal Environ. Int. Conf.* 1985, **2**, 531.
[69] S. Rapsomanikis, O. F. X. Donard, and J. H. Weber, *Appl. Organometall. Chem.*, 1987, **1**, 115.
[70] J. R. Ashby and P. J. Craig, *Appl. Organometall. Chem.*, 1987, **1**, 275.
[71] C. Huey, F. E. Brinckman, S. Grim, and W. P. Iverson, 'Proc. Conf. Trans. Persist. Chem. Aquat. Ecosystem,' NRC, Ottawa, Canada, 1974, p. 11.
[72] F. E. Brinckman, *J. Organometall. Chem. Library*, 1981, **12**, 343.
[73] J. S. Thayer and F. E. Brinckman, *Adv. Organometallic Chem.*, 1981, **20**, 314.
[74] L. E. Hallas, J. C. Means, and J. J. Cooney, *Science*, 1982, **215**, 1505.
[75] H. E. Guard, A. B. Cobet, and W. M. Coleman, III, *Science*, 1981, **213**, 770.

yield was 6% (high for biomethylation experiments) but there was no evidence that the extra methyl group arose *de novo* from a methylating agent in the sediment. Various redistribution reactions, some sulphide mediated, could account for the results without biomethylation being invoked. The extra yield in active sediments could be due to more sulphide being present which would assist the redistribution. Some sulphide might be lost on autoclaving the sediments, leading to less Me_4Sn production in sterile sediments.

In view of the laboratory evidence for environmental methylation of tin further studies to detect methyl tin species in the dispersed environment should be made (mainly North American examples are extant). Particularly in Europe, where less methyl tins are used industrially, would the detection of methyl tins be of great interest. Finally more experiments showing the production of methyl tins in the laboratory from incubation experiments with inorganic tin are urgently required. Present evidence, though compelling, is scarce.

3 ORGANOLEAD COMPOUNDS

3.1 Use and Toxicity

The use of organolead compounds as additives to petrol for spark ignition engines has been much debated recently. This is easily the biggest use for organometallic lead compounds. 'Knocking' is caused by detonation of the petrol–air mixture rather than smooth combustion, and its occurrence may be removed by adding alkyl-lead compounds to petrol. Lead anti-knock additives are Et_4Pb, Me_4Pb, and mixed ethyl–methyl alkyl-leads. In a number of countries, amounts used have been reduced recently following legislation; *e.g.*, in the USA maximum allowable lead concentrations in gasoline fell from 2.5 g US gal^{-1} in 1970 to 0.5 g gal^{-1} in 1979; in the UK the earlier level of 0.4 g dm^{-3} fell to 0.15 g dm^{-3} in 1985. About 200 000 tonnes of lead ethyl anti-knock compounds are presently produced annually throughout the world. This has fallen from 317 000 tonnes in 1974.[76]

The toxic effects of Et_4Pb arise by conversion to Et_3Pb^+ in the liver.[77] This is more soluble and attacks the central nervous system. Et_4Pb is readily absorbed through the skin, following which conversion to the trialkyl form takes place, perhaps through initial hydroxylation at a β-carbon position (*q.v.* organotin compounds). In severe poisoning with Et_4Pb, highest Et_3Pb^+ concentrations are found in the liver. Conversion of Et_4Pb to Et_3Pb^+ in the liver is so rapid that a half-life of minutes for Et_4Pb has been assumed. Elsewhere it is longer. The half-life in the liver or kidneys of rats is around 40 days for Me_3Pb^+ and 15 days for Et_3Pb^+.

Trimethyl lead, triethyl lead, and diethyl lead are also accumulated by aquatic organisms. Concentration factors are generally much lower than for tetraalkyl lead species with concentration factors up to 375 after exposure to

[76] I. M. Robinson, in Ref. 2, p. 99.
[77] J. W. Robinson, E. L. Kiesel, and J. A. L. Rhodes, *Environ. Sci. Health*, 1979, **A14**, 65.

diethyl lead.[78] Generally, however, concentration factors were in the range 1–24 after 96 hours to 14 days exposure times at concentrations of 0.1 mg dm^{-3} to 25 mg dm^{-3}.[78]

Poisoning with organic lead compounds differs somewhat from that with inorganic lead which generally leads to colic, neurological symptoms, and anaemia. The critical organ for organic lead poisoning appears to be the brain. Incipient anaemia is indicated as a first symptom of chronic exposure, and this parallels the case with inorganic lead exposure.

3.2 Detection and Transformations in the Environment

Most of the alkyl-lead compounds in petrol are converted to inorganic species during use. These are emitted to the atmosphere, but up to about 10% of the lead released from vehicles (from the exhaust or from evaporation) may be in the organic form. Reactions in aqueous systems will be discussed here though atmospheric levels will be noted. Because most of the organic lead is converted to inorganic lead in use, the environmental problem in the main reduces to the inorganic problem of lead contamination of air, water, food, or organisms. The main point of controversy is the extent to which human problems are caused by petroleum—derived lead rather than natural (geological) lead or other industrial inorganic lead emission. The general problem of inorganic lead in the environment will not be covered here but the behaviour of the organometallic species will be discussed.

Organolead species can usually be detected in urban atmospheres but it has been concluded that the lifetimes of the tetraalkyl species in street air are short[79] and this has been shown for Et_4Pb.[80] Higher concentrations are present near garages and cold-choked vehicles, but in general tetraalkyl lead compounds usually account for about 1–4% of the total airborne lead.[81] These decay to inorganic lead(II) via R_3Pb^+ and R_2Pb^{2+} ions (probably co-ordinated to airborne particulate matter),[112,113] but the major route for decay is photolytic homogeneous reaction in the atmosphere with hydroxyl radical attack being the chief initiation route.[82,83] Photolysis and reaction with ozone also occur. The rate of decay in the atmosphere during daylight has been estimated at up to 21% per hour for Me_4Pb and 88% for Et_4Pb. These pathways do not account for much decomposition at night. Half-life estimates for the persistence of the ionic trimethyl and triethyl lead species have recently been estimated as 5 and 1.5 days respectively.[84,85] Decay routes of this kind seem likely to occur in view of the low concentrations of alkyl-leads found in air despite years of steady emission in certain locations. From various atmospheric samples the proportion of tetra-alkyl-lead present has ranged from 0.4% to over 15.0% of the total lead

[78] Y. K. Chau, P. T. S. Wong, G. A. Bengert, and J. Wassien, *Appl. Organometall. Chem.*, 1988, **2**, 427.
[79] R. M. Harrison, R. Perry, and D. H. Slater, *Atmos. Environ*, 1974, **8**, 1187.
[80] B. Radziuk, Y. Thomassen, J. C. Van Loon, and Y. K. Chau, *Anal. Chim. Acta*, 1979, **105**, 255.
[81] R. M. Harrison and R. Perry, *Atmos. Environ.*, 1977, **11**, 847.
[82] R. M. Harrison, *J. Environ. Sci. Health*, 1976, **A11**, 419.
[83] R. M. Harrison and D. P. H. Laxen, *Atmos. Environ.*, 1977, **11**, 201.

present[84,85] with total tetraalkyl-lead concentrations generally in the range 10 to 200 ng dm^{-3},[55] although[86-89] a recent case has reported up to 400 ng m^{-3}.[90] In one case it was claimed that up to 62% of the total airborne lead was organic.[85] This is unusual. GLC–MS analysis was used and metal–organic complexes rather than organometallic lead compounds may have been measured. In general greater concentrations of organolead species are found near garages, indoor car parks, road tunnels, and in urban air. A monograph in which this question is reviewed has been published.[19] Recent work has shown that in urban and semi-rural UK sites the predominant gas-phase alkyl-lead species was tetramethyl-lead. Ionic Me_3Pb^+ was also present. The alkyl:organic lead ratio was in the range 1.3–26.9% at the urban site; 0.6–20% at the semi-rural site implying the proximity of emission sources of alkyl-lead in this region (Colchester, UK).[91]

Some workers have produced results for 'molecular' lead which are rather higher than these but the results may be due to collection problems leading to too high ratios for the real alkyl to total lead content of air. Some of these methods might measure lead organic complexes also.[91,92]

Interestingly analysis of urban air in a region where only Et_4Pb (and not Me_4Pb) was used in petrol has shown that methyl-lead species were also present in the atmosphere presumably by chemical rearrangement reactions.[79] The source of the methyl groups is unknown.

Whilst most lead release to the atmosphere is derived from lead in petrol, there is more uncertainty about the origin of the lead contents of soils. Sources of contamination might include agricultural sprays (lead arsenate), lead smelters, power stations (lead in coal) as well as fall out from petrol-derived lead. Recent work with lead isotopes has suggested that, in one location at least, lead from petrol is the major source of the lead found in soils.[93]

The proportion of blood lead concentrations due to alkyl-lead in petrol is actively disputed at present. One suggestion is 10%; other groups have suggested higher figures.[94] Reduction of the amount of organolead compounds in gasoline does lead to a reduction in the concentration of lead in air: *e.g.*, in some German cities this concentration was reduced by 60% following the reduction in 1976 of the lead content of petrol from 0.4 g dm^{-3} to 0.15 g dm^{-3}.[95] The extent to which such aerial lead reductions eventually lower blood lead levels is disputed. Further, even the correlation between lead concentrations in blood and the

[84] C. N. Hewitt and R. M. Harrison, *Environ. Sci. Technol.*, 1986, **20**, 799.
[85] C. N. Hewitt and R. M. Harrison, *Atmos. Environ.*, 1985, **19**, 545.
[86] L. J. Snyder, *Anal. Chim. Acta*, 1967, **39**, 591.
[87] L. J. Purdue, R. E. Enrione, R. J. Thompson, and B. A. Bonfield, *Anal. Chem.*, 1973, **45**, 527.
[88] S. Hancock and A. Slater, *Analyst (London)*, 1975, **100**, 422.
[89] T. Nielsen, H. Egsgaard, E. Larsen, and G. Scholl, *Anal. Chim. Acta*, 1981, **124**, 1.
[90] W. R. A. de Jonghe, D. Chakraborti, and F. C. Adams, *Environ. Sci. Technol.*, 1981, **15**, 1217.
[91] A. Allen, M. Radojevic, and R. M. Harrison., *Environ. Sci. Technol.*, 1988, **20**, 517.
[92] J. W. Robinson, L. Rhodes, and D. K. Wolcott, *Anal. Chim. Acta*, 1975, **78**, 78.
[93] E. L. Gulson, K. G. Tiller, K. J. Mizon, and R. H. Merry, *Environ. Sci. Technol.*, 1981, **15**, 691.
[94] Lead and Health (The Lawther Report DHSS), HMSO, London 1980.
[95] D. Turner, *Chem. Br.*, 1980, **16**, 312.

results of behavioural and cognitive tests in humans is disputed. This topic is also of political interest at present.

The decay of organolead compounds in the aqueous environment has been studied.[96] Sunlight, surfaces, and certain ions accelerate the decay rate. Suspensions of Et_4Pb and Me_4Pb were quite stable in water in darkness (2% decomposition over 77 days, 16% over 22 days for ethyl and methyl respectively). In the presence of sunlight, decay is rapid (99% after 15 days for ethyl, 59% after 22 days for methyl). The rates were catalysed in darkness by copper(II) or iron(II). The products were Et_3Pb^+ and Me_3Pb^+. Only traces of Et_2Pb^{2+} were detected. Absorption of Et_4Pb and Me_4Pb onto silica was complete from aqueous solution and led to decomposition to the trialkyl-lead ions (after 30 days 97% of Et_4Pb and 55% of Me_4Pb had reacted). This suggests that absorption onto sediments promotes the decay of tetraalkyl-lead species in the environment. Some groups have reported little absorption but others have detected these compounds on sediments.[97] In darkness, solutions of Me_3PbCl, Et_3PbCl, and n-Bu_3PbCl are stable.[72] Only Me_3PbCl shows decay (1% reaction after 220 days). Metal cations had no effect. Sulphide anion promoted the production of R_4Pb. Sunlight increased the decomposition rate; in 15 days there was 4% loss of Me_3Pb^+, 25% loss of n-Bu_3Pb^+, and 99% loss of Et_3Pb^+. The main detectable product was inorganic lead. Presumably R_4Pb was also formed. Again silica absorbed the organolead cations completely from water and the breakdown rates were promoted slightly.

Dialkyl-lead compounds disproportionate slowly in darkness (after 30 days 10% of Me_2Pb^{2+}, 6% of Et_2Pb^{2+}, and 4% of n-Bu_2Pb^{2+} had reacted). Trialkyl-lead products and lead(II) were formed.

In sunlight after 40 days, 70% of n-Bu_2PbCl_2, 25% of Et_2PbCl_2, and 5% of Me_2PbCl_2 had disproportionated. This and other work suggests that alkyl-lead compounds emitted from vehicles and deposited in waterways undergo fairly rapid decomposition in the presence of light, surfaces, or ions. In theory any inorganic lead formed might undergo biomethylation to Me_4Pb. This is considered later. Decomposition of alkyl-lead cations in water has also been studied by other groups with rather similar conclusions.[98,99] In sea water it has been suggested that Et_4Pb would lie on the seabed as a separate phase, slowly dissolving into seawater. Some would evaporate, but most would form Et_3Pb^+.

Et_4Pb has been analysed at a 30 $\mu g\ g^{-1}$ level in mussels near a sunken ship which had been carrying Et_4Pb.[100] These results suggest that alkyl-lead compounds are not quickly metabolized by organisms and may remain in the organic form in tissue for some time. The occurrence of tetraalkyl-lead in aquatic biota is significant because of the possibility of food chain effects.[101] However, in

[96] A. W. P. Jarvie, R. N. Markall, and H. R. Potter, *Environ. Res.*, 1981, **25**, 241.
[97] Y. K. Chau and P. T. S. Wong, in Ref. 8, p. 39.
[98] F. Huber, U. Schmidt, and H. Kirchmann, in Ref. 8, p. 65.
[99] J. R. Grove, 'Intern. Experts Disc. Lead-Occurrence, Fate, and Pollution in the Marine Environ.,' Rovinj, Yugoslavia, 18–22 Oct., 1977.
[100] G. F. Harrison, 'Intern. Experts Disc. Lead-Occurrence, Fate, and Pollution in the Marine Environ.,' Rovinj, Yugoslavia, 18–22 Oct., 1977.
[101] P. J. Craig, in Ref. 1, and reference therein.

one example only one out of 50 Canadian fish was observed to contain Me_4Pb, $(0.25 \mu g\ g^{-1})$ and the source was unknown.[102]

Where Me_4Pb is present it may be accumulated by fish either through water or food.[103,104] One work suggests a rate of $0.4 \mu g\ g^{-1}$ to $2.5 \mu g\ g^{-1}$ daily from a concentration of Me_4Pb in water of $3.5 \mu g\ dm^{-3}$, giving a daily accumulation factor from water of 100–700. Most Me_4Pb is accumulated in fatty tissue and the concentration factor in lipids in the intestine was calculated at 16 000 from water containing $25 \mu g\ dm^{-3}$ of Me_4Pb. The gills, air bladder, and liver also accumulate Me_4Pb. The half-lives for loss of Me_4Pb from intestinal fat and skin were 30 and 45 hours to lead-free water. This presumably occurs by metabolism to Me_3Pb^+ derivatives, the basic toxic substance.[104,105] Accumulation of alkyl lead has been found in cod, lobster, and mackerel tissue where levels of tetraalkyl-lead of between 0.1 and $4.79 \mu g\ g^{-1}$ were found. In lobster digestive gland the tetraalkyl-lead was 81% of the total lead. Percentages of organic lead in various matrices range from 9.5 to 89.7 (flounder meal). The analytical method was extraction with benzene/aqueous EDTA solution followed by flameless AA,[106] which is not as specific as the methods used in more recent work.

Tetramethyl-lead has been shown to be non-toxic to algae, the apparent toxicity being again due to the R_3Pb^+ breakdown products. For a range of alkyl-lead species the trialkyl-leads were most toxic and within the trialkyl series toxicity increased with alkyl chain length. In further work conventional toxicity tests with an estuarine bivalve, *Sorobiculara plana* (Da Costa) were carried out for triethyl-lead and trimethyl-lead species.[107] Acute toxicity was observed at concentrations below $0.14\ mg\ dm^{-3}$ and $0.08\ mg\ dm^{-3}$ respectively with mortality continuing to occur after 35 and 60 days respectively. Approximate incipient lethal concentrations fall in the range $0.07–0.1\ mg\ dm^{-3}$ for both compounds. In algae Me_4Pb accumulates in the cytoplasm.[107]

An analysis of 107 fish showed 17 samples containing tetraalkyl-lead compounds. In this study water, vegetation, algae, weeds, and sediments showed no organic lead to be present. Some of the fish were caught in locations far from likely contamination use (*viz.* parts of Ontario, Canada) and biomethylation was mentioned as a possible cause for the presence of tetraalkyl-lead compounds. In general the alkyl-leads were less than ten per cent of the total lead in the fish.[108,109]

Much analytical development work for alkyl-lead analysis has been carried out. Recent methods used include GLC–AA with electrothermal atomization,[79] GLC with microwave plasma detection,[110] and flameless AA with heated

[102] Y. K. Chau, P. T. S. Wong, G. A. Bengert, and O. Kramar, *Anal. Chem.*, 1979, **51**, 186.
[103] B. G. Maddock and D. Taylor, 'Intern. Experts Disc. on Lead-Occurrence, Fate, and Pollution in the Marine Environ.' Rovinj, Yugoslavia, 18–22 Oct., 1977.
[104] P. T. S. Wong, Y. K. Chau, O. Kramar, and G. A. Bengert, *Water Res.*, 1981, **51**, 621.
[105] P. Grandjean and T. Nielson, *Residue Rev.*, 1979, **72**, 97.
[106] G. R. Sirota and J. F. Uthe, *Anal. Chem.*, 1977, **49**, 823.
[107] S. J. Marshall and A. W. P. Jarvie, *Appl. Organometall. Chem.*, 1988, **2**, 143.
[108] Y. K. Chau, P. T. S. Wong, O. Kramer, G. A. Bengert, R. B. Crux, J. O. Kinrade, J. Lye, and J. C. Van Loon, *Bull. Environ. Contam. Toxicol.*, 1980, **24**, 265.
[109] B. A. Silverberg, P. T. S. Wong, and Y. K. Chau, *Arch. Environ. Contam. Toxicol.*, 1977, **5**, 305.
[110] D. C. Reamer, W. H. Zoller, and T. C. O'Haver, *Anal. Chem.*, 1978, **50**, 1448.

graphite atomization.[110] Other analyses have also been performed by similar methods.[111-116]

Development of a derivatization method using sodium tetraethyl borate, to produce the corresponding ethyl lead compounds has recently been used. Dimethyl-lead (Me_2Pb^{2+}) and trimethyl-lead (Me_3Pb^+) have both been successfully determined in standard solutions in this way.[117]

Tetraalkyl- or other alkyl-lead species have been measured frequently in the environment and their origin is usually ascribed to evaporation of unused petrol or to incomplete combustion of the organoleads in petrol but sometimes biomethylation of inorganic lead is cited.

Detection of methyl-lead species in the environment is not, in itself, evidence of biomethylation and some of the sediments used in lead methylation studies already contained tetramethyl-lead.[118] There are a number of reports that inorganic lead may be methylated in the environment. A mixture of Great Lakes, Canada, water and sediments with nutrients has produced Me_4Pb without any addition of lead in the laboratory; these sediments already contained lead. Addition of Me_3PbOAc greatly increased the amounts either through disporportionation of biomethylation.[118] Addition of some inorganic lead(II) salts also caused an increase in the amount of Me_4Pb present. Using pure species of various bacteria up to 6% of Me_3Pb^+ present was converted to the tetramethyl form in one week. Inorganic lead was not converted under these conditions. It is conceivable that some of these observations could be due to Lewis acid displacement of weakly bound pre-existing Me_4Pb, as lead was already present in the sediment. It was not demonstrated that the lead added was the same as the lead later analysed as Me_4Pb.[118] This possibility might be unlikely but it should be borne in mind. Analysis was by GLC–AA.

Me_3Pb^+ salts can also be converted to Me_4Pb by chemical disproportionation. One route involves sulphide ions or hydrogen sulphide in an analogous route to that existing for tin or mercury.[119] Sulphide mediation may be of general importance in the conversion of partially substituted alkyl metals in the environment to the fully methylated species. There is agreement that Me_3Pb^+ salts are converted to Me_4Pb in the environment, but there is debate about the proportion of the conversion arising from redistribution of the existing methyl groups (disproportionation) and the proportion arising from biological methylation.[119-121] This latter has been calculated to be up to 20% of the whole

[111] R. M. Harrison and D. P. H. Laxen, *Nature*, 1978, **275**, 738.
[112] J. W. Robinson, E. L. Kiesel, J. P. Goodbread, R. Bliss, and R. Marshall, *Anal. Chim. Acta*, 1977, **92**, 321.
[113] S. A. Estes, P. C. Uden, and R. M. Barnes, *Anal. Chem.*, 1981, **53**, 1336.
[114] Y. K. Chau, P. T. S. Wong, and P. D. Goulden, *Anal. Chim. Acta*, 1976, **85**, 421.
[115] J. D. Messman and T. C. Rains, *Anal. Chem.*, 1981, **53**, 1632.
[116] M. D. Dupuis and H. H. Hill, Jr., *Anal. Chem.*, 1979, **51**, 292.
[117] S. Rapsomanikis, O. F. X. Donard, and J. H. Weber, *Anal. Chem.*, 1986, **58**, 35.
[118] P. T. S. Wong, Y. K. Chau, and P. L. Luxon, *Nature*, 1975, **253**, 263.
[119] A. W. P. Jarvie, R. N. Markall, and H. R. Potter, *Nature*, 1975, **255**, 217.
[120] A. P. Whitmore, 'A Study of Lead Alkylation in Natural Systems,' Ph.D. Thesis, University of Aston, 1981.
[121] P. J. Craig, *Environ. Tech. Lett.*, 1980, **1**, 17.

methylation.[122] However different experimental conditions between sterile and biological experiments might have changed the rate of the disproportionation mechanism on which the calculation was based. One group has suggested a biological methylation occurring ten times faster than the sulphide promoted route,[123] but other workers have not invoked a biological component. One group has found that the conversion was up to 4%[124] a result in agreement with previous work by other groups using sediments from widely differing locations.[118,121] It has been suggested that where the concentration of lead is insufficient to stop growth of micro-organisms, the proportion of Me_3Pb^+ methylated biologically is over 80% with less than 20% converted chemically.[97] In a different system a biological methylation of 50–76% of the total was claimed [125] but the yield was only 0.009%. Et_3PbCl has been methylated as well as disproportionated in a micro-organism culture; 13% of the methylated derivative was found,[122] but another group found no methyl group was added to the Et_3Pb moiety.[126]

There are other examples of the conversion of lead(II) salts to Me_4Pb. In a series of water samples seeded with micro-organisms from an aquarium, incubation of lead(II) acetate produced Me_4Pb (NB mercuric acetate may produce methyl mercury by methylation from decomposition of the acetate methyl group alone, but under more extreme conditions).[127] It has also been reported that lead(II) salts added to St. Lawrence River sediment will produce Me_4Pb. Two out of three sediment sites produced this from lead(II) nitrate. No control experiments were reported and analysis was by GLC retention times.[123] Results of incubations with anoxic British Columbia sediments containing added lead(II) nitrate suggested a 0.03% conversion to Me_4Pb.[128] Control experiments were reported and identification of the product was by a GLC–MS method. Although these sediments also contained lead, Me_4Pb was not observed in unspiked samples suggesting that the results were not due to displacement of pre-existing organic lead. In this study lead(II) acetate did not methylate. However, a more recent report by these workers has placed doubt on the existence of lead(II) methylation here.[124] Using ^{14}C-labelled methyl donors (L-serine, methanol, $MeCoB_{12}$ and D-glucose) no ^{14}C-methyl was found in the Me_4Pb arising from incubation experiments with Me_3Pb^+, suggesting chemical methylation only.[129,130] Some groups, however, have been unable to detect lead(II) methylation in micro-organism or sediment media.[119,121,129] There is evidence

[122] F. Huber, U. Schmidt, and H. Kirchmann in Ref. 9, p. 65.
[123] J. P. Dumas, L. Pazdernik, and S. Belloncik, 'Proc. 12th Canad. Symp. Water Pollution, Res. Canada,' 1977, p. 91.
[124] J. A. J. Thompson, 'Abstracts Intern. Conf. Heavy Metals Environ.,' Amsterdam, Sept. 1981, p. 653.
[125] M. D. Baker, P. T. S. Wong, Y. K. Chau, C. A. Mayfield, and W. E. Innes, 'Abstracts Intern. Conf. Heavy Metals Environ.,' Amsterdam, Sept. 1981, p. 645.
[126] F. Huber, U. Schmidt, and H. Kirchman, in Ref. 8, p. 66.
[127] H. Agaki, Y. Fujita, and E. Takabatake, *Chem. Lett.*, 1975, **1**, 171.
[128] J. A. J. Thompson and J. A. Crerar, *Mar. Pollut. Bull.*, 1980, **11**, 251.
[129] K. Reisinger, M. Stoeppler, and H. W. Nurnburg, 'Abstracts Intern., Conf. Heavy Metals Environ.,' Amsterdam, Sept. 1981, p. 649.
[130] C. N. Hewitt and R. M. Harrison, *Environ. Sci. Technol.*, 1987, **21**, 260.

of lead methylation based on air movement analyses. The ratio of alkyl to total lead in the atmosphere, in the absence of a nearby source, lies in the general range 0.5–8%.[111] However, samples of air taken from various rural sites were found to have ratios up to a maximum of 33% and these high ratios were connected with air that had passed over the open sea, estuarine, and coastal areas.

Variation in the alkyl total lead ratio depended critically on wind direction and was highest for air that had not passed over possible sources of anthropogenic lead. For example, in the Outer Hebrides islands (UK) oceanic western oriented air continued higher alkyl:total ratios than air originating over the UK mainland. This is good evidence for a maritime source of volatile alkyl-lead, presumably in the methyl form as westerly air reaching the Hebrides from the Atlantic will not have crossed land surfaces.[130,131]

More recent work has also reinforced the evidence that incubation of organic lead(II) compounds produce methyl lead(IV) products in low yields. Recent culture experiments with lead(II) nitrate (where there is no possibility of the methyl group arising from the salt anion) and various marine phytoplankton and macrophyte cultures led to the productions of the tri- and dimethyl lead products.[131]

Inclusion of ^{210}Pb labelled lead(II) nitrate with various sediments produced ^{210}Pb containing alkyl lead over a 14 day period.[130] This direct demonstration of a methyl group being attached to **added** lead overcomes some theoretical difficulties with earlier incubation of pre-existing organolead compounds by the added inorganic lead. Non-biotic methylation of the ^{210}Pb by any already present methyl-*lead* seems unlikely and it must be concluded that the ^{210}Pb was subject to a real environmental methylation process. This process could account for the alkyl:total lead ratios found in the UK rural sites discussed above.

Organolead compounds have not been detected from the reaction of $MeCoB_{12}$ with lead(II) salts.[119,131–134] The naturally occurring methylating agent iodomethane (MeI) is variously reported to a) react and b) not react with lead(II) salts in aqueous media to produce low yields of Me_4Pb by oxidative addition.

Low yields of dimethyl-, trimethyl-, and tetramethyl lead, were observed for the reaction of lead(II) and $Me_2Co(N_4)^+$ (a methylcobalamin model), after mixing with sediment taken from the Great Bay Estuary, USA.[69] Methyl-lead species were also observed for lead(II) in the presence of $Me_2Co(N_4)^+$ and $MeCoB_{12}$ together, but not for $MeCoB_{12}$ alone. It was thought that methyl lead products form via disproportionation of MePb(II) and $Me_2Pb(II)$ followed by successive methylation of di- and trimethyl-lead.

Abiotic chemical alkylation of lead(II) in water to give Me_4Pb is not

[131] R. M. Harrison and A. G. Allen, *Appl. Organometall. Chem.*, 1989, **3**, 49.
[132] C. Agnes, S. Bemdle, H. A. O. Hill, F. R. Williams, and R. J. P. Williams, *J. Chem. Soc., Chem. Commun.*, 1971, 850.
[133] R. T. Taylor and M. L. Hanna, *J. Environ. Sci. Health*, 1976, **A11**, 201.
[134] J. Lewis, R. H. Prince and D. A. Stotter, *J. Inorg. Nucl. Chem.*, 1973, **35**, 341.

inherently impossible. It has been carried out with boron alkyls[135,136] by methyl carbanion alkylation (*q.v.* $MeCoB_{12}$). There is, then, little theoretical reason why lead(II) might not be alkylated under environmental conditions; although monomethyl-lead species are unstable, if the rate of further methylation to the more stable R_2Pb^{2+} or R_3Pb^+ derivatives is greater than decomposition, then Me_4Pb may be formed—as was observed for the boron alkylating agents. It is also possible that monomethyl-lead species might be stabilized in the environment by co-ordination to natural ligands.

The question of lead(II) methylation remains open. There is a need for isotopic work which, in a methylating system, would clearly demonstrate that the lead added is that which is methylated. Ideally analysis should be in an absolute method (*e.g.*, GLC–MS).

4 ORGANOMERCURY COMPOUNDS

4.1 Uses and Toxicity

Much has been said about the environmental role of organomercury compounds and much research is still being carried out. A number of detailed accounts exist and the discussion that will appear here will be closely defined. As for lead, increases in total metal concentrations today compared to the pristine environment are not solely caused by use of organomercury products. The main use for mercury has been as the metal as the cathode in chlor-alkali cells and in electrical apparatus, and most mercury introduced to the environment has probably come from these sources. Only organomercury concentrations and uses will be discussed (See Table 5).

Restrictions on organomercury compounds have reduced the quantities released to the environment. Uses of organomercury compounds are covered in publications by the Organization for Economic Co-operation and Development[137] and the International Atomic Energy Agency,[138] but these are rather historic in terms of present use. Use of alkyl mercury compounds has declined and other organomercurials are generally being further restricted in use.

The chief toxic effects of inorganic mercury poisoning are tremor, psychological disturbance, gingivitis, and occasionally proteinuria. Acute doses of mercury (II) lead to kidney injury and perhaps death. The symptoms of methyl mercury poisoning are well known. The critical organ is the brain and penetration of the blood-brain barrier leads to sensory disturbance, tremor, ataxia, constriction of the visual fields, and impaired hearing.[139] There is often a long latent period after exposure and the effects are frequently irreversible. At Minamata pre-natal exposure occurred, giving symptoms in the infants of cerebral palsy characterized by mental retardation and motor disturbance. The mothers may or may not show symptoms.

[135] J. B. Honeycutt, Jr. and J. M. Riddle, *J. Am. Chem. Soc.*, 1960, **82**, 3051.
[136] J. B. Honeycutt, Jr. and J. M. Riddle, *J. Am. Chem. Soc.*, 1961, **83**, 369.
[137] Ref. 17, p. 23.
[138] Ref. 18, p. 35.
[139] K. Beijer and A. Jernelov, in Ref. 3, p. 203.

Table 5 *Uses of organomercury compounds*

Compound	Use	Notes
MeHgX	Seed dressings	Banned Sweden, 1966,
EtHgX	(fungicides)	USA, 1970, Canada, 1970, *etc.*
RHgX (ROAc,py)	Catalysts for urethane, vinyl acetate production	
	Slimicides	Little used.
PhHgX	Seed dressings (fungicides)	Banned as slimicide USA, 1970, banned for rice
	Bactericides, Slimicides	Japan, 1968, *etc.*
p-MePhHgX	Spermicide	
MeOCH$_2$CH$_2$HgX	Seed dressings (Fungicides)	Banned Japan, 1968.
EtOCH$_2$CH$_2$HgX	Seed dressings	
Thiomersal (EtHg derivative)	Antiseptic	Reducing in usage.
Mercurochrome (organomercury fluorescein derivative)	Antiseptic	Reducing in usage.
Mersalyl (methoxyalkyl Hg derivative)	Diuretic	Reducing in usage.
Chlormerodrin (Alkoxyalkyl Hg derivative)	Diuretic	Reducing in usage.

X = inorganic or organic anion

Elimination of methyl mercury is slow, and this species produces more mercury in the brain than intake of other forms of mercury. Excretion eventually takes place to about fifty per cent inorganic form. It is the symptomless build-up coupled with the acute effects which is so serious. The half-life in the human body for methyl mercury is about 70 days, compared to 4 or 5 for mercury(II) and 10 for MeOCH$_2$CH$_2$Hg$^+$ in the rat.[18,140]

Excretion rates of methyl mercury are slow. After 10 days for chicks 20% of methyl mercury, 60% of inorganic mercury(II), 80% of phenyl, and 90% of MeOCH$_2$CH$_2$Hg$^+$ had been eliminated. Half-lives for methyl mercury in chickens are 70 days, osprey 2–3 months, mallard drakes 84 days; similar to man.[141] From fish half-lives of up to 1 000 days have been measured but as low as 8–23 days in goldfish.[142]

A daily intake of methyl mercury of 0.3 mg for the average human can lead to the appearance of toxic symptoms. Acceptable weekly intake levels of 0.3 mg of total mercury were defined by the World Health Organization and the Food and Agriculture Organization. This leads to allowable maximum levels of about 0.5 to 1.0 µg g^{-1} in fish and shellfish.[11,17,89,143]

[140] Ref. 17, p. 39.
[141] N. Fimreite, in Ref. 3, p. 601.
[142] J. W. Huckabee, J. W. Elwood, and S. C. Hildebrand in Ref. 3, p. 277.
[143] T. Takizawa, in Ref. 3. p. 359.

It is estimated that the lowest whole blood mercury level assumed capable of producing neurological symptoms is about 0.2 µg g^{-1}. To prevent this level being reached limits for mercury concentration in foods have been set in various countries (*e.g.*, for fish these are 1.0 (Sweden), 0.5 (Canada), 0.5 (USA), 1.0 (Finland), and 0.4 (Japan)—all in µg g^{-1}).[142] Phenyl and alkoxyalkyl mercury compounds break down to organic materials (*e.g.*, ethane) and inorganic mercury and with their shorter half-lives are more akin to inorganic mercury in their effects. For these compounds the kidney seems to be the critical organ.[144] Methyl mercury also decays in soils or sediments to mercury(0) but again at a slower rate[144-146] than phenyl or alkoxyalkyl derivatives.[147-149]

In cases of prolonged exposure toxicity becomes chronic at lower concentration, and metal levels in the tissues can be higher than those found in cases of acute poisoning.[150] A number of studies of the toxicity of organomercury compounds have been made.[3,17,18] In one case the fish-eating population near the Agano river in Japan who consumed 300–1500 g of fish daily, with the methyl mercury level in the fish at 3–4 µg g^{-1}, suffered poisoning. The rate of methyl mercury uptake by fish from water may be 10–100 times faster than for mercury(II) ion but its elimination is slower. Hence there is particular concern about methyl mercury levels in fish which is a main pathway exposing man to mercury.

4.2 Detection and Transformation in the Environment

Methyl mercury concentrations have been measured in many environmental matrices during the past fifteen years. Much data is available in the work edited by Nriagu.[3] The emphasis here is to compare measured methyl mercury levels with calculated pre-man environmental levels, to compare the ratios of organic to inorganic mercury in various media and also to assess bioconcentration effects in various food nets. Reported methyl mercury levels are a measurement of an equilibrium between methylation and demethylation. It should not be assumed that any methyl mercury detected is necessarily a man made pollution problem; there is evidence to show that levels in certain marine species are much the same today as they were centuries ago. Environmental levels of organomercury species are available.[3,6,11]

The organomercury content for fish and freshwater biota has been investigated in detail. Most freshwater fish have analysable levels of mercury in their tissues and usually more than 80% is in the methyl form. All forms of mercury seem to be accumulated by fish from water and food and methyl mercury is

[144] T. Sazuki, in Ref. 3. p. 399.
[145] Y. Kimura and V. L. Miller, *J. Agric. Food Chem.*, 1964, **12**, 253.
[146] K. Furukawa, T. Suzuki, and K. Tonamura, *Agric. Biol. Chem.*, 1969, **13**, 128.
[147] Th. M. Lexmond, F. A. M. de Haan, and M. J. Frisel, *Neth. Agric. Sci.*, 1976, **24**, 79.
[148] K. Tonamura and F. Kanzaki, *Biochim. Biophys. Acta*, 1969, **184**, 227.
[149] G. Billen, C. Joiris, and R. Wollast, *Water Res.*, 1974, **8**, 219.
[150] F. Ribeyre, A. Delarche, and A. Boudou, *Environ. Pollut. (Ser. B)*, 1980, **1**, 259.

absorbed faster than inorganic mercury and retained longer. In other aquatic species the percentage of the methyl form seems to vary between about 50% and 80%. The concentration of methyl mercury in fish tends to be proportional to the mercury concentration in water and accumulation is correlated with temperature to a certain value for each species which is connected with oxygen consumption rates.

Methyl mercury is absorbed by fish from bottom and suspended sediments and food and possibly as the dissolved species from water. There may be no actual methylation in fish though *in vitro* methylation from homogenates has been demonstrated,[151,152] but demethylation occurs slowly and so methyl mercury accumulates.

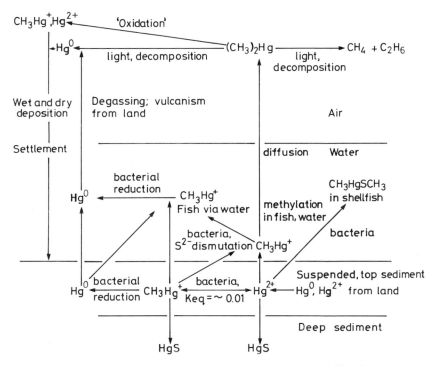

Figure 1   Adapted from Ref. 1

[151] N. Imura, S-K. Pan, M. Shimitzu, and T. Ukita, in 'New Methods Environ. Chem. Toxicol., Collected Papers Res. Conf. New Methods Ecol. Chem.,' ed. F. Coulston, Inst. Acad. Print. Co. Ltd., Tokyo, Japan, 1973, p. 211.
[152] N. Imura, S-K. Pan, M. Shimitzu, T. Ukita, and K. Tonamura, *Ecotoxicol. Environ. Safety*, 1977, **1**, 235.

Freshwater biota, even from remote areas can accumulate detectable quantities of mercury from natural and crustal sources. Fish from such areas still contain ng g^{-1} concentration, of which up to 90% is in the methyl form.[153] Acid mobilization of mercury in remote areas and deposition of mercury from the atmosphere may ultimately be man-derived as these processes provide more mercury for methylation.

Data on dissolved methyl mercury may not really refer to actual dissolved species, rather to mercury complexed to suspended particulate matter. A sample of interstitial water has been reported to contain 1.4 µg dm^{-3} of methyl mercury. Dissolved mercury has been reported at 0.01–0.03 µg dm^{-3} from one source, while some Canadian lakes contained 0.5–1.7 ng g^{-1}.[154-156]

The proportion of mercury in the methyl form in birds varies considerably depending on the mercury source and may vary between 10 and 100%.[1]

The best indicators of methyl mercury exposure in humans are hair, urine, or blood. The average methyl mercury level in hair in Japan was reported to be 2.07 µg g^{-1}; people living in a mercury polluted district had 6.69 µg g^{-1}, heavy fish eaters from Oyabe had 11.21 µg g^{-1}, and tuna fishermen had 12.82 µg g^{-1}. Pre-industrial examples of this latter group may have had similar levels.[157]

As most mercury in fish is in the methyl form, some total mercury levels will be discussed. Marine fish having longer life spans have the highest levels of mercury. These are from about 0.1 to 0.3 µg g^{-1} dry weight for small pelagic fish whereas larger species such as tuna and swordfish are between 0.5 and 1.5 µg g^{-1}, mostly as methyl mercury. Predators accumulate higher levels of methyl mercury than fish or other species lower in the food net.[158] There seems little difference in the percentage of mercury in the methyl form whether or not the fish comes from a contaminated location. Apart from the methyl mercury in fish, there is a report of ethyl mercury being found in fish downstream of phenylmercury effluent.[159]

Maximum background levels of mercury in freshwater fish are about 0.2 µg g^{-1} (muscle) although more than 1.0 µg g^{-1} can be found from geological mercury sources alone. The minimum amount for freshwater fish is about 0.035 µg g^{-1}.[127] Fish from polluted rivers show higher total mercury levels, *e.g.*, in 1965 fish from the Agano River (Niigata, Japan) contained an average of 2.34 µg g^{-1} of mercury. Interestingly this has decreased in more recent years (to 0.60 µg g^{-1} in 1968, and to 0.32 in 1971).[156] A similar decrease in methyl mercury levels in sediments from the River Mersey, UK, has been observed recently.[160-163]

[153] J. W. Huckabee, J. W. Elwood, and S. G. Hildebrand, in Ref. 3, p. 283.

[154] F. Frimmel and H. A. Winkler, *Vom Wasser*. 1975, **45**, 285.

[155] Y. K. Chau and H. Saito, *Int. J. Environ. Anal. Chem.*, 1973, **3**, 133.

[156] J. Stary, B. Havlik, J. Prasilova, K. Kratzer, and J. Hanusova, *Int. J. Environ. Anal. Chem.*, 1978, **5**, 84.

[157] Y. Takizawa, in Ref. 3. p. 325.

[158] H. L. Windom and D. R. Kendall, in Ref. 3, p. 303.

[159] L. Frieberg, Report Expert Group, Nord. Hyg. Tidskr. Suppl. 4, US Andersons Tryckeri, Stockholm 1971.

[160] P. D. Bartlett, P. J. Craig and S. F. Norton, *Sci. Total. Environ.*, 1978, **10**, 245.

[161] P. D. Bartlett and P. J. Craig. *Water Res.*, 1981, **15**, 37.

[162] P. J. Craig and P. M. Moreton, *Environ. Pollut. (B)*, 1985, **10**, 141.

[163] P. J. Craig and S. F. Morton, *Nature*, 1976, **261**, 125.

Analysis of methyl mercury is usually carried out by the Westoo technique or variations. Samples are extracted with toluene or benzene, and methyl mercury is extracted to an aqueous phase followed by re-extraction into benzene or toluene. Analysis is usually by electron capture GLC.[161-166]

Biota in oligotrophic lakes often show higher methyl mercury levels that those in otherwise similar eutrophic lakes. This may be due to dilution in the higher biomass of the latter, and also to the higher pH favouring volatile dimethyl mercury formation. In eutrophic lakes the sediments are largely reduced, forming mercuric sulphide which is little available for methylation.[167] High sulphide levels tend to promote conversion of methyl to dimethyl mercury,[168] *viz.* $2CH_3Hg^+ + S^{2-} \rightarrow (CH_3Hg)_2S \rightarrow (CH_3)_2Hg + HgS$. Oligotrophic lakes subject to acidification (*e.g.*, acid rainfall) have less productivity, lower biomass, and greater availability of mercury. At low pH methyl mercury is favoured and this is absorbed by the reduced biomass—hence the higher concentrations. Reduced complexation to sediments also leads to greater bioavailability, which may, however, be opposed by lower microbial activity.[169] Acid precipitation, methylation, and mobilization have been reviewed recently.[170]

Several observations suggested that inorganic mercury could be methylated in the environment; (i) most of the mercury present in fish is present in the methyl form,[171] (ii) inorganic mercury when added to aquarium sediments is partly converted to methyl mercury[172] and (iii) $MeCoB_{12}$—utilizing methanogenic bacteria can methylate mercury in sediments.[173]

It has now been demonstrated on many occasions that inorganic mercury added to sediments may be converted to methyl mercury. In addition various pure bacterial strains with the capacity for mercury methylation have been isolated.[173-176] However, several authors report an inability to methylate mercury by various systems.[128,129,177,178] This must be a property of the particular system or strain studied; mercury methylation in general terms is firmly established, and the main parameters seem to be as follows.

Both methylation and demethylation,[179-181] occur in sediments and

[164] F. Glocking, *Anal. Proc.*, 1980, 417.
[165] P. D. Bartlett, P. J. Craig, and S. F. Morton, *Nature*, 1977, **267**, 606.
[166] J. F. Uthe, J. Solomons, and B. Grift, *J. Assoc. Official Anal. Chem.*, 1972, **55**, 583.
[167] T. Fagerstrom and A. Jernelov, *Water. Res.*, 1971, **5**, 121.
[168] P. J. Craig and P. D. Bartlett, *Nature*, 1978, **275**, 635.
[169] K. Beijer and A. Jernelov in Ref. 3. p. 208.
[170] J. M. Wood, *Environ. Sci. Res.*, 1980, **17**, 223.
[171] G. Westoo, *Acta Chem. Scand.*, 1966, **20**, 2131.
[172] S. Jenson and A. Jernelov, *Nature*, 1969, **223**, 753.
[173] J. M. Wood, P. S. Kennedy, and C. G. Rosen, *Nature*, 1968, **220**, 173.
[174] J. W. Vonk and A. K. Sijpesteij, *J. Ant. Van Leeuwenhoek*. 1973, **39**, 505.
[175] B. H. Olsen and R. C. Cooper, *Nature*, 1974, **252**, 682.
[176] B. H. Olsen and R. C. Cooper, *Water Res.*, 1976, **10**, 113.
[177] M. K. Handy, O. R. Noyes, and S. R. Wheeler, in 'Biol. Implics. Metals Environ.,' ERDA, Symp. Ser. No. 42, 1977, p. 20.
[178] M. K. Hamdy and O. R. Noyes, *Appl. Microbiol.*, 1975, **30**, 424.
[179] K. L. Jewett, F. E. Brinckman, and J. M. Bellama, in 'Marine Chem. in the Coastal Environ.,' ed. T. C. Church, ACS Symp. Ser. No. 18, ACS, Washington, DC, 1975, p. 304.
[180] W. J. Spangler, J. L. Spigarelli, J. M. Rose, R. S. Filippin, and H. M. Miller, *Appl. Microbiol.*, 1973, **25**, 488.
[181] W. J. Spangler, J. L. Spigarelli, J. M. Rose, and H. M. Miller, *Science*, 1973, **180**, 192.

measurements of methyl mercury are normally measurements of an equilibrium process. Aerobic and anaerobic methylation and demethylation occur.[181-189] Methylation is mainly to monomethyl mercury under normal conditions and demethylation is to methane and mercury(0). Workers have on occasions found faster *net* methylation rates under aerobic,[174,190-192] or anaerobic conditions. Factors which influence the rate of methylation include total mercury concentration, silt and organic content of a sediment, pH/Eh temperature, concentration of methanogenic bacteria, sulphide content, and complexation. It has been concluded that maximum rates of methylation are found in the oxidizing anaerobic zone where redox potential (Eh) ranges from about -100 to $+150$ V.[161] There is an environmental mercury cycle by which mercury compounds may be inter-converted (see Figure 1). This explains why methyl mercury compounds may be inter-converted. Such a cycle shows why methyl mercury was found in fish downstream from pulp mills (phenyl mercury in the effluent) and chlor-alkali plants (inorganic mercury). Methyl mercury is preferentially formed under neutral and acidic conditions and dimethyl mercury under basic conditions.[193] The presence of sulphide ion in natural systems reduces methylation by formation of the largely unavailable mercuric sulphide. Any methyl mercury present may be converted at higher sulphide levels to the dimethyl form by formation of an organo mercury sulphide intermediate (see above).[194,195] Mercuric sulphide may be microbially converted to the sulphate so sulphide formation is not irreversible. Methyl mercury does not seem to build up on sediments beyond about 1.5% compared to total mercury.[161] This is a rough equilibrium level between formation and removal by various processes. Percentage levels in fish and other biota may be much higher. Methyl mercury is also produced in the water column[196-198] in soil both biologically[182,184] and abiotically,[197,198] MeHgSMe has been found in shellfish.[199,200] In soils dimethyl mercury may be an important product.[200-203] It is not established

[182] M. Yamada and K. Tonamura, *J. Ferment. Technol.*, 1972, **50**, 159.

[183] M. Yamada and K. Tonamura, *J. Ferment. Technol.*, 1972, **50**, 893.

[184] M. Yamada and K. Tonamura, *J. Ferment. Technol.*, 1972, **50**, 901.

[185] I. R. Rowland, M. J. Davies, and P. Grasso, *Arch. Environ. Health*, 1977, **32**, 24.

[186] S. E. Lindberg and R. C. Harris, *J. Water Pollut. Control Fed.*, 1977, **49**, 2479.

[187] J. Gavis and J. F. Ferguson, *Water Res.*, 1972, **6**, 989.

[188] D. C. Gillespie, *J. Fish. Res. Board Can.*, 1972, **29**, 1035.

[189] R. J. Pentreath, *Int. J. Exp. Mar. Biol. Ecol.*, 1976, **24**, 103.

[190] J. J. Bisogni, Jr., in Ref. 3, p. 211.

[191] J. J. Bisogni, Jr. and A. H. Lawrence, *J. Water Pollut. Control Fed.*, 1975, **47**, 135.

[192] B. H. Olsen and R. C. Copper, in Ref. 150, p. 111.

[193] K. Beijer and A. Jernelov, in Ref. 3, p. 205.

[194] P. J. Craig and S. Rapsomanikis, in 'Environ. Spec. and Monitoring Needs Trace Metal-Containing Substances from Energy-Related Processes,' Proc. DoE, NBS Spec. Publ. No. 618, p. 54., Washington DC, 1981.

[195] L. R. Rowland, M. J. Davies, and P. Grasso, *Nature*, 1978, **265**, 718.

[196] G. Topping, *Nature*, 1981, **290**, 243.

[197] R. D. Rogers, US Environ. Proc. Agency, Ecol. Res. Ser. Dept.,' 1977, EPA-600/3-77-007.

[198] R. D. Rogers, *J. Environ. Qual.*, 1977, **66**, 463.

[199] G. Lofroth, 'Ecol. Res. Bull. No. 4,' Swed. Nat. Sci. Res. Council, 1969.

[200] S. Kitamura, *Jumamoto Igk.*, 1963, **37**, 494.

[201] R. D. Rogers, 'Int. Conf. Heavy Metals Environ., Toronto,' Oct. 1975, p. C218.

[202] D. L. Johnson and R. S. Braman, *Environ. Sci. Technol.*, 1974, **8**, 1003.

that fish themselves can methylate mercury, although *in vitro* extracts may do so. It may be that absorption of methyl mercury from outside is the main cause of methyl mercury concentration in fish. Net methylation rates from sediments and rivers have been measured and range from 17 to 690 ng g^{-1} day^{-1} (river),[204] 15–40 ng g^{-1} day^{-1} (sediment),[171] 25 ng g^{-1} day^{-1} (sediment),[156] and 137 ng g^{-1} day^{-1} (marsh sediment).[205] A saline environment produces less methyl mercury than equivalent non-saline regions, suggesting methylation may be a methyl carbanion transfer[205,206] (*i.e.*, invoking $MeCoB_{12}$). Further discussions on mercury methylation may be found in several recent reviews.[5,190,207]

5 OTHER ORGANOMETALLIC COMPOUNDS

5.1 Use and Toxicity

Release of organoarsenic compounds to the environment is declining. The main organometallic uses are as pesticides and herbicides. Pentavalent arsenic compounds, particularly the aryls, have in the past been much used as bactericides and pest control agents.

A number of derivatives of phenyl arsonic acid ($PhAsO(OH)_2$) have bactericidal, biocidal, or pharmaceutical properties. Arsenilic acid (p-NH_2PhAsO-$(OH)_2$) and its sodium salt are listed in the British Pharmacopoeia (Veterinary).[16] Agricultural or veterinary uses of compounds are less controlled than medical use and there is greater potential for release to the environment and for toxicity problems. Inorganic arsenic may be methylated in the natural environment. The input of organoarsenic compounds into the environment by man may be insignificant in comparison with natural arsenic movements, particularly as some of the compounds are common to both systems. In fact the marine arsenic cycle appears to be a detoxification process by the organisms concerned. This is discussed later. Where arsenic compounds are found their origin may be from anthropogenic or natural processes, or, in certain locations, from both.

The use of arsenic compounds in pesticides has been the subject of a monograph.[208] Although of minor importance now in many countries, the organoarsenic drugs are still available. For herbicidal use the organoarsenicals and other pesticides mainly replaced inorganic arsenic pesticides or defoliants (*e.g.*, lead, calcium arsenate, sodium arsenite, arsenic trioxide). High rates of application of the inorganic compounds were required and this caused arsenic build-up in the soil, severely reducing plant growth. The organic arsenical herbicides that have been in common use are shown in Table 6. They are methyl arsenic and dimethyl arsenic (cacodylic) acids or salts. The most extensive use in the USA has been for control of annual grass weeds in cotton. Since 1945

[203] R. D. Rogers, *J. Environ. Qual.*, 1976, **5**, 454.
[204] D. G. Langley, *J. Water Pollut. Control Fed.*, 1973, **45**, 44.
[205] H. Windom, W. Gardner, J. Stephens, and F. Taylor, *Estuarine Coastal Mar. Sci.*, 1976, **4**, 579.
[206] J. E. Blum and R. Bartha, *Bull. Environ. Chem. Toxicol.*, 1980, **25**, 404.
[207] 'The Chemistry of Mercury,' ed. N. McAuliffe, Macmillan, London, 1977.
[208] 'Arsenical Pesticides,' ed. E. A. Woolson, ACS Symposium Series No. 7, ACS Washington, DC, 1975.

organoarsenic compounds have been shown to have both therapeutic and growth properties as feed additives for poultry and swine.

The toxicity of arsenic compounds is different from those of the heavy metals. For arsenic the toxicity to the rat declines with arsenite > arsenate > methyl arsenate = dimethyl arsenate (cacodylate). The toxic dose increases by about fifty between arsenite and the methyl acid salts.[208] The toxic effects appear to be caused by binding to sulphydryl lipid groups by trivalent arsenic, and pentavalent arsenic appears to be reduced to the trivalent form. Lower doses produce liver and kidney damage, while acute administration leads to enteritis and death. Arsine poisoning results in anaemia and renal damage. The possible role of arsenic as a cause of cancer is still not clear.[209,210]

Table 6   *Uses for organoarsenic compounds*

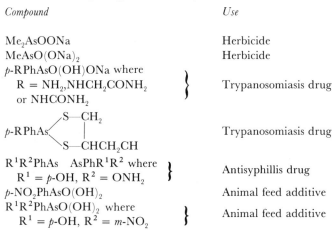

Compound	Use
$Me_2AsOONa$	Herbicide
$MeAsO(ONa)_2$	Herbicide
p-$RPhAsO(OH)ONa$ where $R = NH_2, NHCH_2CONH_2$ or $NHCONH_2$	Trypanosomiasis drug
p-$RPhAs$ (S—CH₂ / S—CHCH₂CH)	Trypanosomiasis drug
R^1R^2PhAs $AsPhR^1R^2$ where $R^1 = p$-OH, $R^2 = ONH_2$	Antisyphillis drug
p-$NO_2PhAsO(OH)_2$	Animal feed additive
$R^1R^2PhAsO(OH)_2$ where $R^1 = p$-OH, $R^2 = m$-NO_2	Animal feed additive

5.2   Detection and Formation in the Environment

As a ubiquitous element in water, air, or soil and living tissue, the detection of arsenic in a sample is not in itself indicative of man-derived contamination. The existence of a natural biomethylation of arsenic in the natural environment reinforces this position, particularly as it increases transport possibilities for arsenic.[211] Arsenic concentrations are given in Table 7. Natural processes, when taken together with the use of both organic and inorganic forms of arsenic by man, can make it difficult to assess the true cause of a particular arsenic level.

The chemistry of arsenic in the marine development is complex and fascinating. Although arsenate is the dominant species in waters of the photic zone, arsenite $MeAsO(OH)_2$ and Me_2AsOOH are found in significant concentration.[212] There is a correlation between the methyl arsenic species and

[209] S. A. Peoples, in Ref. 182, p. 1.
[210] W. R. Penrose, 'C.R.C. Critical Revs. in Environ. Control,' 1974, 465.
[211] J. S. Edmonds and K. A. Francesconi, *Mar. Poll. Bull.*, 1981, **12**, 92.
[212] M. O. Andreae, *Deep Sea Res.*, 1978, **25**, 391.

Table 7 *Some organoarsenic concentrations**

Matrix	MeAs ($ng\,g^{-1}$)	Me_2As ($ng\,g^{-1}$)	Me_nAs as (%) of Total As
Pond	7.4		50
Saline lake	0.7	6.75	10–90
Lake water	0.2	0.25	20
Sea water	0.004	0.02	10–90
California (Sea)	0.1–0.2	1	
Marine biota[†]	0–0.05	0–0.24	10–90
Kelp			4
Macroscopic algae	0.4–8.0×10^3		
Shark muscle			40–80
Shale process waters	200		
Fly ash slurry	24	109	

* Relatively few organoarsenic levels have been determined for environmental matrices and more data are required before these values can be considered as typical.
[†] Diatoms, Coccolithophorids, Dinoflagellates, and Prasmophyceae.

$MeAsO(OH)_2$

Me_2AsOOH

Me_2AsO

Me_2As

Me_2AsH

$Me_3AsCH_2CH_2CH$

Me_3AsCH_2COO

$PhAsO(OH)_2$

$Me_3AsCH_2CH(COO)OPO_2OCCH_2CH(OCOR)CH_2OCOR^1$

Figure 2 *Organoarsenic compounds found in the environment*

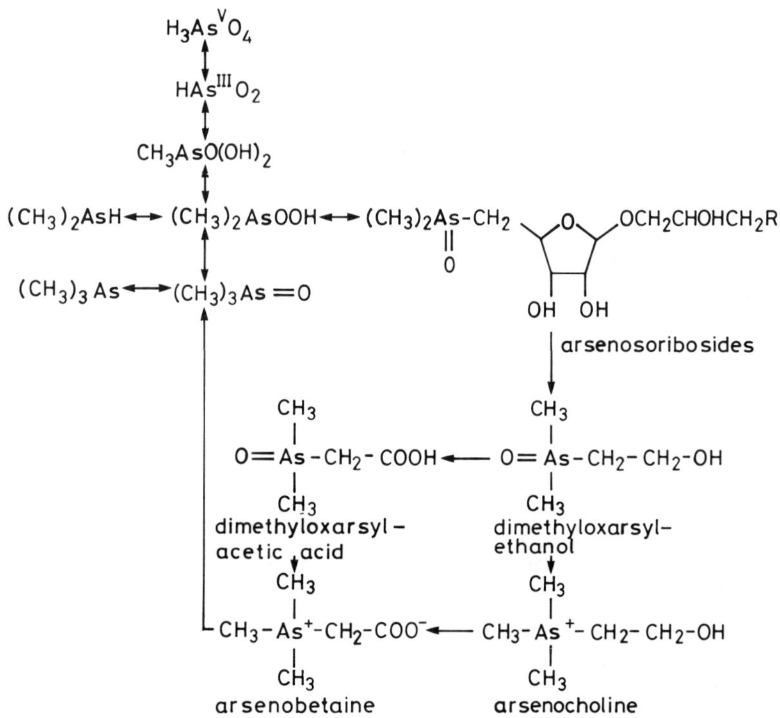

Figure 3 *From Ref. 1*

biological productivity (*e.g.*, chlorophyll concentration and carbon uptake) which suggests the importance of biological activity in arsenic speciation. The methyl arsenicals appear to be formed from and return to the main form, arsenate, and a marine arsenic cycle has been derived. The formation and cycling of these species occurs with natural levels of arsenic and is not a result of man's activities. The same methyl arsenic species are also observed in macroscopic algae. A number of other methylarsenic compounds are formed in the marine environment. Biosynthetic organoarsenical compounds are shown in Figure 2, together with the arsenic cycle.

$MeAsO(OH)_2$ and Me_2AsOOH can also be found in freshwater rivers and lakes.[2] Reduction of arsenate to arsenite in seawater by various species has been demonstrated and the reverse process has been known for some time. Methylation seems to proceed initially by reduction of arsenate as demonstrated in laboratory systems.[213-216]

Below the photic zone the methylated arsenicals rapidly decrease to levels below detection limits (pg g^{-1}). Arsenic is concentrated by aquatic organisms but not a great deal is known about accumulation ratios for organic arsenic compounds. Algae accumulate more arsenic than do fish, with crustacae accumulating intermediate amounts. The chemistry of arsenic within plant and other aquatic species has been discussed above.[217] Arsenobetaine ($Me_3As^+CH_2$-COO^-) has been found as 80% of the arsenic present in shark muscle and this derivative is thought to be the major form of arsenic in higher animals. For shark liver and lipid 40% and 45% respectively were in this form.[218] The origin of the methylarsenic compounds is in the substitution of arsenate for phosphate in metabolism in the oceans at low phosphate levels. By conversion to organic forms of arsenic the algae detoxifies the arsenate originally taken in and a critical point is an organo arsenic phospholipid.[219,220] The final sequence is postulated to be degradation to the arsonium lactate, ($Me_3As^+CH_2CHOHCOO^-$), which probably degrades further to Me_2AsOOH salts.[220] Reductive methylation of arsenate to Me_3As and formation of the lactate is also believed to take place in terrestrial plant roots and may account for the arsenic levels found in various plants and trees. We can therefore consider arsenic methylation as a detoxification process, but as arsenic serves no known biological function[209,210] in living organisms, arsenic metabolism must still be viewed as a contamination rather than as a part of an inherent metabolic process. However, such contamination is usually a consequence of the natural cycling of this element and it is not particularly man-derived.

Hence most of the levels of organoarsenic compounds shown in Table 7 though superfluous to the organisms concerned, are natural levels. It should be

[213] D. L. Johnson, *Nature*, 1972, **240**, 44.
[214] N. J. Blake and D. L. Johnson, *Deep Sea Res.*, 1976, **23**, 773.
[215] D. L. Johnson and M. E. Q. Pilson, *Environ. Lett.*, 1975, **8**, 157.
[216] A. W. Turner and J. W. Legge, *Aust. J. Biol. Sci.*, 1954, **7**, 452, 496, 504.
[217] E. A. Woolson in Ref. 182, p. 97.
[218] S. Kurosawa, K. Yasunda, M. Taguchyi, S. Yamazaki, S. Toda, M. Morita, T. Uehiro, and K. Fuwa, 1980, *Agric. Biol. Chem.*, 1980, **44**, 1983.
[219] R. V. Cooney, R. O. Mumma and A. A. Benson, *Proc. Nat. Acad. Sci. USA*, 1978, **75**, 4262.
[220] A. A. Benson, R. V. Cooney and J. M. Herrera-Lasso, *J. Plant Nutrition*, 1981, **3**, 285.

noted that most of the inorganic arsenic ingested by humans is excreted in the urine as the methyl arsenic compounds.[221-223] Methylation is presumably by intestinal flora but little work has been done yet on arsenic methylation in higher animals.

Organoarsenical pesticides are applied at lower concentrations and consequently accumulate less in the soil than the inorganic arsenate or arsenite derivatives used on a large scale previously.

At these lower concentrations (*viz.* about 2–4 kg ha^{-1} compared to 10–1000 kg ha^{-1}) organic arsenic compounds appear to persist as such less in the environment than the inorganic derivatives. However, as they partly decay through arsenate this is equivocal. Decay to volatile products and the reduced concentrations at which they are applied minimizes their effect on the soil compared to inorganic arsenic compounds. If no losses from the soil are allowed for such application would lead to excess arsenic in the soil of 1.4 µg g^{-1} to 4.6 µg g^{-1} dependent on the number of applications and the half-life.[221] Losses due to methylation and reductive volatilization, erosion, and crop removal may however reach 40%–60% of applied arsenic.

As a consequence of agricultural use, soil absorption studies of organoarsenics show absorption is rapid, and further changes to less soluble forms then occur. Applied arsenate and dimethyl arsenate (cacodylates) became gradually less extractable over six months after addition. Although the arsenic–carbon bond is stable in plants, microbiological oxidation takes place in soils producing up to ten per cent yields of carbon dioxide over 30 days, together with arsenate. A series of experiments has shown that over 24 weeks aerobic conditions about 41% of applied Me_2AsOOH is oxidized to carbon dioxide and arsenate while 35% is lost from the soil as a volatile organoarsenic compound.[224] Under anaerobic conditions there is no production of carbon dioxide but 61% was volatilized. These appear to be biomethylation processes by fungi and methanogenic bacteria respectively.

Biomethylation also takes place for phenyl arsenic compounds in the natural environment. Rate of oxidation of organoarsenicals to arsenate in various soils seems to be between two and ten per cent per month. This leads to estimates of 0.78 and 0.28 respectively, remaining one year after application at unit concentration, giving half-lives of 34.3 and 6.6 months in each case[151] ignoring volatilization. Both organic and inorganic forms of arsenic in plants seem to absorb and translocate in relatively small amount.[225] Experiments with $MeAs(O)(OH)ONa$ over a six year period showed no arsenic residues in the cottonseed.[226] Most of the arsenic losses were not to the plants but by methylation and adsorption *in situ* to soil. Greater absorption may occur in roots. In general, little arsenic is found in plants after arsenic treatment. To achieve

[221] T. J. Smith, E. A. Crecelius, and J. C. Reading, *Environ. Health Perspect.*, 1977, **19**, 89.

[222] E. A. Crecelius, *Environ. Health Perspect.*, 1977, **19**, 149.

[223] A. E. Hiltbold, in Ref. 182, p. 53.

[224] E. A. Woolson and U. C. Kearney, *Environ. Sci. Technol.*, 1973, **7**, 47.

[225] L. R. Johnson and A. E. Hiltbold, *Soil Sci. Soc., Amer. Proc.*, 1969, **33**, 279.

[226] A. E. Hiltbold, B. F. Hajek, and G. A. Buchanan, *Weed Sci.*, 1974, **22**, 272.

significant levels, $Me_2AsO(ONa)$ at 35 kg ha^{-1} had to be applied to the soil, ten times the normal rate. Most arsenic is volatilized, eroded, or retained as insoluble forms with aluminium or iron compounds in the soil.[223,227]

At normal rates of use, organoarsenic compounds fed to animals do accumulate in tissues but to amounts less than the elevation of arsenic in the animal food over natural levels. In poultry and swine a high proportion of ingested arsenic is rapidly excreted in soils described above. There is no evidence for conversion to inorganic forms of arsenic within the animal. $MeAsO(OH)_2$ and $Me_2AsO(OH)$ have been found in shells and human urine but not in rainwater.[228,229]

Use of organic arsenicals in animal feed (permissible levels, 50–375 µg g^{-1}) does lead to arsenic residues in tissues. Feeding the approved levels gives 1–2 µg g^{-1} in chicken liver; feeding at ten times the approved level gives 6–7 µg g^{-1} in chicken liver; (permissible level, 0.5–2.0 µg g^{-1}). As noted, these levels rapidly decline after arsenic-containing feed is withdrawn.[228] Feeding of arsenilic acid (p-$NH_2PhAsO(OH)_2$) at the 273 µg g^{-1} level (nearly three times the approved for poultry or swine) produced arsenic levels of 29.2 (liver), 23.6 (kidney), and 1.2 µg g^{-1} (muscle) respectively. Original levels were less than 0.01 µg g^{-1}. The wholeblood level increased from 0.01 to 0.54 µg g^{-1}. All weights are dry weight.[228] The depletion period was examined following withdrawal of arsenic and it was found that after six days liver arsenic levels had fallen from 29.2 to 5.0 µg g^{-1}. Most arsenic compounds are excreted unchanged from animals. The use of arsenic-containing animal waste as fertilizer has been considered from the point of view of increase in the arsenic burden of the soil; fortunately it would take more than 2 000 tonnes of waste containing 15–50 µg g^{-1} arsenic applied each year per acre to raise the soil arsenic to the 500 µg g^{-1} level found in some arsenic-treated soils.

Analysis of arsenic in environmental matrices is by a number of methods.[153,229,230–234] Conversion to arsine derivatives by zinc and hydrochloric acid, electrolysis, or by borohydride is commonly used to speciate arsenite, arsenate, $MeAsO(OH)_2$, and Me_2AsOOH. The latter two evolve methyl arsines. The volatile arsines are trapped and measured by AA or atomic emission. Photometric measurement of arsenic complexes is also used.[235–237] Historically, methylation of arsenic(III) compounds to Me_3As by cultures of the mould *S. brevicaulis* by Challenger was the first demonstration of biological methylation

[227] P. J. Elman, *Proc. S. Weed Conf.*, 1965, **18**, 685.

[228] C. C. Calvert, in Ref. 182, p. 70.

[229] R. S. Braman and C. C. Foreback, *Science*, 1973, **182**, 1247.

[230] M. O. Andreae, *Anal. Chem.*, 1977, **49**, 820.

[231] R. H. Fish, F. E. Brinckman, and K. L. Jewett, *Environ. Sci. Technol.*, 1982, **16**, 174.

[232] R. S. Braman, D. L. Johnson, C. C. Foreback, J. M. Ammons, and J. I. Bricker, *Anal. Chem.*, 1977, **49**, 621.

[233] M. H. Arbab-Zhaver and A. G. Howard, *Analyst (London)*, 1980, **105**, 744.

[234] A. A. Grabinski, *Anal. Chem.*, 1981, **53**, 966.

[235] 'Am. Pub. Health Assoc.,' Std. Methods Exam. Water Waste water, 13th edn., Washington, DC, 1971.

[236] G. M. George, L. H. Frahm, and J. P. McDonnell, *J. Assoc. Off. Anal. Chem.*, 1973, **56**, 793.

[237] E. B. Sandell, 'Colorimetric Determination of Traces of Metals,' 3rd edn., Interscience, New York, 1959.

of a metalloidal element. $MeAsO(ONa)_2$ and $Me_2AsO(ONa)$ also gave Me_3As, whereas alkyl arsonic acids $(RAsO(OH)_2, R_2AsO(OH))$ produced the appropriate methyl alkyl arsine, confirming the supply of methyl by the mould. The methylations took place in well aerated mould cultures which nevertheless exhibited strong reducing action. The mechanism appears to operate by methyl carbonium ion transfer from *S*-adenosyl methionine (SAM) in the mould, to a lone electron pair on arsenic(III) (*i.e.*, an oxidative addition). Further reduction and subsequent methylation may occur as appropriate. This explanation was put forward by Challenger in two reviews.[238,239] It is possible that methylation of Group IV divalent elements (tin and lead) might also involve methyl carbonium transfer (via SAM) as leading to the observed tetravalent methyl compound. Soil organisms have also been shown to produce Me_3As.[240] Anaerobic synthesis of Me_2AsH has been demonstrated from methanogenic bacteria, particularly *Methanobacterium strain* (Mo.H). Arsenate soils were the substrate and $MeCoB_{12}$ was a requirement for Me_2AsH production. Hydrogen, adenosine triphosphate (partially), and cell extract were also needed. Boiling prevented synthesis of Me_2AsH. This suggests the methylation works by a non-enzymatic transfer from $MeCoB_{12}$ to arsenic. The original mechanism of Challenger was supported by this work. In addition with *C. humicola*, cell extracts and arsenate, $Me_2AsO(OH)$, $MeAsO(OH)_2$, and Me_3AsO are found, *i.e.*, the intermediates in Challenger's mechanism.[241-245] Arsenite, methylarsonate $(MeAsO_3^{2-})$, and dimethylarsinate $(Me_2AsO_2^-)$ may also be methylated to Me_3As by *C. humicola*.

Such methylation under environmental conditions sheds light on to the organic arsenic derivatives found in seawater, marine biota, and other waters. No free arsine derivatives seem to have been found in sea water to date; there is no evidence yet of biological volatilization from surface waters but if there is it is likely that oxidation then takes place.

In a similar fashion to the mercury case, mixed bacterial cultures can biomethylate $Me_2AsO(OH)$ to arsenate.[246] This suggests a balanced arsenic cycle in the marine environment[246] (Figure 3).

A number of organoarsenic compounds have been detected in oil shale retort and process waters, including $MeAsO(OH)_2$ and $PhAsO(OH)_2$. These were ascribed to prehistoric formation and fixing in the shale or they could have been formed chemically during the industrial process.[231]

Arsenic methylation from freshwater and other organisms has also been

[238] F. Challenger, 'Aspects of the Organic Chemistry of Sulphur,' Butterworths, London, 1959.

[239] F. Challenger, *Quart. Rev. Chem. Soc.*, 1955, **9**, 255.

[240] D. P. Cox and M. Alexander, *Appl. Microbiol*, 1973, **25**, 408.

[241] B. C. McBride and R.S. Wolfe, *Biochemistry*, 1971, **10**, 4312

[242] B. C. McBride and T. L. Edwards, 'Biol. Implic. Metals. Environ.,' ERDA Symp. Ser. 42, Proc. Ann. Hanford Life Sci. 15th Symp., 1977, 1.

[243] B. C. McBride, H. Merilees, and W. Pickett, in Ref. 8, p. 94.

[244] W. R. Cullen, B. C. McBride, and W. Pickett, *Can. J. Microbiol.*, 1979, **25**, 1201.

[245] W. R. Cullen, B. C. McBride, and M. Reimer, *Bull. Environ. Contam. Toxicol.*, 1979, **21**, 157.

[246] J. G. Sanders, *Chemosphere*, 1979, **3**, 135.

demonstrated recently with arsine or arsonic acid derivatives being identified.[125,247,248]

A number of other metals have been postulated to undergo environmental
methylation. Some of the experimental evidence has been somewhat remote
from environmental conditions (*e.g.*, non-aqueous solvents). Here only those
cases where reasonable environmental evidence exists are mentioned. Aqueous
or sediment incubations leading to observable stable methyl metal species have
been made for thallium(I), producing Me_2Tl^+ species,[249] selenium, and tellur-
ium species being methylated, to the dimethyl compound.[250] The role of the
anti-knock additive $MeCpMn(CO)_3$ in the environment has also been discussed.[251] Methyl antimony species have recently been detected in the environment.[252] There is some early laboratory evidence for the methylation of this
element.

Detection was by sodium borohydride reduction to methyl stibine derivatives
followed by AA analysis. From natural waters methylstibonic acid [MeSbO-
$(OH)_2$] and dimethyl stibinic acid [$Me_2SbO(OH)$] were detected between the 1
and 13 ng dm^{-3} level. The origin of the methyl species was ascribed to biological
methylation by algae.[252]

Let it not be inferred that the environmental impact of organometallic
compounds in general, and biomethylation in particular has not been of historic
significance. We should not forget the case of Napoleon I, Emperor of France.
It has been suggested that Napoleon was probably poisoned by gaseous Me_3As
emission from arsenic biomethylation of the wallpaper in his bedroom on St.
Helena.[253] The green wallpaper has been shown to contain arsenic and the
damp climate of St. Helena is suitable for the growth of moulds capable of
biomethylation of arsenic to Me_3As. However, this interesting hypothesis has
also been disputed.[254]

The role of arsenic in the environment has been reviewed in particular detail
recently, first in a review and also in a monograph issue of a Journal.[255,256]

Acknowledgements. Dr. P. J. Craig gratefully acknowledges funding over the past
few years from the Natural Environment Research Council and the Science and
Engineering Research Council UK. The receipt of fieldwork and travel funding
from the Royal Society, the British Council, and Leicester Polytechnic is also
acknowledged with thanks. Janet R. Ashby acknowledges a grant from NERC.

[247] W. R. Cullen, C. L. Froese, A. Lui, B. C. McBride, D. J. Patmore, and M. Reimer, *J.
Organometall. Chem.*, 1977, **139**, 61.
[248] P. T. S. Wong, Y. K. Chau, L. Luxon, and G. A. Bengert, Proc. 11th Ann. Conf. Trace Subst.
Environ. Health, Columbia, Mo., USA, June 1977.
[249] F. Huber, U. Schmidt, and H. Kirchmann, in Ref. 8, p. 73.
[250] Y. K. Chau, P. T. S. Wong, B. A. Silverberg, P. L. Luxon, and G. A. Bengert, *Science*, 1976, **192**,
1130.
[251] M. D. Dupuis and H. H. Hill Jr., *Anal. Chem.*, 1979, **51**, 292.
[252] M. O. Andreae, J-F. Asmode, P. Foster, and L. Van'tdack, *Anal. Chem.*, 1981, **53**, 1766.
[253] D. E. H. Jones and K. W. D. Ledingham, *Nature*, 1982, **299**, 626.
[254] P. K. Lewin, R. G. Hancock, and P. Voynovitch, *Nature*, 1982, **299**, 627.
[255] W. Maher and E. Butler, *Appl. Organometall. Chem.*, 1988, **2**, 191.
[256] Various, *Appl. Organometall. Chem.*, 1988, **2**, Issue No. 4.

CHAPTER 17

Radioactivity in the Environment

C. N. HEWITT

1 INTRODUCTION

Pollution of the natural environment by radioactive substances is of concern because of the considerable potential that ionizing radiation has for damaging biological material. Although both the benefits of controlled exposure for medical purposes and the catastrophic effects of large doses of radiation (for example, those received by the inhabitants of Hiroshima and Nagasaki in 1945) are well understood what is less clear are the effects of small doses on the general population. For this reason it is necessary to evaluate in detail the exposure received from each of the multiplicity of natural and man made sources of radioactivity in the environment. In order to do this an understanding is required of the actual and potential source strengths, the pathways and cycling of radioactivity through the environment and their flux rates, and the possible routes of exposure for man.

2 RADIATION AND RADIOACTIVITY

2.1 Types of Radiation

Radiation arises from a spontaneous rearrangement of the nucleus of an atom. Whilst some nuclei are stable many are not and these can undergo a change, losing mass or energy in the form of radiation. Some unstable nuclei are naturally occurring whilst others are produced synthetically. The most common forms of these radiations are alpha particles, beta particles, and gamma rays, the physical properties of which are shown in Table 1 and are described below.

    (a) Alpha (α) particles consist of two protons and two neutrons bound together and so are identical to helium nuclei. They have a mass number

Table 1   *Types of radioactive emissions*

Radiation	Symbol	Composition	Charge	Mass Number	Approximate Tissue Penetration (cm)
Alpha	α	particle containing two protons and two neutrons	2+	4	0.01
Beta	β	particle of one electron	1−	0	1
Gamma	γ	very short wavelength electromagnetic radiation	0	0	100

of 4 and a charge of $+2$. Because they are so large and heavy, alpha particles travel slowly compared with other types of radiation (at maximum, about 10% the speed of light) and can be stopped relatively easily. Their ability to penetrate into living tissue is therefore limited and damage occurs only when alpha-emitting isotopes are ingested or inhaled. However, their considerable kinetic energy and the double positive charge which attracts and pulls away electrons from atoms belonging to tissue means that alpha particles can cause the formation of ions and free radicals and hence severe chemical change along their path. The emission of alpha particles is common only for nuclides of mass number greater than 209 and atomic number 82, nuclides of this size having too many protons for stability. An example of α-decay is:

$$\ce{^{210}_{84}Po} \rightarrow \ce{^{206}_{82}Pb} + \ce{^{4}_{2}He}$$

Note that the sum of the superscripts (mass numbers or sum of neutrons and protons) and the sum of the subscripts (atomic numbers or sum of protons) remain unchanged during the decay.

(b) Beta (β) particles are simply electrons emitted by the nucleus during the change of a neutron into a proton. They have minimal mass (1.36×10^{-4} times that of an alpha particle) but high velocity (typically 40% the speed of light) and a charge of -1. They may penetrate through skin or surface cells into tissue and may then pass close to the orbital electrons of tissue atoms where the repulsion of the two negative particles may force the orbital electron out of the atom, ionizing the tissue, and forming radicals. Because beta decay involves the change of a neutron into a proton the atomic number of the nuclide increases by one, but the mass number does not change. Beta decay is a common mode of radioactive disintegration and is observed for both natural and synthetic nuclides. Examples are:

$$\ce{^{90}_{38}Sr} \rightarrow \ce{^{90}_{39}Y} + \ce{^{0}_{-1}e}$$

(c) Gamma (γ) radiation is very short wavelength electromagnetic radiation. It travels at the speed of light, is uncharged, but is highly energetic and so has considerable penetration power. As it passes through biological tissue the electric field surrounding a gamma ray may eject orbital electrons from atoms and so can cause ionization of the tissue and formation of radicals along its path. The emission of gamma radiation does not lead to changes in mass or atomic numbers, and may occur either on its own from an electronically excited nucleus or may accompany other types of radioactive decay. Examples of these two processes are:

$$_{52}^{125}\text{Te*} \rightarrow _{52}^{125}\text{Te} + \gamma$$

the asterisk signifying the excited state of the nucleus and:

$$_{55}^{137}\text{Cs} \rightarrow _{56}^{137\text{m}}\text{Ba} + _{-1}^{0}\text{e} \rightarrow _{56}^{137}\text{Ba} + \gamma$$

2.2   The Energy Changes of Nuclear Reactions

The energy changes associated with nuclear reactions are considerably greater than those associated with ordinary chemical reactions. The sum of the mass of the products of the nuclear reaction is invariably less than the sum of the mass of the reactants and the amount of energy released (ΔE) is equivalent to this difference in mass (Δm). Most of this energy is released as kinetic energy, although some may be used to promote the nucleus to an excited state from where it will lose energy in the form of γ-radiation and return to the ground state.

The energy equivalent of a given mass can be calculated by means of Einstein's equation:

$$\Delta E = (\Delta m)c^2$$

The energy equivalent of one atomic mass unit ($1\,\text{u} = 1.660566 \times 10^{-27}\,\text{kg}$) is:

$$\Delta E = (1.660566 \times 10^{-27}\,\text{kg})\,(2.99792 \times 10^8\,\text{m s}^{-1})^2$$
$$= 1.49244 \times 10^{-10}\,\text{J}$$

This is usually expressed in units of electron volts (eV) where:

$$1\,\text{eV} = 1.60219 \times 10^{-19}\,\text{J}$$
$$\text{and} \qquad 1\,\text{MeV} = 1.60219 \times 10^{-13}\,\text{J}$$

The energy equivalent of 1 u is therefore 931.5 MeV.

The amount of energy released by a decay process can now be calculated. For example, in the alpha decay:

$$_{84}^{210}\text{Po} \rightarrow _{82}^{206}\text{Pb} + _{2}^{4}\text{He}$$
$$\Delta m = (\text{mass}\ _{84}^{210}\text{Po}) - (\text{mass}\ _{82}^{206}\text{Pb} + \text{mass}\ _{2}^{4}\text{He})$$
$$= 209.9829\,\text{u} - (205.9745\,\text{u} + 4.0026\,\text{u})$$
$$= 0.0058\,\text{u}$$

The energy released by the decay is therefore:

$$\Delta E = 0.0058\,\text{u} \times 931.5\,\text{MeV/u} = 5.4\,\text{MeV}$$

2.3   Rates of Radioactive Decay

The rates of decay of radioactive nuclides are first-order and independent of temperature. This implies that the activation energy of radioactive decay is zero and that the rate of decay depends only on the amount of radioactive substance present. If N is the number of atoms present at time t the rate of change of N is given by

$$dN/dt = -\lambda N$$

where λ is the characteristic (or disintegration or decay) constant for that radionuclide. Integrating between times t_1 and t_2 give:

$$N_2 = N_1 \exp\left[-\lambda(t_2 - t_1)\right]$$

where N_1 and N_2 are the number of atoms of the radionuclide present at times t_1 and t_2 respectively. If t_1 is set to zero then:

$$N = N_0 \exp\left(-\lambda t\right) \qquad \text{(equation a)}$$

where t is the elapsed time and N_0 is the number of atoms of the radionuclide present when $t_1 = 0$.

When N/N_0 is equal to 0.5 (*i.e.*, half the atoms have decayed away) t is defined as being the half-life, $t_{\frac{1}{2}}$. Then:

$$N/N_0 = 0.5 = \exp\left(-\lambda t_{\frac{1}{2}}\right)$$
$$t_{\frac{1}{2}} = 0.693/\lambda$$

Equation (a) can now be written in terms of the more readily available $t_{\frac{1}{2}}$, rather than λ, to give

$$N = N_0 \exp(-0.693 \, t/t_{\frac{1}{2}}).$$

The half-lives of some selected radionuclides are given in Table 2.

Table 2   *Half-lives of some environmentally important radionuclides*

Radionuclide	*Half-life*
^{131}I	8.1 d
^{85}Kr	10.8 y
^{3}H	12.3 y
^{90}Sr	28 y
^{137}Cs	30 y
^{239}Pu	2.4×10^4 y
^{238}U	4.5×10^9 y

2.4   Activity

The amount of radiation emitted by a source per unit time is known as the activity of the source, expressed in terms of the number of disintegrations per second. The unit of activity, the becquerel (Bq), is defined as being one disintegration per second. The activity of a source is proportional to the number of radioactive atoms present and so diminishes with time according to first-order kinetics.

2.5 Radioactive Decay Series

Some radioactive decay processes lead in one step to a stable product but frequently a disintegration leads to the formation of another unstable nucleus. This can be repeated several times producing a radioactive decay series which only terminates on the formation of a stable nuclide. There are three naturally-occurring decay series, headed by ^{232}Th, ^{238}U, and ^{235}U. Each of these nuclides has a half-life which is long in relation to the age of the earth and each finally produces stable isotopes of lead, ^{208}Pb, ^{206}Pb, and ^{207}Pb respectively. The 14 steps of the ^{238}U series are shown in Figure 1 and it will be seen that at several points branching occurs as the series proceeds by two different routes which rejoin at a later point.

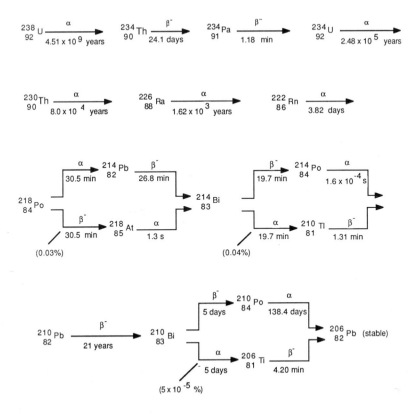

Figure 1 *Disintegration series of $^{238}_{92}$U (half-lives of isotopes are indicated)*

2.6 Production of Artificial Radionuclides

The first artificial transmutation of one element into another was achieved in 1919 when Ernest Rutherford passed α-particles (produced by the radioactive decay of ^{214}Po) through nitrogen:

$$^{14}_{7}N + ^{4}_{2}He \rightarrow ^{17}_{8}O + ^{1}_{1}H$$

Subsequently the first artificial radioactive nuclide, ^{30}P, was produced:

$$^{27}_{13}N + ^{4}_{2}He \rightarrow ^{30}_{15}P + ^{1}_{0}n \text{ (neutron)}$$
$$^{30}_{15}P \rightarrow ^{30}_{14}Si + ^{0}_{1}e \text{ (positron)}$$

Bombardment reactions of this type have been used to produce isotopes of elements which do not exist in nature and which are particularly important as pollutants of the environment, for example:

$$^{238}_{92}U + ^{1}_{0}n \rightarrow ^{239}_{92}U + \gamma$$
$$^{239}_{92}U \rightarrow ^{239}_{93}Np + ^{0}_{-1}e \qquad \text{neptunium}$$
$$^{239}_{93}Np \rightarrow ^{239}_{94}Pu + ^{0}_{-1}e \qquad \text{plutonium}$$
$$^{239}_{94}Pu + ^{2}_{1}H \rightarrow ^{240}_{95}Am + ^{1}_{0}n \qquad \text{americium}$$

2.7 Nuclear Fission

Many heavy nuclei with mass numbers greater than 230 are susceptible to spontaneous fission, or splitting into lighter fragments, as a result of the forces of repulsion between their large number of protons. Fission can also be induced by bombarding heavy nuclei with projectiles such as neutrons, alpha particles, or protons. When the fission of a particular nuclide takes place the nuclei may split in a variety of ways, producing a number of products. For example, some of the possible fission reactions of ^{235}U are:

$$^{235}_{92}U + ^{1}_{0}n \rightarrow ^{95}_{39}Y + ^{138}_{53}I + 3^{1}_{0}n$$
$$^{235}_{92}U + ^{1}_{0}n \rightarrow ^{97}_{39}Y + ^{137}_{53}I + 2^{1}_{0}n$$
$$^{235}_{92}U + ^{1}_{0}n \rightarrow ^{90}_{36}Kr + ^{144}_{56}Ba + 2^{1}_{0}n$$
$$^{235}_{92}U + ^{1}_{0}n \rightarrow ^{90}_{35}Br + ^{143}_{57}La + 3^{1}_{0}n$$

In each of these reactions neutrons are formed as primary products and these neutrons can, in turn, cause the fission of other ^{235}U nuclei, so causing a chain reaction of the fissile uranium.

The number of nuclei of a particular daughter product formed by the fission of 100 parent nuclei is defined as the yield of the process. Fission yields vary, fission induced by slow neutrons (as in a nuclear power plant) having a different set of yields to that induced by fast neutrons (as in many nuclear weapons). Table 3 shows examples of the percentage yields of the fast-neutron induced fission of ^{239}Pu. It also shows some of the yields of ^{235}U fission, which are similar for both slow and fast neutron induced processes.

Table 3 *Yields of some long-lived radionuclides following uranium and plutonium fission*

		Yield (%)	
Radioisotope	*Half-life*	^{235}U	^{239}Pu
Strontium-90	28 y	5.8	2.2
Iodine-131	8 d	3.1	3.8
Caesium-137	30 y	6.1	5.2
Krypton-85	10.3 y	0	—
Cerium-141	33 d	6.0	5.2

2.8 Beta Decay of Fission Products

The primary daughter products of nuclear fission are almost always β-radio-active and often these also quickly produce β-radioactive products. Only after several such decays are products with long half-lives formed. For example the fission of U-235 produces Br-90:

$$^{235}_{92}\text{U} + ^{1}_{0}\text{n} \rightarrow ^{90}_{35}\text{Br} + ^{143}_{57}\text{La} + 3^{1}_{0}\text{n}$$

which quickly decays to ^{90}Kr with a half-life of 1.4 seconds. This in turn has a half-life of 33 seconds and decays to ^{90}Rb ($t_{\frac{1}{2}} = 2.7$ minutes), and this to ^{90}Sr:

$$^{90}_{35}\text{Br} \rightarrow ^{90}_{36}\text{Kr} + _{-1}^{0}\text{e} \ (t_{\frac{1}{2}} = 1.4 \text{ s})$$
$$^{90}_{36}\text{Kr} \rightarrow ^{90}_{37}\text{Rb} + _{-1}^{0}\text{e} \ (t_{\frac{1}{2}} = 33 \text{ s})$$
$$^{90}_{37}\text{Rb} \rightarrow ^{90}_{38}\text{Sr} + _{-1}^{0}\text{e} \ (t_{\frac{1}{2}} = 2.7 \text{ min})$$

Strontium-90 decays to yttrium-90:

$$^{90}_{38}\text{Sr} \rightarrow ^{90}_{39}\text{Y} + _{-1}^{0}\text{e} \ (t_{\frac{1}{2}} = 28 \text{ years})$$

and has a half-life of 28 years, which is sufficiently long for it to be widely circulated through the environment, in contrast to its short-lived precursors. After one more beta decay a stable product, ^{90}Zr, is formed. It is therefore radioactive strontium rather than its precursors that is the important pollutant in this series.

Examples of other environmentally-important nuclides formed by beta decay chains are iodine-131 and caesium-137. ^{131}I is formed from ^{131}Sn, itself produced by the fission of ^{235}U, and forms the stable nuclide ^{131}Xe:

$$^{131}_{50}\text{Sn} \rightarrow ^{131}_{51}\text{Sb} + _{-1}^{0}\text{e} \ (t_{\frac{1}{2}} = 3.4 \text{ min})$$
$$^{131}_{51}\text{Sb} \rightarrow ^{131}_{52}\text{Te} + _{-1}^{0}\text{e} \ (t_{\frac{1}{2}} = 23 \text{ min})$$
$$^{131}_{52}\text{Te} \rightarrow ^{131}_{53}\text{I} + _{-1}^{0}\text{e} \ (t_{\frac{1}{2}} = 24 \text{ mins})$$
$$^{131}_{53}\text{I} \rightarrow ^{131}_{54}\text{Xe} + _{-1}^{0}\text{e} \ (t_{\frac{1}{2}} = 8 \text{ days})$$

^{137}Cs begins as the uranium fission daughter ^{137}I and ends as the stable barium isotope, ^{137}Ba:

$$^{137}_{53}\text{I} \rightarrow ^{137}_{54}\text{Xe} + _{-1}^{0}\text{e} \ (t_{\frac{1}{2}} = 24 \text{ s})$$
$$^{137}_{54}\text{Xe} \rightarrow ^{137}_{55}\text{Cs} + _{-1}^{0}\text{e} \ (t_{\frac{1}{2}} = 3.9 \text{ min})$$
$$^{137}_{55}\text{Cs} \rightarrow ^{137}_{56}\text{Ba} + _{-1}^{0}\text{e} \ (t_{\frac{1}{2}} = 30 \text{ years})$$

2.9 Units of Radiation Dose

The amount of biological damage caused by radiation and the probability of the occurrence of damage are directly related to dosage and so before discussing the effects of radiation it is necessary to consider the units of dose.

2.9.1 Absorbed Dose. The amount of energy actually absorbed by tissue or other material from radiation is known as the *absorbed dose* and is expressed in a unit called the *gray* (Gy). One gray is equal to the transfer of one joule of energy to one kilogram of material. Table 4 shows the relationship of this and other SI units with the older units they replace.

Table 4 *SI and old radiation units*

Quantity	SI Unit	Old Unit	Relationship
Activity	becquerel	curie	$1\ Ci = 3.7 \times 10^{10}\ Bq$
Absorbed dose	gray	rad	$1\ rad = 0.01\ Gy$
Dose equivalent	sievert	rem	$1\ rem = 0.01\ Sv$

2.9.2 Dose Equivalent. Because the physical and chemical properties of α, β, and γ radiations vary, equal absorbed doses of radiation do not necessarily have the same biological effects. In order to equate the biological effects of one type of radiation with another it is therefore necessary to multiply the absorbed dose by a quality factor that accounts for the differences in biological damage caused by radioactive particles having the same energy. This is known as the *dose equivalent* and is expressed in units of *sieverts* (Sv). For γ-rays, X-rays, and β-particles the quality factor equals 1, whereas for α-particles it is 20. The dose equivalent resulting from an absorbed dose of 1 Gy of alpha radiation therefore equals 20 Sv.

2.9.3 Effective Dose Equivalents. Having established the dose equivalent for each tissue (H_T) it is now necessary to weight this to take account of the differing susceptibilities of different organs and types of tissue to damage. For example the testes and ovaries are more easily damaged than are the lungs or bones. The risk weighting factors (W_T) currently recommended by the International Commission for Radiological Protection (ICRP) are shown in Table 5. The effective dose equivalent (H_E) for the body is then expressed as the sum of the weighted dose equivalents:

$$H_E = \Sigma_T H_T W_T$$

2.9.4 Collective Effective Dose Equivalents. As well as quantifying the effective dose equivalent (or *dose*) received by individuals it is also important to have a measure of the total radiation dose by a group of people or population. This is the *collective effective dose equivalent* (or *collective dose*) and is obtained by multiplying the mean effective dose equivalent to the group from a particular source by the number of people in that group, to give units of man Sieverts (man Sv).

Table 5 *ICRP risk weighting factors*

Tissue or organ	Weighting factor
Testes and ovaries	0.25
Breast	0.15
Red bone marrow	0.12
Lung	0.12
Thyroid	0.03
Bone surfaces	0.03
Remainder	0.30
Whole body total	1.00

3 BIOLOGICAL EFFECTS OF RADIATION

3.1 General Effects

The formation of ions and free radicals in tissue by radiation and the subsequent chemical reactions between these reactive species and the tissue molecules causes a range of short-term and long-term biological effects. As a frame of reference acute exposure can cause death, an instantaneously absorbed dose of 5 Gy probably being lethal. The range of effects include cataracts, gastrointestinal disorders, blood disorders including leukaemia, damage to the central nervous system, impaired fertility, cancer, genetic damage, and changes to chromosomes producing mutations in later generations. Of the long-term effects of radiation to the exposed generation cancer and especially leukaemia is probably the most important. The number of excess (or extra) cancers observed in an exposed group compared with a non-exposed control group divided by the product of the exposed group size and the mean individual dose gives the *risk factor*. This is the risk of the effect occurring per unit dose equivalent.

Risk factors have been estimated by the United Nations Scientific Committee on the Effects of Atomic Radiation (UNSCEAR) by studying various groups of exposed people. The risk factor found for fatal leukaemia, for example, is about $20 \times 10^{-3} \, \text{Sv}^{-1}$, or about a 1 in 500 chance of dying of leukaemia following an exposure of 1 Sv. When irradiation of the testes or ovaries occurs before the conception of children there is a risk that damage to the DNA may cause hereditary defects in future generations. The current UNSCEAR estimate of the risk factor for serious hereditary damage to humans is about $2 \times 10^{-2} \, \text{Sv}^{-1}$, or about 1 in 50 per Sv, with about half of this damage manifesting itself in the first two generations. When fatal cancers and serious hereditary defects in the first two generations are considered together the current ICRP estimate of the risk factor is $1.65 \times 10^{-2} \, \text{Sv}^{-1}$. This is based on the assumption that there is no lower threshold of dose below which the probability of effect is zero. In other words it is assumed that any exposure to radiation carries some risk and that the probability of the effect occurring is proportional to the dose. It is worth noting that the historical trend has been for estimates of risk to increase with time, and hence the estimates of acceptable dose have decreased with time.

3.2 Biological Availability and Residence Times

The rate of uptake of a radionuclide into the body depends upon its concentrations in the various environmental media (air, water, food, dust, *etc.*), the rates of intake of these media and the efficiency with which the body absorbs the nuclide from the media. This latter parameter largely depends upon the physical and chemical properties of the radioactive substance in question. For example, the chemical properties of strontium are similar to those of another Group II element, calcium, which is of vital biological significance. Calcium enters the body to form bones and to carry on other physiological functions and ^{90}Sr can readily follow calcium into the bones where it will remain. In the same way the chemical similarity of caesium to potassium, which is present in all cells in the body, allows caesium to be readily transported throughout the entire body. On the other hand the inert gases krypton, xenon, and radon, which are all radioactive air pollutants, are not absorbed by the body (although they can still contribute to the exposure of an individual by external irradiation and by forming radioactive daughters with different physico-chemical properties).

In any given organism there exists a balance between the intake of an element and its excretion and this controls the concentration of the element in the organism. For radioactive isotopes this balance will include on the debit side the radioactive decay of the isotope. The effective half-life, t_{eff}, of a nuclide in an organism, *i.e.*, the time required to reduce the activity in the body by half, is a function of both the biological half-life, t_{biol}, and the radioactive half-life, t_{rad}:

$$t_{eff} = t_{rad} \, t_{biol} / (t_{rad} + t_{biol})$$

Whilst t_{rad} is fixed for any given nuclide, t_{biol} will vary from species to species and between individuals with age, sex, physical condition, and metabolic rate. For ^{90}Sr, which has $t_{rad} = 28$ y and $t_{biol} = \sim 35$ y the effective half life for man is ~ 15.5 y. Obviously those nuclides of most concern as pollutants are those present in the environment in the highest concentrations, with the most energetic emissions and with the longest effective biological half-lives.

4 NATURAL RADIOACTIVITY

4.1 Cosmic Rays

The earth's atmosphere is continuously bombarded by highly energetic protons and alpha particles emitted by the sun and of galactic origin. Their energies range from about 1 MeV to about 10^4 MeV with a flux rate at the outer edge of the atmosphere of about 2×10^7 MeV m^{-2}s^{-1}. These primary particles have two effects. They cause some radiation exposure directly to man, the magnitude of which varies with altitude and latitude, and they interact with stable components of the atmosphere causing the formation of radionuclides. Of these, ^{3}H and especially ^{14}C are important to the biosphere while several of the shorter-lived nuclides (*e.g.*, ^{7}Be, $t_{\frac{1}{2}} = 53$ days; ^{39}Cl, $t_{\frac{1}{2}} = 55$ min) have applications as tracers in the study of atmospheric dispersion and deposition processes.

Carbon-14 is formed from atmospheric nitrogen by:

$$^{14}_{7}\text{N} + ^{1}_{0}\text{n} \rightarrow ^{15}_{7}\text{N}$$
$$^{15}_{7}\text{N} \rightarrow ^{14}_{6}\text{C} + ^{1}_{1}\text{p}$$

The ^{14}C is then oxidized to CO_2 and enters the global biogeochemical cycle of carbon, being incorporated into plants by photosynthesis and into the oceans by absorption. Most natural production of tritium is also by the reaction of cosmic ray neutrons (with energy $>4.4\,\text{MeV}$) with nitrogen:

$$^{14}_{7}\text{N} + ^{1}_{0}\text{n} \rightarrow ^{12}_{6}\text{C} + ^{3}_{1}\text{H}$$

Tritium is then oxidized or exchanges with ordinary hydrogen to form tritiated water and enters the global hydrological cycle. The National Radiological Protection Board (NRPB) estimate that the annual effective dose equivalent from cosmic rays in the UK is about $300\,\mu\text{Sv y}^{-1}$ on average. However, people that live at high altitudes and high latitudes can receive rather greater doses.

4.2 Terrestrial Gamma Radiation

The earth's crust contains three elements with radioactive isotopes which contribute significantly to man's exposure to radiation. These are ^{40}K, with an average concentration in the upper crust of ~ 3 ppm, ^{232}Th, which is present in granitic rocks at 10–15 ppm, and the three isotopes of uranium which total an average 3–4 ppm in granite. These latter nuclides have relative abundances of 99.274% ^{238}U, 0.7205% ^{235}U, and 0.0056% ^{234}U in the crust.

The NRPB estimate that the gamma rays emitted by these radionuclides and their daughters in soil, sediments, and rocks and in building materials give an average annual effective dose equivalent in the UK of about $400\,\mu\text{Sv y}^{-1}$, with the whole body being irradiated more or less equally. The local geology and types of building material used lead to considerable variation about this figure.

4.3 Radon and its Decay Products

The decay series of ^{238}U and ^{232}Th both contain radioactive isotopes of the element radon, ^{222}Rn in the former case and ^{220}Rn (sometimes known as thoron) in the latter. Radon, being a noble gas, readily escapes from soil and porous rock and diffuses into the lower atmosphere. There the nuclides decay with half-lives of 3.8 days (^{222}Rn) and 55 s (^{220}Rn) producing a series of short-lived daughter products. The full ^{238}U decay series is shown in Figure 1. These daughters attach themselves to aerosol particles in the atmosphere which are efficiently deposited in the lungs if inhaled. Their subsequent α and β emissions can then irradiate and damage the lung tissue.

Ambient outdoor concentrations of radon are low, typically about $3\,\text{Bq m}^{-3}$ in the UK. However when these gases enter a building, either through the floor following soil emissions or from the building's construction materials or by desorption from the water supply, the low air exchange rates in modern buildings causes a substantial increase in concentrations. Considerable variations in indoor concentrations have been observed, a positively-skewed distribution with a mean of $\sim 25\,\text{Bq m}^{-3}$ being typical. The concentrations in the outside air, in the soil, and in soil pore water, the rate of emanation from building materials and the building ventilation rate will all affect the indoor concentration, and in a large scale measurement programme a few individual dwellings with concentrations two or three orders of magnitude above the mean might be observed. Because of the large variations in concentration and differences in people's activity patterns the calculation of an individual's exposure to radiation from radon is rather uncertain. The average annual effective dose equivalent in the UK is estimated by NRPB to be $\sim 800\,\mu\text{Sv y}^{-1}$, although some individuals may receive very much higher doses than this. In any case radon and its daughters generally give a greater dose to man than any other sources of radiation. In the UK two 'Action Levels' have been set. They are $20\,\text{mSv y}^{-1}$ for existing houses and $5\,\text{mSv y}^{-1}$ for new houses, corresponding to 400 and $100\,\text{Bq m}^{-3}$ respectively.

In the USA in particular attempts are made to reduce indoor radon exposure in the worst affected areas and buildings. This can be done by preventing radon entering the building, by increasing the building ventilation rate or by removing the radon decay products from the air, for example, using an electrostatic precipitator. Since most radon in buildings comes from the ground below the best method is to reduce the ingress of radon into the building by sealing the floor or by creating or increasing underfloor ventilation.

4.4 Radioactivity in Food and Water

The most important naturally occurring radionuclides in food and water are ^{226}Ra, formed from ^{238}U, its α-emitting daughter products, ^{222}Rn, ^{218}Po, and ^{214}K. There is a wide variation in the ^{226}Ra content of public water supplies, it usually being very low and representing a minor source of intake, but some well and spring water may contain 0.04–$0.4\,\text{Bq L}^{-1}$. As previously mentioned the domestic water supply, particularly water used in showers, may act as a source of ^{222}Rn into the indoor environment. For the population in general the main source of radium intake is from food. Wide variations in concentration are found but a typical daily intake from the average diet is about $0.05\,\text{Bq}$ per day. Some foods, for example Brazil nuts and Pacific salmon, accumulate radium in preference to calcium and can have very much higher concentrations than the average for food. Unbalanced diets based on these foods could lead to enhanced intakes of radium. Both ^{210}Pb and ^{210}Po can enter food, both from the soil but also by wet and dry deposition from the atmosphere where they are present as daughters of ^{222}Rn. For the average individual in the UK the NRPB estimate of the total effective dose equivalent of alpha activity from the diet is about $200\,\mu\text{Sv y}^{-1}$.

An additional source of α-activity for some individuals which should be mentioned is cigarette smoke. The decay products of radon emitted from the ground under tobacco plants can adsorb onto the growing plant leaves and hence be incorporated into cigarettes. Dose rates have been estimated to be as high as 6–7 mSv per year from this source for some individuals.

Most of the naturally occurring β activity in food is due to ^{40}K. The availability of potassium to plants is subject to wide variations and hence the ^{40}K activity of foods also varies. However potassium is an essential element for plants and animals, constituting about 0.2% of the soft tissue of the body. This leads to a total ^{40}K content of ~ 4 kBq for the average person although the individual value depends upon age, weight, sex, and proportion of fat. The NRPB estimate that this gives an average effective dose equivalent of about $170\,\mu\text{Sv}\,\text{y}^{-1}$ to individuals in the UK.

5 MEDICAL APPLICATIONS OF RADIOACTIVITY

Although not generally regarded as being pollutants, radiations used for medical purposes make a significant contribution to man's exposure. Indeed they provide the major artificial route of exposure to man and so warrant mention here. X-rays are used for a wide range of diagnostic purposes, a typical chest X-ray giving an effective dose equivalent of about $20\,\mu\text{Sv}$. Short-lived radionuclides are also used diagnostically, for example, $^{99\text{m}}$Tc is used for bone and brain scans. The use of radiation for diagnostic purposes gives an average effective dose equivalent in the UK of about $250\,\mu\text{Sv}$ per year. Of this about $120\,\mu\text{Sv}\,\text{y}^{-1}$ is considered to be genetically significant, compared with about $1000\,\mu\text{Sv}\,\text{y}^{-1}$ genetically significant dose from natural sources. Restrictions on the use of X-rays during pregnancy, limiting the X-ray beam area to the minimum needed, reducing radiation leakage from the source and improving the sensitivity and reliability of the measuring methods used are all aimed at reducing the genetically significant dose to reduce the probability of future genetic effects.

Externally administered beams of X-rays, gamma rays (from ^{60}Co sources) and neutrons and radiations from internally administered radionuclides (*e.g.*, ^{131}I in the thyroid) are all used for therapeutic purposes. Some of the doses involved are very high but the potential adverse effects have to be weighed against the benefits accrued to the individual patient. Regular and stringent calibration and testing of medical radiation sources is vital if patients are to receive the correct dose in the correct place.

6 POLLUTION FROM NUCLEAR WEAPONS EXPLOSIONS

Since 1945 the products of nuclear weapons explosions, both of fission and fusion devices, have caused considerable pollution of the globe. The major bomb testing programmes were held in 1954–8 and 1961–2, with the number of documented test explosions now approaching 1 000 and totalling about 1 000 megatons of explosive force. Both underground and atmospheric tests have been used, with most of the explosive yield being from explosions in the atmosphere. Following an atmospheric explosion there will be some local deposition (or fallout) of activity but the majority of the products are injected into the upper troposphere and stratosphere, allowing their dispersion and subsequent deposition to the earth's surface on a global scale.

Fission bombs, such as those used against Japan, depend upon the rapid formation of a critical mass of ^{235}U or ^{239}Pu from several components of subcritical size. The critical mass for ^{235}U is about 10 kg which on fission produces a very large amount of radioactivity: the total activity released by the 14 kiloton device at Hiroshima was about 8×10^{24} Bq. Most of the released activity is in the form of short-lived nuclides, but considerable amounts of environmentally important fission products, including ^{85}Kr, ^{89}Sr, ^{90}Sr, ^{99}Tc, ^{106}Ru, ^{131}K, ^{137}Cs, ^{140}Ba, and ^{144}Ce, are also produced. Of particular biological significance are ^{89,90}Sr, ^{131}I, and ^{137}Cs.

As mentioned above strontium is of importance because it has similar chemical properties to those of calcium. It enters the body by inhalation and ingestion, particularly through milk and vegetables. Iodine differs from strontium in that it is present in the atmosphere in both the gas and aerosol phases. Exposure may be by inhalation and ingestion, but of particular importance is the concentration of iodine in milk by cows grazing on contaminated grass and its subsequent consumption by man. Estimates of the amount of ^{137}Cs released into the atmosphere by bomb tests vary but it is probably of the order of 10^{18} Bq giving rise to human exposure mainly through grain, meat, and milk.

Fusion bombs (also known as thermonuclear or hydrogen bombs) use the reaction of lithium hydride with slow neutrons to generate tritium:

$$^{6}_{3}\text{Li} + ^{1}_{0}\text{n} \rightarrow ^{3}_{1}\text{H} + ^{4}_{2}\text{He}$$

which then reacts with deuterium releasing energy:

$$^{2}_{1}\text{H} + ^{3}_{1}\text{H} \rightarrow ^{4}_{2}\text{He} + ^{1}_{0}\text{n}$$

The slow neutrons are initially supplied by the fission of ^{235}U or ^{239}Pu and the deuterium by the use of lithium-6 deuteride. Fusion bombs therefore release fission products as well as large amounts of tritium.

Tritium is a pure beta emitter with a half-life of 12.3 y. It is readily oxidized in the environment forming tritiated water and hence enters the hydrological cycle. The amount of tritium injected into the atmosphere by weapons tests probably exceeds 10^{20} Bq giving rise to estimated dose commitments over an average lifetime of 2×10^{-5} Gy in the northern hemisphere and 2×10^{-6} Gy in the southern hemisphere.

Various other radionuclides are produced by weapons explosions by the neutron activation of other elements in the soil or surface rocks, the air and the bomb casings. Of these ^{14}C is the most significant pollutant being formed from stable nitrogen in the air. It is a pure beta emitter of mean energy 49.5 keV and a half-life of 5 730 years. The rate of natural production of ^{14}C by cosmic ray interactions is about 1×10^{15} Bq y^{-1}, compared with a mean of $\sim 5 \times 10^{15}$ Bq y^{-1} produced by weapons testing since 1945.

The other important group of pollutant nuclides produced by weapons testing are the transuranics, including plutonium. Of these ^{239}Pu is the most significant as it has been produced in large quantities ($\sim 1.5 \times 10^{16}$ Bq) and has a long half-life (24 360 y). It is formed by capture of a neutron by ^{238}U:

$$^{238}_{92}\text{U} + ^{1}_{0}\text{n} \rightarrow ^{239}_{92}\text{U} \xrightarrow[\rightarrow 93]{\beta-} \text{Np} \xrightarrow[\rightarrow 94]{\beta-} ^{239}\text{Pu}$$

The most important route to man for plutonium from weapons tests is by inhalation and the UNSCEAR estimates for the population—weighted dose, up to 2000 AD from tests carried out to 1977, are 1×10^{-5} Gy in the northern hemisphere and 3×10^{-6} Gy in the southern.

The current NRPB estimate for the average effective dose equivalent in the UK from weapon test activity is about 10 μSv per year at present, compared with about 80 μSv per year in the early 1960's.

7 POLLUTION FROM ELECTRIC POWER GENERATION PLANT AND OTHER NUCLEAR REACTORS

7.1 Emissions Resulting from Normal Reactor Operation

There are a large number of nuclear reactors in operation worldwide of many different sizes, designs, and applications. As well as about 580 reactors operational, under construction, or shut down for electricity production (total for 1987) there are many other reactors used for research, isotope production, education, materials testing, and military purposes. Included in this latter category are reactors used to power submarines and other naval vessels. However the greatest use remains the generation of electricity.

The production of electric power by using thermal energy derived from nuclear fission involves a chain of steps from the mining and preparation of fissionable fuel through to the disposal of radioactive wastes. There are actual and potential emissions of radioactivity into the environment at each of these steps.

7.1.1 Uranium Mining and Concentration. Uranium ores contain, typically, about 0.15% U_2O_3 and require milling, extraction, and concentration before shipping to the consumer. At several stages in this process dusts bearing radioactivity are produced which may lead to contamination of the environment in the vicinity of the plant. Large volumes of mill tailings are also produced and as these contain virtually all the radium and thorium isotopes present in the ore they give rise to emissions of radon. Also of environmental significance are liquid releases from the mills which, if uncontrolled, may lead to contamination of surface and ground waters. It should also be mentioned that the mining and milling of uranium ore inevitably leads to the occupational exposure of workers to uranium and its daughters, especially via inhaled ^{222}Rn.

7.1.2 Purification, Enrichment, and Fuel Fabrication. The concentrated and purified ore extract, known as yellowcake, contains about 75% uranium and is converted to usable forms of uranium metal and uranium oxide by several processes. First uranium tetrafluoride, UF_4, is produced. The yellowcake is digested in nitric acid, insoluble impurities removed by filtration and soluble impurities extracted with an organic solvent. The uranyl nitrate solution is concentrated, oxidized to UO_2 and reacted with hydrogen fluoride to form uranium tetrafluoride, UF_4. The UF_4 so produced contains the relative natural abundances of the various uranium isotopes, including about 0.7% of ^{235}U. This is adequate for some reactors, such as the first generation British Magnox reactors, and unenriched uranium metal is produced for them by reaction of the UF_4 with magnesium. The metal is cast and machined into fuel rods, heat treated, and then inserted into magnesium aluminium alloy ('Magnox') cans

Other designs of reactors, including the Advanced Gas-cooled and Pressurized Water Reactors, require fuel in the form of enriched uranium oxide containing 2–3% of ^{235}U. This is produced by first reacting UF_4 with fluorine gas to form uranium hexafluoride, UF_6. The UF_6 gas is then repeatedly centrifuged or diffused through porous membranes, the small mass differences between the isotopes resulting in their eventual fractionation. Following enrichment the UF_6 is hydrolysed and reacted with hydrogen to form uranium oxide, UO_2, in powder form. This is pressed into pellets and packed in helium filled stainless steel or zirconium–tin alloy cans.

Radioactive releases into the environment during all these steps should be minimal, although as with all chemical and mechanical processes some emissions are inevitable. The current NRPB estimates of the maximum annual effective dose equivalents arising from fuel preparation for members of the public living near to the relevant plants in the UK are less than 5 μSv from discharges to the air and about 50 μSv from discharges to water. These each give estimated *collective* effective dose equivalents for the UK of about 0.1 man Sv per year.

7.1.3 Reactor Operation. An operational nuclear reactor contains, and potentially may emit, radioactivity from three sources, from the fuel, from the products of fission reactions, and from the products of activation reactions. The amount of activity present as fuel itself is of course variable but a Pressurized Water Reactor with 100 tonnes of 3.5% enriched uranium would contain ~ 0.25 TBq of ^{235}U and 1.1 TBq of ^{238}U. Unless a catastrophic accident occurs, releases of fuel into the environment should not occur.

Fission of the fuel produces a large number of primary products which in turn decay producing a large number of secondary decay products. Because of their very wide range of half-lives the relative composition and amounts of fission products present in a reactor varies with time in a complex manner. As each nuclide has a different rate of decay the rate of accumulation will vary. However there reaches a point when the rate of production of a nuclide equals its rate of removal by decay and an equilibrium arises between parent and daughter. If the rate of formation is constant then for all practical purposes equilibrium may be assumed to have been reached after about seven half-lives. For a short-lived nuclide such as ^{85}Kr equilibrium will be reached about 10 weeks after the start of the reactor, whereas for ^{137}Cs with a half-life of 30.1 y it will take more than 200 years of continuous operation to reach equilibrium. As the reactor continues to operate with a given set of fuel rods the long-lived fission products become relatively more important and eventually the reactor might contain 10^{20} Bq of activity in total.

Most of the fission products are contained within the fuel cans themselves but total or partial failure of the fuel cladding will allow contamination of the coolant. In addition contamination of the outer surfaces of the fuel rods by uranium fuel will allow fission products to form in the coolant. Most of the fission product activity is in the form of gaseous elements and although various scrubbing systems are used to remove them from the coolant leakage into the environment can and does occur. The amount of gaseous fission products released will depend upon the number of fuel cladding failures, the design of the ventilation, cooling, and coolant purification systems and the length of operation of the reactor. Of the fission products ^{85}Kr, which has a half life of 10.8 y, makes the greatest contribution to the global dose commitment from reactor operation.

The third source of activity in the reactor results from the neutron activation of elements present in the fuel and its casing, the moderator, the coolant, and other components of the reactor itself. Those areas most subjected to neutron activation are those where the incident neutron flux is highest, and include the fuel casings and the coolant system. From an environmental point of view the most important activation product is probably tritium. This is formed by various routes, especially by the activation of deuterium present in the water of Heavy Water Reactors but also by the activation of deuterium present in the water of light water-cooled reactors, the activation of boron added as a regulator to the primary coolant of Pressurized Water Reactors and by activation of boron present in the control rods of Boiling Water Reactors. The small size of the tritium nucleus allows it to diffuse through the cladding materials and so be released into the environment where it will be inhaled and absorbed through the skin. A smaller dose will also be received from drinking and cooking water, following the incorporation of released tritium into the water cycle. Estimates have been made of the annual dose equivalent for individuals living 1, 3, and 10 km from a Canadian CANDU Heavy Water Reactor and are of the order of 8, 3, and 2 μSv respectively.

Other important activation products are ^{58,60}Co, ^{65}Zn, ^{59}Fe, ^{14}C, and the actinides. This latter group, which includes the three long lived isotopes of plutonium ^{238}Pu, ^{239}Pu, and ^{240}Pu (with half-lives of 86.4, 24 360, and 6 580 y respectively), are produced by the neutron activation of fuel uranium. Plutonium has a very low volatility and is not significantly released during normal reactor operation. However it can potentially enter the environment following a reactor accident and as a result of fuel reprocessing.

Various estimates are available for the total dose arising from these various discharges of radioactivity to the environment during the normal operation of power generation reactors. The current NRPB estimate of the maximum effective dose equivalents resulting from discharges to the atmosphere for the most exposed individuals living near to power stations in the UK is about 100 μSv per year, giving a collective effective dose equivalent in the UK of about 4 man Sv per year. In addition to discharges to the atmosphere there are also discharges of radioactivity from power stations to natural waters. These result from the temporary storage of used fuel under water in specially constructed ponds. NRPB estimate that they give a maximum effective dose equivalent to local individuals of less than 350 μSv per year. The collective effective dose equivalent in the UK, mainly through the eating of contaminated seafood, is about 0.1 man Sv per year.

7.2 Pollution Following Reactor Accidents

With the large number of reactors operational world-wide it is inevitable that accidents with environmental consequences should occur. To date four such incidents have taken place in the UK, USA, and USSR.

7.2.1 Windscale, 1957. In October, 1957, a graphite moderated reactor at Windscale, UK, overheated, rupturing at least one fuel can and causing the release of fission products into the environment. An estimated 740 TBq of ^{131}I, 22 TBq of ^{137}Cs, 3 TBq of ^{89}Sr, and 3 TBq of ^{90}Sr were released. The cause of the accident was the sudden release of a huge amount of Wigner energy from the graphite in the core. When graphite is bombarded by fast neutrons, as in a reactor, an increase in volume and a decrease in thermal and electrical conductivity results. The graphite can then efficiently store energy and, if heated above 300 °C, this is released in the form of further heat. Normally this Wigner energy is released by slowly heating the graphite, allowing the stored energy to be released under control and reversing the radiation damage. During 7–8 October, 1957, however, an unsuccessful attempt was made to release the Wigner energy at Windscale, following which the reactor rapidly overheated releasing radioactive material into the atmosphere.

Following the accident the inhalation of aerosol and gaseous nuclides probably gave the largest source of exposure to the general population, although the drinking of milk contaminated by ^{131}I was also significant, despite a restriction on the sale of milk from farms in the worst affected area. The total committed effective dose equivalent in the UK has been estimated at about 2 000 man Sv, giving about 30 excess deaths from cancer in the UK during the first 40 y after the accident.

7.2.2 Idaho Falls, 1961. In January, 1961, during maintenance work on an enriched uranium-fuelled boiling water reactor at the National Reactor Testing Station, Idaho Springs, USA, the central rod was accidentally removed causing a large explosion, killing three men. Most of the coolant water was ejected, together with an estimated 5–10% of the total fission products in the core. About 4 TBq of activity escaped from the reactor building.

7.2.3 Three Mile Island, 1979. Whilst operating at full power on 28 March, 1979, the cooling water system failed on the pressurized water reactor at Three Mile Island reactor 2 at Harrisburg, Pennsylvania. The reactor automatically shut down and three auxiliary pumps started to provide cooling water. However valves in this circuit had inadvertently been left closed and the resultant increase in temperature and pressure in the primary circuit caused an automatic relief valve to open, allowing about 30% of the primary coolant to escape. Fresh cooling water was injected into the primary circuit by the emergency cooling system but following an error in their analysis of the accident this and the main coolant pumps were deactivated by the operators. It was only when cooling water was later added that the overheating of the core ceased.

Despite the reactor core sustaining serious damage the amount of activity released into the environment was limited to about 10^{17} Bq, almost entirely as short-lived noble gases, especially ^{133}Xe (which has a half-life of 5.2 days). The resultant total committed effective dose equivalent to the population was estimated to be about 20 man Sv with the maximum individual dose equivalent being about 1 mSv.

7.2.4 Chernobyl, 1986. The most serious reactor accident to date began on 26 April, 1986, when an explosion and fire occurred at the Chernobyl number 4 reactor near Kiev in the Ukraine. The reactor came into service in 1984 and was one of 14 RBMK boiling water pressure tube reactors operational in the USSR. These are graphite-moderated reactors of 950–1450 MW electrical power generation capacity fuelled with 2% enriched uranium dioxide encased in zirconium alloy tubing. The normal maximum temperature of the graphite is about 700 °C and in order to prevent this being oxidized the core is surrounded by a thin-walled steel jacket containing an inert helium/nitrogen mixture.

Prior to the accident the reactor had completed a period of full power operation and was being progressively shut down for maintenance when an experiment was begun to see whether the mechanical inertia of one of the turbogenerators could be used to generate electricity for a short period in the event of a power failure. The reactor core contained water at just below the boiling point but when the experiment began some of the main coolant pumps slowed down causing the core water to boil vigorously. The bubbles of steam so formed displaced the water in the core and, because steam absorbs neutrons much less efficiently than water, the number of neutrons in the core began to rise. This increased the power output of the reactor, so increasing the heat output and the amount of steam in the core which in turn led to a further rise in the neutron density. This positive feedback mechanism led to a rapid surge in power causing the fuel to melt and disintegrate. As the fuel came into contact with the surrounding water, steam explosions occurred destroying the structure of the core and the pile cap causing radioactive material to be ejected into the atmosphere. The core fires allowed a continuing release of activity which was slowly reduced by the dumping of clay and other materials onto the core debris. However the core temperature again began to rise and a second peak in activity release occurred on 5 May. After this the core was progressively buried and finally sealed in a concrete sarcophagus.

Estimates of the amount of activity released from the reactor vary but Soviet measurements suggest that all of the noble gases, 10–20% of the volatile fission products (mainly iodine and caesium), and 3–4% of the fuel activity, giving a total of 1.85×10^{18} Bq, escaped into the environment. On the basis of air concentration and deposition measurements the UKAEA's Harwell Laboratory estimate that about 7×10^{16} Bq of ^{137}Cs was released.

The immediate casualties of the accident were 31 killed and about two hundred diagnosed as suffering from acute radiation effects. About 135 000 people and a large number of animals were evacuated from a 30 km radius area surrounding the plant.

The meteorological conditions prevailing over Europe at the time of the accident were rather complex, leading to the dispersion of activity over a very wide area. In the UK peak air concentrations occurred on 2 May when ~ 0.5 Bq m^{-3} of ^{137}Cs was recorded at Harwell. No Chernobyl radioactivity was immediately detected in the southern hemisphere. In the year following the accident the annual mean ^{137}Cs concentration in the northern hemisphere was about the same as that of 1963 when weapons testing activity was at its highest. The amount of ^{137}Cs deposited on the ground surface obviously varied with air concentration, rainfall, and other parameters but close to Chernobyl (within 30 km) as much as 10^4 kBq m^{-2} of ^{137}Cs was deposited. The average for Austria was 23 kBq m^{-2}, for the UK 1.4 kBq m^{-2}, and for the USA 0.04 kBq m^{-2}. The very heavy but localized rainfall which occurred in parts of Europe and the UK during the time when the plume was overhead led to a very patchy distribution of deposited activity on the ground. In the UK for example it varied from > 10 kBq m^{-2} of ^{137}Cs in parts of Cumbria to < 0.3 kBq m^{-2} in parts of Suffolk.

Following the wet and dry deposition of activity from the atmosphere some contamination of foodstuffs was inevitable and outside of the immediate area around Chernobyl this was the major consequence of the accident. In general the most vulnerable foods in the UK were lamb and milk products. In the UK a ban on the movement and slaughter of lambs was imposed within specified areas until the meat consistently contained less than 1000 Bq kg^{-1} of radio-caesium. No restrictions were placed on the sale of milk.

The main exposure pathways to man following the accident were direct gamma irradiation from the cloud and from activity deposited on the ground (groundshine), inhalation of gaseous and particulate activity from the air and, most importantly, the ingestion of contaminated foods. The Organization for Economic Co-operation and Development Nuclear Energy Agency estimate that the average individual effective dose equivalents received in the first year after the accident range from a few microsieverts or less for Spain, Portugal, and most countries outside of Europe to about 0.7 mSv for Austria. However, hidden within these averages are the higher peak doses received by the most exposed individuals, or critical group, in each country. These vary from a few micro-sieverts outside Europe to an upper extreme of 2–3 mSv for the Nordic countries and Italy. The total collective dose in Eastern and Western Europe has been estimated to be about 1.8×10^5 man Sv. The NRPB estimate that the average effective dose equivalent received in the UK during the first year was about 40 μSv with a total of about 20 μSv being received over subsequent years. The total committed effective dose equivalent for the UK population is estimated to be about 3 000 man Sv. When compared with the other doses of radioactivity normally received by individuals, for example the 2 mSv per year individual effective dose equivalent received on average in Europe from natural back-ground radiation, these doses are small and probably insignificant. Using the ICRP estimate of risk it is possible to calculate the additional mortality likely to result from Chernobyl. In the 40 years following the accident about 50 excess deaths are likely in the UK (compared with about 145 000 non-radiogenic cancers per year in the UK), indicating that outside the immediate vicinity of the plant the impact of the accident will be completely undetectable among future mortality statistics.

The Chernobyl accident presented a unique opportunity for experiments and studies in a wide range of environmental studies. It allowed the validation and refinement of atmospheric dispersion models, the calculation of washout ratios and deposition velocities, and the study of the behaviour of caesium and iodine in food chains, natural waters, sediments, and in the urban environment. It demonstrated the necessity of better international harmonization of scientific databases and public health protection policies and it provided a vivid example of the transboundary long-range transport of pollutants.

7.3 Radioactive Waste Treatment and Disposal

An inevitable consequence of man's use of radioactivity is that radioactive waste material is produced which must then be disposed of. Although there is no universal scheme for the classification of such waste it is usual for it to be categorized in terms of its activity content as low, intermediate and high level.

7.3.1 Low Level Waste. Low level wastes are produced in large volumes by all the various medical, industrial, scientific, and military applications of radioactivity. They include contaminated solutions and solids, protective, cleaning and decontamination materials, laboratory ware, and other equipment. They also include gases and liquids operationally discharged from power stations and other facilities. It has been estimated that during the period 1980–2000 about 3.6×10^6 m^3 of such wastes will be produced, some of which is sufficiently low in activity to be directly discharged into the environment, either with or without prior dilution or chemical treatment. Typical maximum activity concentrations in low level waste are 4 GBq t^{-1} (alpha) and 12 GBq t^{-1} (beta and gamma). Much low level waste is currently disposed of by shallow burial in landfill sites, often with the co-disposal of other, non-radioactive, controlled wastes, or by discharge into surface waters in rivers, lakes, estuaries, or coastal seas or by discharge into the atmosphere. If the environmental biophysicochemical behaviour of the radionuclides in question and their possible pathways back to man are well understood then it is possible to make reliable estimates of the likely resultant doses of radioactivity to those most exposed in the population. If these doses are suitably low then the disposal methods may be deemed to be acceptable. However at the present time there are many uncertainties in the understanding of such behaviour, pathways, and doses but nevertheless the very large volumes of low level waste being generated will, through lack of economically and environmentally viable alternatives, continue to be disposed of in these relatively uncontrolled ways.

7.3.2 Intermediate Level Waste. Intermediate level wastes are sufficiently active to prevent their direct discharge into the environment, with maximum specific activities of typically 2×10^{12} Bq m^{-3} (α) and 2×10^{-14} Bq m^{-3} (β and γ). They comprise much of the solid and liquid wastes generated during fuel reprocessing, residues from power station effluent plants, and wastes produced by the decommissioning of nuclear facilities. Very large quantities ($\sim 2\,000\,\mathrm{t\,y}^{-1}$) of intermediate and low level wastes have been disposed of by dumping in deep ocean waters in the NE Atlantic. Although some authorities still consider this method of disposal to be the best practicable environmental option for these categories of waste it is no longer practised and intermediate level waste produced in the UK is now stored on land, mainly at Sellafield, awaiting further policy decisions.

7.3.3 High Level Waste. High level wastes mainly consist of spent fuel and its residues and very active liquids generated during fuel reprocessing. Typical maximum activities are $4 \times 10^{14}\,\mathrm{Bq\,m}^{-3}$ (α) and $8 \times 10^{16}\,\mathrm{Bq\,m}^{-3}$ (β and γ). At present such wastes generated in the UK are stored at Sellafield in storage ponds where it is proposed they will be vitrified prior to further storage (to allow the decay of shorter lived nuclides) and finally disposal in deep repositories. No such repository yet exists but deep mines and boreholes on land and sea as well as other more exotic solutions including extraterrestrial disposal have all been proposed.

7.4 Fuel Reprocessing

In most nuclear reactors the economic lifetime of the fuel in the core is determined not by the depletion of fissile material but by the production and accumulation of fission products which progressively reduce the efficiency of the reactor. In a typical reactor the fuel is changed on a three-year cycle, generating large amounts of partially spent fuel contaminated with fission products. Apart from the economic considerations, which may or may not be in favour of recovering the unreacted fissile material from the spent fuel depending upon the relative costs of reprocessing and importing further uranium ore, there are several reasons why such reprocessing is carried out: it allows the production of plutonium for military purposes, it reduces dependence on imported ores and it reduces the volume of high-level waste produced by a reactor. As a commercial operation it also successfully attracts overseas currency to the UK.

In the reprocessing method currently used in the UK the short lived activity is allowed to decay by storage for several months after which the spent fuel is dissolved in nitric acid and a sequential extraction procedure used successively to remove the uranium, plutonium, and fission products. The uranium is then re-enriched and fabricated into fuel rods and the plutonium used in the mixed oxide fuel or for military purposes. About 25 000 t of Magnox fuel has been reprocessed at Sellafield, yielding about 10 000 t of uranium for re-enrichment. The THORP thermal oxide reprocessing plant currently under construction at Sellafield will process fuel from advanced gas-cooled and pressurized water reactors at a rate of about $600\,\mathrm{t\,y}^{-1}$ and will handle the 2 500 t of waste currently being stored at Sellafield and that being produced now and in the future in the UK and overseas. The current NRPB estimates of the annual effective dose equivalents arising from fuel reprocessing to the UK population are $1\,\mathrm{mSv\,y}^{-1}$ to the most exposed individuals and a collective dose of $80\,\mathrm{man\,Sv\,y}^{-1}$

8 POLLUTION FROM NON-NUCLEAR PROCESSES

Two non-nuclear industrial processes, the burning of fossil fuels and the smelting of non-ferrous metals, release non-trivial quantities of radioactivity into the environment and require brief consideration here. Coal contains uranium and thorium in varying concentrations, typically 1–2 ppm of both ^{238}U and ^{232}Th but at much higher concentrations (100–300 ppm) in some areas, *e.g.*, the western USA. It also contains significant quantities of ^{14}C and ^{40}K. Similarly oil and natural gas both contain members of the naturally-occurring U and Th decay series. When the fuel is burned, as in a conventional power station, these nuclides and any daughters present are either released into the atmosphere in the flue gas and fly ash or retained in the bottom ash. One current estimate of the amount of ^{226}Ra emitted by a typical 1 000 MW coal-fired station is 10^9–10^{10} Bq y^{-1} with a larger amount being retained in the ash. These emissions undoubtedly give rise to elevated environmental concentrations with the whole body dose equivalents for those most exposed living in the vicinity of large coal-fired plants being possibly as high as 1 mSv per year.

The second non-nuclear industrial source of radioactivity, the smelting of non-ferrous metals, arises because of the natural occurrence of radioactive isotopes of lead. The geochemistry of lead is intimately associated with that of uranium and thorium, there being four radioactive isotopes of lead: ^{210}Pb and ^{214}Pb in the ^{238}U decay chain, ^{211}Pb in the ^{235}U decay chain, and ^{212}Pb in the ^{232}Th decay chain. All except ^{210}Pb have half-lives of less than 12 hours, but the 22 y half-life of ^{210}Pb and its subsequent decay to form ^{210}Po, itself an alpha-emitter of half-life 138 days, makes it of environmental significance. During the primary and secondary smelting of lead and other non-ferrous metals and their ores some lead is released into the atmosphere. A fraction of this will be ^{210}Pb together with a similar quantity of ^{210}Po. This aerosol may then be inhaled, giving rise to an exposure of the lung to alpha particles. Investigation and understanding of such 'non-nuclear' pathways of radioactivity to man is at an early stage but it is possible that those living close to large smelters and other sources of ^{210}Po may receive measurable and non-trivial doses of radioactivity.

Acknowledgements. I thank Dr. M. Kelly for his constructive comments and Mrs. J. Dixon for typing the manuscript.

9 BIBLIOGRAPHY

R. S. Cambray, *et al.*, 'Observations on radioactivity from the Chernobyl accident,' *Nuclear Energy*, 1987, **26**, 77–101.

K. D. Cliff, J. C. H. Miles, and K. Brown, 'The Incidence and Origin of Radon and its Decay Products in Buildings,' NRPB–R159, HMSO, London, 1984.

J. H. Gittus, *et al.*, 'The Chernobyl Accident and its Consequences,' United Kingdom Atomic Energy Authority, HMSO, London, 1988.

J. S. Hughes and G. C. Roberts, 'The Radiation Exposure of the UK Population—1984 Review,' NRPB–R173, HMSO, London, 1984.

R. Kathren, 'Radioactivity in the Environment: Sources, Distribution, and Surveillance,' Harwood, Amsterdam, 1984.

National Radiological Protection Board, 'Living with Radiation,' HMSO, London, 1986.

Organization for Economic Co-operation and Development Nuclear Energy Agency, 'The Radiological Impact of the Chernobyl Accident in OECD Countries,' OECD, Paris, 1987.

CHAPTER 18

Quantitative Systems Methods in the Evaluation of Environmental Pollution Problems

P. C. YOUNG

1 INTRODUCTION

One of the first steps in the scientific method is the formulation of a hypothesis, usually in the form of a mathematical model of some kind. The confidence that the scientist has in this model will vary considerably, depending on the nature of the application. In the case of environmental pollution problems, however, the degree of confidence is often low: whilst the scientist may be able to specify numerous mechanisms which *could* be operative in the system, he often finds it difficult to define which of these mechanisms will be dominant under the specific circumstances encountered in the problem under study. And to include all the mechanisms which he feels *may* affect the system will lead to large over-parameterized models which, as we shall see, present many problems of statistical identification, estimation, and validation.

This lack of confidence is compounded by the problems associated with the planning and execution of *in situ* experiments on the system. Planned laboratory experiments may help to clarify certain aspects of the system behaviour but they can rarely define unambiguously the exact nature of this behaviour in the natural environment. Faced with such a dilemma, the scientist may resort to passive observation or 'monitoring' of the system during its normal operation. At the very best, such monitoring will include the acquisition of data from partially planned experiments, as when tracer studies are used to evaluate the transportation and dispersion of pollutants in a river or estuarine system. At the worst, monitoring will be restricted by logistic and cost considerations, so that the resultant data set can only reveal limited aspects of the system behaviour.

Such difficulties are exacerbated if it is the *dynamic* behaviour of the system which is important to the resolution of the problem under consideration. The objective of this Chapter is to introduce new methods of time-series and systems analysis that can help in the solution of these kind of 'badly defined' dynamic system problems.[1-6] Needless to say, the time-series analysis can be no better than the data set on which it is based; and it may well be that the information/ database is so meagre that little sensible data analysis, as such, is possible. In these situations, which we do not consider here, the model builder must resort to some form of more speculative *simulation modelling*. Such modelling has become quite straightforward with the development of advanced computer systems and it has a very real role to play in environmental systems analysis. But it cannot be considered as an end in itself and must be carried out with due caution, as we discuss in the last Section of this Chapter.

The rather safer, yet less flexible, *data-based modelling* considered in detail here, is intended to provide a system of 'checks and balances' which can help the model builder at least to identify difficulties that may be inherent in the limited data set at his disposal. In this manner it is possible to avoid problems such as 'over-parameterization,' 'surplus unvalidated content,' and unjustified confidence in the resulting model.[1,2]

2 MODELLING THE DISPERSION OF POLLUTION IN A RIVER SYSTEM

Before discussing a general approach to the analysis and modelling of poorly defined systems, let us consider a simple, yet practical, example: namely, the characterization of solute dispersion in a non-tidal river system. Here, perusal of most text books on the subject would suggest that the behaviour is relatively well understood. However, difficulties of model formulation and parameter estimation can still arise, particularly if conventional modelling methods are used without taking note of certain important results in systems theory and time-series analysis.

2.1 Conventional Modelling *vs.* the Systems Approach

Classical hydrodynamic analysis associated with dispersion in flowing media, such as the seminal work of G. I. Taylor[7] on flow in pipes, often provides the

[1] P. C. Young, in 'Modelling, Identification, and Control in Environmental Systems,' ed. G. C. Vansteenkiste, North Holland, Amsterdam, 1978, p. 103.
[2] P. C. Young, Chapter 2 in 'Uncertainty and Forecasting of Water Quality,' ed. M. B. Beck and G. Van Straten, Springer-Verlag, Berlin, 1983.
[3] P. C. Young, 'Recursive Estimation and Time-Series Analysis,' Springer-Verlag, Berlin, 1984.
[4] P. C. Young, Chapter 6 in 'River Flow Modelling and Forecasting,' ed. D. A. Kraijenhoff and J. R. Moll, D. Reidel, Dordrecht, 1986.
[5] P. C. Young, Chapter to appear in 'Control and Dynamic Systems, Vol. XXXII,' ed. C. T. Leondes, Academic Press, Florida, 1989.
[6] P. C. Young and S. Benner, 'MicroCAPTAIN Handbook: Version 2.0,' Centre for Research on Environmental Systems, University of Lancaster, 1988.
[7] G. I. Taylor, *Proc. Roy. Soc.*, 1954, **A223**, 446.

starting point for modelling studies on the longitudinal transportation and dispersion of solutes in stream channels. This involves a mathematical representation in the form of the following single dimensional, partial differential equation (see, *e.g.*, references 8–11), usually known as the *Fickian Diffusion Equation* or the *Advection Dispersion Equation* (ADE),

$$\frac{\partial x(s,t)}{\partial t} + U\frac{\partial x(s,t)}{\partial s} = D\frac{\partial^2 x(s,t)}{\partial s^2} \tag{1}$$

where $x(s,t)$ is the concentration of the solute at spatial location s and time t; U is the cross-sectional average longitudinal velocity; and D is the longitudinal dispersion coefficient. The solution to equation (1) for an impulsive ('gulp') input of solute with finite mass M is, with suitable choice for the origins of s and t, given by,

$$x(s,t) = \frac{M}{2A(\pi \, Dt)^{0.5}} \exp\left\{-\left[\frac{(s - ut)^2}{4Dt}\right]\right\} \tag{2}$$

Although it is well known that this represents a Gaussian distribution in space, it is not always appreciated that, when viewed in the time domain at a fixed longitudinal location, it yields a skewed distribution, because the solute cloud continues to evolve as it passes the fixed observation point. It is also not often acknowledged that neither equation (1) nor (2) is even theoretically valid until an asymptotic state is achieved; *i.e.*, until the solute cloud has been evolving for a sufficiently long period of time (see, *e.g.*, reference 12).

In practice, observations of dispersing solute clouds in rivers have revealed persistent deviations from the behaviour predicted by the Fickian theory and the measured concentration profiles rarely, if ever, attain a Gaussian distribution, except at very large distances from the injection point. Partly this is due to the fact that it is usual, for logistic reasons, to measure temporal rather than spatial profiles. But even when this is taken into account, the skewness of the profiles still persists and the Gaussian shape is rarely achieved in practice, as we see in Figures 1 and 2, which show the results of tracer experiments conducted on streams in the USA and the UK. The reasons for this are essentially twofold: firstly, observations are rarely made at long enough times after injection for the Gaussian distribution to evolve; and secondly, real stream channels are non-uniform, and this non-uniformity leads to rather different patterns of mixing than those predicted by the pure Fickian theory.

It is inevitable that most measurements taken during pollution incidents are carried out close to the pollution sources, because it is here where the highest and potentially most environmentally damaging concentrations are likely to occur.

[8] H. B. Fischer, E. J. List, R. C. Y. Koh, J. Imberger, and N. H. Brooks, 'Mixing in Inland and Coastal Waters,' Academic Press, New York, 1979.

[9] G. T. Orlob, 'Mathematical Modeling of Water Quality,' Wiley, Chichester, 1983.

[10] B. Henderson Sellers, P. C. Young, and J. Ribeiro da Costa, to appear in Proc. Int. Symp. on Water Quality Modelling of Agricultural Non- Point Sources, Utah State University, Utah, 1988.

[11] P. C. Young and S. G. Wallis, to appear in Channel Networks, ed. K. J. Beven and M. J. Kirby, Wiley, Chichester, 1989.

[12] P. C. Chatwin and C. M. Allen, *Ann. Rev. Fluid Mech.*, 1985, **17**, 119.

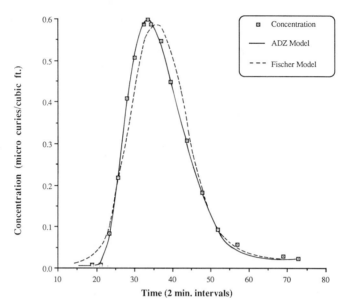

Figure 1 *Modelling the Copper Creek dispersion data: comparison of ADZ model fit with the
results obtained by Fischer using the ADE model*

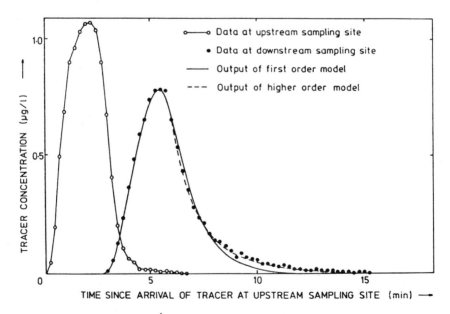

Figure 2 *ADZ modelling results for the River Conder dispersion data: comparison of first and
second order model fits*

At these short time scales, the most striking deviation between observations and Fickian theory is the additional skew in the measured concentration profiles. For the temporal profiles, this manifests itself in a sharper rise to the peak, followed by a longer and more slowly decaying tail, as revealed in Figures 1 and 2. As a result, although most textbooks still subscribe to the use of the ADE model in a river context, reasonable arguments can be presented to suggest that it is not necessarily an appropriate model for pollutant transportation in naturally occurring channels and streams.

The limitations of the ADE have stimulated scientists either to modify the Fickian inspired theory or to develop alternatives. Of these, one approach which shows particular promise depends on the assumption that 'dead zones' play an important role in the dispersion process. These zones are traditionally associated with side pockets, bed irregularities, and other 'roughness' elements in the stream channel; elements which tend to create relatively slower moving regions or even areas of 'backflow'.* Solute which is entrained within these areas is then released back into the main flow relatively slowly and at diluted concentrations, due to the 'eddy-like' flow structure which tends to characterize such regions. This induces a distributed time-delay effect and tends to raise significantly the tail of the concentration profile.

Various dead zone analyses[13,14] have shown that dead zones can account for some of the observed deviations from the traditional ADE model behaviour. Until recently, however, most of these dead zone models have been based on an extension of the ADE model, in which the dead zones are characterized explicitly as a first order differential equation for the dead zone effect, adjoined to the classical partial differential equation (1) of the ADE. In the past few years, however, some of the concepts of systems theory have been brought to bear on the problem and this has led to a rather different model formulation.

One of the most important aspects of the systems approach is the idea that the modelling should be **objective orientated** and **data-based**; *i.e.*, a model should be constructed to satisfy some user-defined objective and should be obtained, wherever possible, from the objective analysis of experimental data. In practice, such objectives normally demand both that the model is able to characterize the behaviour of the environmental process *as it occurs in the real world* (rather than as it *might occur* within a more limited and abstract scientific setting); and that the model should be in a form which can be used within the real world for purposes such as forecasting, operational control, and management.

It was this systems concept which inspired a more radical innovation in dead zone dispersion modelling which began in the early 1970's[15,16] and came to

[13] E. M. Valentine and I. R. Wood, *Proc. ASCE, J. Hydrol. Eng. Div.*, 1979, **105**, 999.
[14] K. A. Bencala and R. A. Walters, *Water Resour. Res.*, 1983, **19**, 718.
[15] M. B. Beck and P. C. Young, *Water Res.*, 1975, **9**, 769.
[16] M. B. Beck and P. C. Young, *Proc. ASCE, J. Environ. Eng. Div.*, 1976, **102**, 909.

* The casual observer who throws a stick into a river may often observe that, near to the bank, the stick can get caught in such a dead zone, and may often be seen to move in an opposite direction to the main stream flow.

fruition in the 1980's.[17-19] Here, the direct analysis of experimental tracer data, in the form of a special recursive approach to time-series analysis,[3] is utilized to statistically identify and estimate a simple time-series model, of low or minimal dimension, which is able to explain the measured concentration profile much better than previous conventional models, such as the ADE. This data-based approach can be contrasted to more conventional **mechanistic** modelling procedures, such as the ADE, where a model structure—usually of pre-ordained order—is derived by physico-chemical reasoning and then utilized as the basis for exercises in model 'fitting' or 'optimization.' Whereas the data-based approach allows the data to expose the possible model structure, usually within a given larger general class of time-series models, the conventional approach tends to rely on the adequacy of the analyst's preconceptions about the behavioural mechanisms of the system under study.

Since the systems model of Beer and Young is intended primarily for use in the operational control and management of pollution in river systems, usually on the basis of monitored sampled data, it is selected in the form of a lumped parameter, discrete-time (sampled data) equation, which is able to describe the temporal variations in concentration at a finite number of selected spatial locations in the river. This avoids the alternative, and more complicated, distributed parameter formulation of the ADE, which tends to provide more information than is really required in the operational control and management context. Having made use of an objective, data-based method to define a low order dynamic model of this type, however, the systems approach demands, if at all possible, that the analyst should provide a reasonable physical explanation for the model structure. This stage in the analysis can be considered as an exercise in **model validation**, which is aimed at establishing the model's **credibility** within the wider scientific community.[2]

In Beer and Young's case, for example, this physical explanation depends upon a rather novel dead zone interpretation of the dispersive mechanism, in which the *aggregative* effect of the many individual dead zone regions in a given reach of the river is considered as being equivalent to a *single* dead zone, with a defined volume and associated residence time. The ratio of this **Aggregated Dead Zone** (ADZ) volume to the total volume of water in the reach, or the **dispersive fraction**, then provides a physically meaningful measure of the river's dispersive properties. In fact, as we shall see, the detailed experimental study of river channels in the north west of England[18,19] suggests that this ADZ dispersive fraction is an important *flow-invariant* property of the river channel, which is able to define the dispersive nature of the stream channel over most of the normal flow régime. In this sense, the dispersive fraction may eventually replace the dispersion coefficient D of the more conventional ADE model (1), as the primary measure of riverine dispersion.

[17] T. Beer and P. C. Young, *Proc. ASCE, J. Environ. Eng. Div.*, 1983, **109**, 1049.

[18] P. C. Young and S. G. Wallis, in Proc. BHRA Int. Conf. on Water Quality Modelling in the Inland Natural Environment, Bournemouth, 1986.

[19] S. G. Wallis, P. C. Young, and K. J. Beven, *Proc. Inst. Civ. Engrs.: Part 2*, 1989, **87**, 1.

2.2 Mathematical Development of the ADZ Model

Although the ADZ model is obtained from a rigorous exercise in statistical identification and parameter estimation applied to a class of linear, discrete time (*i.e.*, sampled data) models, it can be interpreted best, in physical terms, from its formulation as a continuous-time, first order, differential-delay equation. In order to facilitate understanding of the model, therefore, we will consider first the model in this continuous time form and then proceed later to discuss the manner in which the model is objectively identified and estimated in its equivalent discrete-time form.

We have discussed the ADZ model so far in relation to flow processes occurring in the river system. In alternative, chemical engineering terms, the model can be considered as a combination of 'plug flow,' to account for the translational effect, or advection, introduced by the main river flow processes; and a 'continuous stirred tank reactor' (CSTR) mechanism to describe the dispersive properties inherent in the ADZ. If the flow through the reach is denoted by $Q\,\mathrm{m^3\,s^{-1}}$ and the volume of the ADZ is $V\,\mathrm{m^3}$ then, under the assumption of complete mixing in the ADZ, the relationship between the measured concentration $x(t)$ of the pollutant at the output (or downstream) location in the reach to the measured concentration $u(t)$ at the input (or upstream) location can be obtained from dynamic mass conservation considerations. Assuming, for simplicity, a conservative solute, this mass balance equation takes the form,*

change of mass per unit time		mass flow in per unit time		mass flow out per unit time	
$\dfrac{\mathrm{d}[Vx(t)]}{\mathrm{d}t}$	$=$	$Qu(t-\tau)$	$-$	$Qx(t)$	(3)

or if V and Q are assumed constant (*i.e.*, steady flow conditions),

$$\frac{\mathrm{d}x(t)}{\mathrm{d}t} = -\frac{1}{T}x(t) + \frac{1}{T}u(t-\tau) \qquad (4)$$

where τ is the advective time-delay introduced to allow for the translational effects of the uni-directional river flow; and $T = V/Q$ seconds is the 'residence time' of the ADZ. Clearly, a non-conservative solute could be handled in a similar manner but the losses occurring within the reach (*i.e.*, both in the ADZ and during the advective time delay) would need to be accounted for in some manner. However, we will consider this later in Section 5, when we apply the basic ADZ principles discussed here to the modelling of Dissolved Oxygen and Biochemical Oxygen Demand in a river system.

The discrete time equivalent of equation (4), for a specified sampling interval Δt, can be obtained in various ways. If we assume, however, that the input $u(t)$ is approximately constant over the sampling interval, then the discrete-time

---

* The ADZ volume V will normally be much less than the total reach volume V_R; the dipersive fraction V/V_R, for example, is normally in the range 0.3 to 0.4 for medium sized natural streams (see later Section 2.3 and references 18 and 19).

equation relating the sampled value of the output concentration $x(k)$, at the kth sampling instant, to the sampled input concentration $u(k-\delta)$, δ sampling intervals previously, is of the form,

$$x(k) + \alpha_1 x(k-1) = \beta_0 u(k-\delta) \tag{5}$$

where α_1 and β_0 are constant coefficients or parameters; and δ is the advective time delay in sampling intervals as defined by the integral value of $\tau/\Delta t$. If $\tau/\Delta t$ is not an exact integral number, then this definition of δ obviously implies an approximation, the magnitude of which will depend upon the size of the sampling interval. This can be obviated quite straightforwardly but, for simplicity of presentation, we will assume here that τ is an integral number of sampling intervals. In this situation, T can be related to α_1 by the following expression (see *e.g.*, reference 3),

$$T = \frac{V}{Q} = -\frac{\Delta t}{\log_e (\alpha_1)} \tag{6}$$

so that an estimate of the ADZ residence time and, given knowledge of the river flow Q, an estimate of the ADZ volume can be derived from a statistical estimate $\hat{\alpha}_1$ of the parameter α_1.

It is convenient, at this point, to introduce the backward shift operator z^{-i}, where $z^{-i}x(k) = x(k-i)$, and then convert equation (5) into the following *transfer function* (TF) form,

$$x(k) = \frac{\beta_0 z^{-\delta}}{1 + \alpha_1 z^{-1}} u(k) \tag{7}$$

The statistical estimation of the parameters α_1 and β_0 in this first order TF is a standard problem in time-series anlaysis[3,20] and so we will not discuss it in detail here. It will suffice to point out that, in general, for an arbitrarily selected stretch of river, we cannot assume that a *single* ADZ element such as this will be able to model fully the dispersive properties; it could well be that several ADZ elements such as this would be required to model a reasonably long length of river. In this higher order context, the TF model representing n ADZ elements in a serial *or* parallel connection (see Figure 3) takes the general form,

$$x(k) + a_1 x(k-1) + a_2 x(k-2) + , \ldots , + a_n x(k-n) = b_0 u(k-\delta_T) + , \ldots , + b_m u(k-\delta_T-m) \tag{8}$$

or, in TF terms,

$$x(k) = \frac{b_0 + b_1 z^{-1} + b_2 z^{-2} + \ldots + b_m z^{-m}}{1 + a_1 z^{-1} + a_2 z^{-2} + \ldots + a_n z^{-n}} u(k-\delta_T) \tag{9}$$

where $a_i, i = 1, 2, \ldots, n$ and $b_j, j = 0, 1, \ldots, m$ are the $n + m + 1$ model parameters, possibly with some b_j coefficients having zero value; m will depend upon the nature of the connection between the elements; and δ_T is the combined effect of the individual advective time delays associated with the ADZ elements that make up the complete model.

[20] G. E. P. Box and G. M. Jenkins, 'Time Series Analysis, Forecasting, and Control,' Holden-Day, San Francisco, 1970.

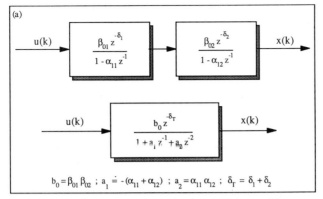

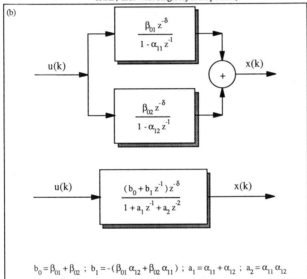

Figure 3 *Typical series and parallel connections of ADZ model elements in river network dispersion models*
(a) $[2,0,\delta_T]$ *model: serial connection of two ADZ elements with different advective time delays and residence times (i.e., two reaches in series, each with single layer dispersion)*
(b) $[2,1,\delta_T]$ *model: parallel connection of two ADZ elements with equal advective time delays but different residence times (i.e., single reach with two layer dispersion)*

The need to specify a general model, such as (9), which allows for the interconnections of first order ADZ elements arises from physical considerations. In a single uniform stretch of river, for instance, we might assume that a serial connection of ADZ elements, each defined by a first order TF such as (7), could describe the dispersive characteristics. However, if we allow for more complicated river networks, or more complex multi-layer structure in each reach of the river channel, then serial and parallel connections are both possible (see later Section 2.4 and reference 11).

In terms of the TF model (9), time series analysis can be considered as a two stage statistical procedure: first, the **identification** of m, n and δ_T (equivalently the determination of how many ADZ elements such as (7) are appropriate to model the given stretch of river and how these are connected); and second, the **estimation** of the $n + m + 1$ parameters that characterize this identified model structure. Such identification and estimation studies need, of course, to be based on the data set $\{x(k); u(k); k = 1, 2, ... , N\}$, where N is the total number of samples collected over the observation interval.

2.3   The Problem of Model Order Identification

For the present purposes, it is the approach to objective model order identification which is most important in the modelling process, since it is this aspect of the analysis which most distinguishes the systems approach from the more conventional methods of mechanistic, simulation modelling. How then, given concentration measurements at two locations in a river, do we identify how many first order ADZ elements are required to provide an adequate dynamic description of the data?; or equivalently, how do we choose the length of river appropriate to a single ADZ description?

In simple terms, the statistical approach to this identification problem reduces to the definition of statistical criteria which indicate to the analyst whether the model is identifiable from the data and if so, what order model (*i.e.*, n, m and δ) is best justified in some statistical sense. The approach to model order identification used in the analysis of the tracer data in Figures 1 and 2 is based on the following identification statistic,[5]

$$\text{YIC} = \log_e\{\sigma_\xi^2/\sigma_x^2\} + \log_e\{\text{NEVN}\} \qquad (10)$$

where,

σ_ξ^2 is the sample variance of the model residuals (*i.e.*, fitting errors)
σ_x^2 is the sample variance of the measured system output $x(k)$ about its mean value.

while NEVN is the 'Normalized Error Variance Norm'[21] defined as,

$$\text{NEVN} = \frac{1}{p} \sum_{i=1}^{i=p} \sigma_\xi^2 \frac{p_{ii}}{[\hat{a}_i]^2}$$

Here, in relation to the general TF model (9), $p = m + n + 1$ is the total number of parameters estimated; $\hat{a}_i$ is the estimate of the ith parameter in the vector a of all the model parameters, *i.e.*,

$$a = [a_1 \, a_2 \ldots \ldots a_n \, b_0 \, b_1 \ldots \ldots b_m]^T$$

[21] P. C. Young, A. J. Jakeman, and R. McMurtrie, *Automatica*, 1980, **16**, 281.

and $\sigma_\xi^2 p_{ii}$ is an estimate, obtained during the analysis, of the error variance associated with the ith parameter estimate.

It can be seen that the first term in the definition of the YIC provides a normalized measure of how well the model explains the data: the smaller the variance of the model residuals in relation to the variance of the measured output, the more negative the first term becomes. Similarly, the second term is a normalized measure of how well the parameter estimates are defined for the pth order model: clearly the smaller the relative error variance, the better defined are the parameter estimates in statistical terms, and this is once more reflected in a more negative value for the term. Thus the model which minimizes the YIC provides a good compromise between model fit and parametric efficiency: as the model order is increased, so the first term tends always to decrease; while the second term tends to decrease at first and then to increase quite markedly when the model becomes over-parameterized and the standard error on its parameter estimates becomes large in relation to the estimated values (in this connection, note that the square root of $\sigma_\xi^2 p_{ii}$ is simply the relative standard error on the ith parameter estimate).

2.4 Analysis of the River Conder Tracer Data

Let us now see how these concepts of time-series modelling can be applied in practice. Figure 1 compares the results obtained by fitting an ADZ model to the Copper Creek data with the equivalent results obtained by Fischer[22] using the ADE model. Here a second order model was identified from the data; *i.e.*, the statistical identification procedure outlined above suggests that the reach model is second order and should consist of two ADZ elements. Since these results are discussed further by Beer and Young,[17] however, we will not consider them in detail here; rather we will discuss the results of the other dye tracer experiment shown in Figure 2.

This experiment was carried out by the author and his colleagues on the River Conder, a medium sized stream in north west England.[18,19,23] The reach of this river under investigation was 5 metres wide and 116 metres in length, characterized by one major pool-riffle structure and a fine cobble bed. A number of different order models were estimated from these data using the Simplified Refined Instrumental Variable (SRIV) method of recursive estimation (see reference 24) in the *MicroCAPTAIN* microcomputer program for the IBM PC–AT. Table 1 compares the results obtained from this exercise and shows that the best identified model in YIC terms is the first order (*i.e.*, [1,0,12] or $n = 1$; $m = 0$; $\delta = 12$) model, which can be interpreted in physical terms as a single ADZ element with ADZ residence time of 65 seconds and an advective time delay of 12 sampling intervals (*i.e.*, 180 seconds). Two separate ADZ model predictions are shown in the figure, one based on this first order model; and the

[22] H. B. Fischer, *Proc. ASCE, J. Sanit. Eng. Div.*, 1968, **94**, 927.
[23] S. G. Wallis, C. Blakeley, and P. C. Young, *J. Inst. Water Eng. Scientists*, 1987, **41**, 122.
[24] P. C. Young, in 'Identification and System Parameter Estimation, 1985.' ed. H. A. Barker and P. C. Young, Pergamon Press, Oxford, 1985, p. 1.

Table 1

Den. Order n	Num. Order m	Advective Delay δ_T(15 sec. ints.)	COD	YIC Order
1	*0*	*12*	*0.9933*	1
2	0	11	0.9923	4
2	*1*	*12*	*0.9978*	2
3	0	11	0.9971	5
3	1	11	0.9981	3

All higher order models rejected on YIC basis

other obtained from a second order model [2,1,12], which can be interpreted as two ADZ elements with ADZ residence times of 46 and 182 seconds respectively. This second order model has a rather larger YIC value but it fits the data rather better, particularly in the important tail of the concentration profile (Coefficient of Determination, COD $= R_T^2 = 0.9978$ compared with 0.9933 for the first order; *i.e.*, 99.78% of data explained).

Clearly, from a purely objective YIC standpoint, we should accept the first order model as the more appropriate representation of the data, with the better fit provided by the higher order models not fully justified by their increased parameterization. However, we see that the second order model is quite well identified, being the second choice in YIC terms, and the improvement in the explanation of the elevated tail seems to be visually significant. Moreover, analysis of the model indicates that the two estimated ADZ elements have considerably different residence times and are arranged in a *parallel* structure, as shown in Figure 3(b). A rather appealing, albeit speculative, interpretation of such a model structure is a two layer concept, with the longer residence time ADZ representing a slower moving layer close to the bed, and the shorter one representing the faster moving layer above this. This seems quite a reasonable conjecture on the basis of the River Conder, where the fine cobble bed could create the right conditions for such multi-layer behaviour.

We have included this example specifically to emphasize that, while objective statistical criteria such as the YIC can be a great help in ensuring that the model is not over-parameterized, they should not provide the *only* criteria on which the model is judged. If the model has a physical interpretation, as here, then physical considerations may well be used to over-ride the purely statistical considerations and add to the model's credibility. This seems quite reasonable from a Bayesian statistical standpoint,* for example, which would suggest that prior knowledge of this type should be allowed to influence the modelling. But we should not allow this influence to be too great without careful thought: in this example, for instance, the reason for the ambiguity arises partly from the poor excitation provided by the impulsive input and so, before reaching a final decision on which model is more appropriate, we might think of conducting other experiments to clarify the situation (see below, Section 5). In practical terms, however,

---

* It is interesting to note that the recursive approach to time-series analysis used here can be considered directly in Bayesian terms, the recursive update at each sample from the *a priori* to *a posteriori* estimate of the parameter vector is the very embodiment of Bayesian estimation.[3]

either of the two best identified models would perform quite well in simulation or predictive applications, since they both explain over 99% of the experimental data and are reasonably efficient in their parameterization.

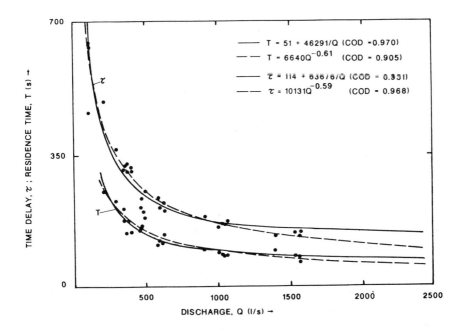

Figure 4 *Relationship between first order ADZ model parameters (advective time delay; residence time) and river flow for the River Conder*

2.5 The Wider Implications of the ADZ Model

The tracer experiment discussed above was one of many conducted on the river Conder and other rivers in north west England. Figure 4 shows a plot of the estimated ADZ residence times and advective time delays for the first order identified models over a wide range of different river flows (discharge) on the Conder; while Figure 5 is a plot of the estimated dispersive fraction (*i.e.*, the estimated ratio of ADZ volume to total reach volume) against flow obtained from the results in Figure 4. From this latter figure, we see that the fractional volume of the reach which appears attributable to ADZ effects is, rather surprisingly at first, approximately constant at a value of 0.372 over the whole range of flows investigated; in other words, only 37.2% of the river volume appears primarily responsible for introducing the pollutant dispersion effects, with the remainder functioning in a largely advective role. This dispersive fraction tends to be river dependent, ranging from values around 40% for fast, turbulent rivers, to 15% for smooth, man made channels.

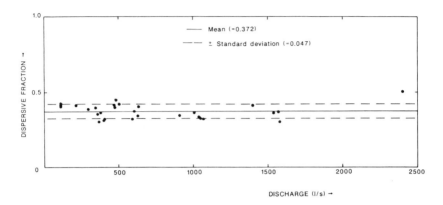

Figure 5   *The Dispersive Fraction; a flow invariant dispersion parameter: relationship between the dispersive fraction and river flow for the River Conder*

These results show the potential value of the time-series approach to modelling proposed in this Chapter. By analysing the experimental tracer data in a careful and largely objective, statistical manner, we have not only been able to model the data efficiently, but also expose an important flow-invariant parameter; namely the dispersive fraction. It is always satisfying if such an invariant characteristic can be discovered in scientific investigations and it is already helping to stimulate new theoretical developments on dispersion modelling (see *e.g.*, reference 25).

2.6   System Theoretic Considerations

Finally, before proceeding further, there are some aspects of the Conder tracer experiment that should concern us from a system theoretic standpoint. For example, we should question whether the single tracer experiment, with such an impulsive type of input, is adequate to define the dispersion model. It is well known[3] that, in a dynamic experiment, the input signal should be 'rich' enough in dynamic terms to allow for complete and unambiguous estimation of the model parameters: for instance, a second order model, such as the one discussed above, is not fully 'identifiable' if the input signal is in the form of a pure, single frequency sinusoid; at least two separate sinusoidal components, at different frequencies, are required for unambiguous estimation of the model parameters. And while an impulsive input allows for identifiability, it is not 'persistently exciting' in a statistical sense. In simple terms, this means that the input does not *continually perturb* the system adequately over the observation interval. As a result, it does not continue to yield information on the dynamic behaviour of the system that can be used to constantly improve the statistical efficiency of the estimates. For this reason, the parameter estimates are not defined as well as they might be

[25] R. Smith, *J. Fluid Mech.*, 1987, **178**, 257.

if a more optimal input signal, in these terms, had been selected. This clearly raises the idea of introducing better or even optimal experimental planning procedures in relation to these and similar experiments (see, *e.g.*, references 26 and 27).

Of course, such ideas of optimum experimental design are only relevant in situations where detailed experimental planning is possible. And since most environmental modelling must be based largely on passively monitored data with little experimental planning, we must ask ourselves whether such data contain sufficient information to allow for complete model identification and estimation. Obviously, the answer to this question will depend upon factors such as: the nature of the system; the model size and form; and the experimental data set. All that can be said in general terms is that the analyst should attempt to evaluate his modelling results with these factors in mind. As a golden rule, he should be extremely wary of using large and complex dynamic models when his database is small, as is so often the case in pollution modelling. Otherwise he may be attaching too much weight to his prior prejudices and may have more confidence in his model predictions than is justified by the analysis.

The present author[1,2] and Kalman[28] have both stressed the importance of limiting the effects of the model builder's 'intuition' or 'prejudice' on model construction. And both have suggested the strong need to go from 'data → model' in as objective a manner as possible. Of course, complete objectivity is difficult in practice; but at least the analyst should try to limit the strength of his prejudice in relation to model construction. This is made easier if he follows a systematic procedure which conforms with the scientific method and provides a series of 'checks and balances' at all stages in the modelling activity. We will consider such a general procedure in Section 4 of this Chapter. First, however, let us see how the ADZ modelling concept can be utilized in the development of more complex multivariable (*i.e.*, multi-input, multi-output) models of pollution in river systems.

3   A TYPICAL MULTIVARIABLE SYSTEM MODEL BASED ON THE ADZ CONCEPT: DO–BOD DYNAMICS IN A NON-TIDAL STREAM

In relation to general problems of pollution modelling, the implications of the example in the last Section are quite far reaching. In specific terms, the ADZ model is simple and yet there is some gathering evidence that it may have potential for wider application in areas such as streamflow modelling,[4] soil-water processes,[29] translocation modelling in plants subjected to air pollution,[30]

[26] A. J. Jakeman and P. C. Young, Proc. 4th Biennial Conf. Simulation Soc. of Australia, 1980, 248.
[27] G. C. Goodwin and R. L. Payne, 'Dynamic System Identification: Experiment Design and Data Analysis,' Academic Press, New York, 1977.
[28] R. E. Kalman, *Int. J. Policy Anal. Inf. Syst.*, 1979, **4**, 3.
[29] K. J. Beven and P. C. Young, *J. Contaminant Hydrol.*, 1988, **3**, 129.
[30] R. Gould, P. Minchin, and P. C. Young, *J. Exp. Bot.*, 1988, **39**, 997.

and modelling water movement dynamics in trees.[31] These are all, however, examples, of linear, **single-input**, **single-output** (SISO) systems: in general, we should expect that most pollution processes in the environment will be multivariable (*i.e.*, **multiple input and multiple output** or MIMO systems).

A typical and well known example of a such a linear, multivariable system is the model of Dissolved Oxygen and Biochemical Oxygen Demand (DO–BOD) in a non-tidal stream. Within an ADZ context, a reasonable formulation for a DO–BOD model is the following two, coupled differential equations, each of which can be considered as a straightforward extension of the single variable ADZ model (3) but allowing for the non-conservativity and interaction between the DO concentration, denoted by $x_1(t)$, and the BOD concentration, denoted by $x_2(t)$,

$$\frac{d(Vx_1(t))}{dt} = Qu_1^*(t) - Qx_1(t) - k_2Vx_2(t) + k_1V\{C_s - x_1(t)\} + \text{other effects}$$

(11)

$$\frac{d(Vx_2(t))}{dt} = Qu_2^*(t) - Qx_2(t) - k_2Vx_2(t) + \text{other effects}$$

Here, $u_1^*(t)$ and $u_2^*(t)$ are, respectively, the advection-delayed input DO and BOD concentrations from the previous reach, defined as follows,

$$u_1^*(k) = \exp(-k_1\tau)u_1(k - \delta) + k_1/(k_2 - k_1)\{\exp(-k_2\tau) - \exp(-k_1\tau)\}u_2(k - \delta)$$
$$+ \{1 - \exp(-k_1\tau)C_s$$

(12)

$$u_2^*(k) = \exp(-k_2\tau)\, u_2(k - \delta)$$

In equations (11) and (12), k_1 is the BOD decay rate; k_2 is the DO re-aeration rate; C_s is the saturation concentration of DO; while the 'other effects' represent any other processes in the reach which affect the DO and BOD concentrations (see references 15 and 16). Note that, while this is quite similar to other 'compartmental' models of DO–BOD dynamics suggested previously, there are significant differences which arise from the ADZ content in which the present model is set. In particular, the volume V is the ADZ volume assessed from tracer studies and *not* the total volume of water in the reach.

This model can be considered in discrete-time terms, as in the case of the single variable equation (3). The exact nature of this discrete-time model will depend upon the many factors affecting the reach dynamics but, if we assume that the 'other effects' are small,* then the general multi-ADZ form of the model is as follows,

$$\mathbf{x}(k) + \mathbf{A}_1\mathbf{x}(k-1) + \mathbf{A}_2\mathbf{x}(k-2) + \dots \mathbf{A}_n\mathbf{x}(k-n) = \mathbf{B}_0\mathbf{u}(k-\delta_T)\,,\, \dots\,,\mathbf{B}_m\mathbf{u}(k-\delta_T-m)$$

which is the multivariable equivalent of equation (8), with the scalar coefficients replaced by appropriate order matrices of parameters (in this example each is

[31] R. Milne and P. C. Young, in 'Identification and System Parameter Estimation 1985,' ed. H. A. Barker and P. C. Young, Pergamon, Oxford, 1985, p. 463.

* This assumption would, of course, need to be tested in practice; see reference 32.

2×2); δ_T denoting the cumulative effect of the advective delays; and with the vectors $\boldsymbol{x}(k)$ and $\boldsymbol{u}(k)$ defined as,

$$\boldsymbol{x}(k) = [x_1(k)x_2(k)]^T; \; \boldsymbol{u}(k) = [u_1(k)u_2(k)]^T$$

The identification and estimation of this model present much more formidable time-series problems than in the SISO case, but computer-based modelling procedures are available for tackling, if not completely solving, these problems (see *e.g.* references 32 and 33). The major principles of identification and estimation enunciated in the SISO context apply equally well to MIMO systems and, not surprisingly, the multivariable model is even more vulnerable to problems of over-parameterization and identifiability than its SISO counterpart.

In a typical modelling exercise based on these DO–BOD equations, DO and BOD measurements logged over a constant flow period could be used to estimate the parameters directly in equation (11). Alternatively, tracer tests could first be conducted on the river reach under investigation in order to estimate or 'calibrate' the river in relation to its ADZ advective and dispersive characteristics over the whole flow régime (*i.e.*, the estimation of the dispersive fraction). Having calibrated the reach for ADZ advection and dispersion in this manner, DO and BOD could then be monitored and the resulting data set used to estimate the remaining unknown parameters rate parameters k_1, k_2, and C_s in the multivariable model.

In general terms, of course, we should assume that most pollution processes in the environment will not only be non-conservative and multivariable but also nonlinear. We will not consider nonlinear systems here, however, since there is as yet no really unified theory for such systems and each example must be treated on its own merits. In any case, the most popular general approach to nonlinear systems analysis is via linearization; a process which yields linear, small perturbation models which can be handled in the manner considered here.

4 A GENERAL SYSTEMS APPROACH TO MODELLING ENVIRONMENTAL POLLUTION PROCESSES

It was the realization of the difficulties associated with modelling from 'normal operational' data, together with development of the advanced time-series techniques discussed in the previous Section of this Chapter, that first led the author[1,2] to suggest a general procedure for modelling 'badly defined' systems. The various steps in this procedure, which has direct relevance to the modelling of pollution in the natural environment, will not be discussed in detail here since they are discussed fully in the quoted references. Nevertheless, it is worth noting that the procedures make extensive use of **recursive** methods of time-series analysis, in which the model parameters are updated sequentially whilst working

[32] P. G. Whitehead and P. C. Young, in 'Computer Simulation of Water Resources Systems,' ed. G. C. Vansteenkiste, North Holland, Amsterdam, 1975, p. 417.
[33] P. C. Young and C. L. Wang, in 'Multivariable Control for Industrial Applications,' ed. J. O'Reilly, Peter Peregrinus, London, 1986, 244.

serially through the time-series data. The value of such estimation is that it allows us to relax the normal assumption that the model parameters are time-invariant. In this manner, we can investigate the possibility of nonstationarity and nonlinearity in the model.[34]

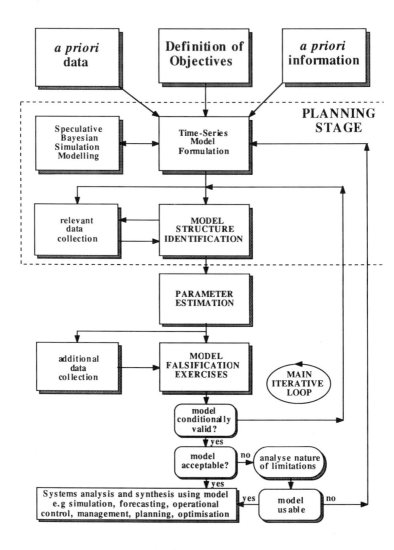

Figure 6 *Simplified block diagram of the system's approach to modelling badly defined dynamic systems described in this chapter*

[34] P. C. Young and D. E. Runkle, plenary session paper, International IFAC Symposium on Adaptive Systems in Control and Signal Processing, Glasgow, 1989.

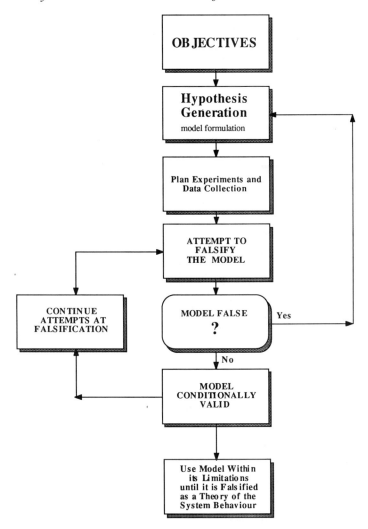

Figure 7 *The approach to model building considered within a Popperian interpretation of the Scientific method*

While the recursive methods of time-series analysis are sophisticated in estimation terms, they are not excessively complicated and are quite easy to use in practice. This is emphasized by the fact that the results discussed earlier in this Chapter were obtained using the *microCAPTAIN* microcomputer program;[6] and simpler versions of this program have been developed for use in the field, based on BBC and EPSON HX20 microcomputers.[23]

A diagram of the overall modelling procedure based in part on these time-series techniques is shown in Figure 6, where it will be seen that, following the careful definition of the objectives of the modelling exercise, the analysis proceeds systematically through various stages, with allowance made for feedback and iteration if satisfactory results are not achieved at first. This is in sympathy with a simple Popperian hypothetico-deductive interpretation of the scientific method,[35] as shown diagrammatically in Figure 7. This approach normally leads to a model which is, at best, 'conditionally valid,' in the sense that the analyst has failed, at that point in time, to falsify it as a theory of the system behaviour. This implies that, while he should use the model in the practical application for which it is intended, he should continue his attempts at falsification and resist the temptation to consider that the model cannot be either improved or even rejected by future studies.

Of course, to use the model in practice, the modeller should be convinced that it is 'acceptable,' in the sense that it describes the system well enough to achieve the defined objectives of the study. If this is not the case, he must proceed to obtain more information about the system until it becomes acceptable in this sense and, therefore, of potential practical utility. The possibility of an 'unacceptable' result may, at first, seem to detract from the proposed approach. But this is, in fact, one of its most important and powerful assets: the alternative, too often encountered in environmental studies, is for the model builder to place too much confidence in his unvalidated model and so use it to reach conclusions, and possibly advise on subsequent decisions, that cannot really be justified on the basis of the limited information at his disposal. Of course, if he consciously decides, in the face of such judgements, to place great trust in his intuition (prejudice?) and so help in reaching some management decision, then this is quite in order. After all most of us have to make decisons everyday on the basis of inadequate information. But at least by following the above procedure, he will not attempt to persuade himself (and others) that the decision has been reached on fully 'scientific' grounds.

5 DISCUSSION AND CONCLUSIONS

Few would dispute that it is difficult to develop reliable mathematical models of large and complex systems. Yet, in areas such as environmental pollution, such models are routinely constructed, usually on the basis of a reductionist philosophy, partial understanding of micro-scale subsystems, and an intuitive feel for the manner in which these subsystems may interact. On this normally inadequate basis, the sub-systems are combined to yield a computer-based, macro-scale simulation model of the complete system. Attempts at model estimation, variously called 'calibration,' 'verification,' and 'validation,' then follow, and the ability of the model to provide a reasonable fit to limited sets of *in situ* data is often considered as a primary justification for assuming its holistic validity.

We do not question these exercises in simulation model building *per se*: if their limitations are acknowledged, simulation models of this type can often provide a useful step in gaining a better insight into the various possible mechanisms which could affect the system behaviour. In this sense, the simulation model can function as a natural extension of the modeller's thought processes and so help to confirm the internal consistency of the model in dynamic systems terms. Bearing

[35] K. R. Popper, 'The Logic of Scientific Discovery,' Hutchinson, London, 1959.

in mind the above comments, however, such simulation modelling should not be considered from a purely deterministic standpoint; rather it should be based on a probabilistic approach, such as the generalized sensitivity method proposed by Hornberger and Spear,[36] which recognizes the many dangers involved in the uncritical development and use of deterministic simulation for complex natural systems analysis (see reference 2).

The dangers of trusting too much in deterministic computer simulation models are very real; and nowhere are these dangers more apparent than in many current attempts at modelling pollution in the natural environment. In a recent study,[37] for instance, it was found that a model intended to describe both longitudinal and lateral dispersion of air pollutants, was very deficient in longitudinal terms, with dynamic behaviour which owed more to the approximate methods of modelling the advective behaviour than to the dispersion coefficients set by the user. Such deficiency only became apparent, however, when time-series analysis of the kind outlined in this Chapter, was applied to the computer generated model outputs obtained with special test inputs that were designed specifically to test this aspect of the model behaviour.

The main purpose of this Chapter has been to question the wisdom and scientific validity of these attempts at complex model building when applied to environmental pollution problems. We have proposed an alternative modelling approach which emphasizes the use of powerful new methods of recursive time-series analysis and the need to avoid, wherever possible, prejudicial judgment during model synthesis. It is an approach in which the model builder should always attempt to build a mathematical model whose size and complexity reflects not so much his unquestioned perception of the system, but more the information content in the experimental data available from the system. It is, in other words, an approach which stresses that, whenever possible, model building should proceed from

Data → Model

without too heavy a reliance on prior conceptions of model structure and parameter values.

Of course, this is not an easy path to follow: it is often much easier and more enjoyable to build speculative simulation models. Furthermore, models constructed in this manner may well not describe the system adequately for all purposes. But surely this merely serves to emphasize the poverty of the data collected in many studies of environmental pollution and our associated lack of real understanding of the *macro-scale* behavioural processes. The main laboratory of the scientist studying environmental pollution is the environment itself. Analytical and research laboratories in universities, research institutions and industry can clearly help in understanding the fundamental processes in environmental pollution. But they cannot hope to define completely these processes as

---

[36] G. M. Hornberger and R. C. Spear, in 'Uncertainty and Forecasting of Water Quality,' ed. M. B. Beck and G. Van Straten, Springer-Verlag, Berlin, 1983.

[37] L. P. Steele, 'Recursive Estimation in the Identification of Air Pollution Models,' Ph.D. Thesis, Australian National University, Canberra, 1981.

they occur in the real world. It is one thing to analyse a complicated chemical engineering process or a sophisticated aerospace vehicle that has been constructed by man and whose basic structure is thus well known. It is quite another to model ill-defined environmental systems where the luxury of the planned *in situ* experiment is the exception rather than the rule.

Finally, the fact that the simple example discussed in this paper has illustrated some of the inherent difficulties of analysing time-series data, even from relatively well planned experiments, should serve to emphasize the enormous problems that face the scientist who attempts to model the environment on the basis of passively monitored data. It is hoped that the caveats presented here will be heeded by new research workers confronted with this difficult but immensely challenging task. The advice may not help them to solve completely their problems but it may prevent them from drawing unwarranted conclusions on the basis of inadequate data.

6 BIBLIOGRAPHY

R. G. Godfrey and B. J. Frederick, *US Geological Survey*, Paper 433-K, 1970.
L. Ljung and T. Soderstrom, 'Theory and Practice of Recursive Estimation,' MIT Press, Cambridge, Mass, 1983.
P. C. Young, *Bull. Inst. Maths. Appl.*, 1974, **10**, 209.
P. C. Young and S. G. Wallis, *J. Appl. Maths. Computation*, 1985, **17**, 299–334.
T. J. Young, 'Recursive Methods in the Analysis of Long Time Series in Meteorology and Climatology,' Ph.D. Thesis, Centre for Research on Environment Systems, Univ. of Lancaster, England, 1987.

Subject Index